# 15th International Scientific Conference of Civil and Environmental Engineering for the PhD. Students and Young Scientists - Young Scientist 2023 (YS23)



MATEC Web of Conferences Volume 385 (2023)

High Tatras, Slovakia
29 - 31 March 2023

**Editors:**

**Peter Mesaros**
**Vinayakam Jothiprakash**
**Kamila Kotrasova**
**Martina Zelenakova**

ISBN: 978-1-7138-8339-5

**Printed from e-media with permission by:**

Curran Associates, Inc.
57 Morehouse Lane
Red Hook, NY 12571

**Some format issues inherent in the e-media version may also appear in this print version.**



For additional information, please contact EDP Sciences – Web of Conferences
at the address below.

EDP Sciences – Web of Conferences
17, Avenue du Hoggar
Parc d'Activité de Courtabœuf
BP 112
F-91944 Les Ulis Cedex A
France

Phone: +33 (0) 1 69 18 75 75
Fax: +33 (0) 1 69 28 84 91

contact-edps@webofconferences.org

**Additional copies of this publication are available from:**

Curran Associates, Inc.
57 Morehouse Lane
Red Hook, NY 12571 USA
Phone: 845-758-0400
Fax: 845-758-2633
Email: curran@proceedings.com
Web: www.proceedings.com

# TABLE OF CONTENTS

# PREFACE

The 15th International Scientific Conference of Civil and Environmental Engineering for the PhD. students and young scientists - Young Scientists 2023 (YS23) under auspices of

- prof. Dušan Katunský, the dean of the Civil Engineering Faculty of the Technical University of Košice, Slovak Republic,
- Ing. Andrej Doležal, the Minister of Transport of the Slovak Republic,
- Mgr. Ján Horecký, the Minister of Education, Science, Research and Sport of the Slovak Republic,
- Ján Budaj, the Minister of Environment of the Slovak Republic, and
- Ing. Martin Hypký, MBA, the Head of Authority for Spatial Planning and Construction of the Slovak Republic,

was held in High Tatras, Slovakia on 29th - 31th March 2023. The traditional meeting of PhD. students, their supervisors and young scientists working in the field of civil and environmental engineering was organised by the Faculty of Civil Engineering, the Technical University of Košice, Slovakia.
The Young Scientist 2023 offered an international platform for the dissemination of the original research results. It provides a pleasant environment to present a new, great ideas of young researchers and discovered advances in the field of civil and environmental engineering and related interdisciplinary topics. The conference provided an opportunity for PhD. students and young researchers to discussed scientific topics and research results, to share their knowledge and experiences, to collaborate in innovative designs and proposals.

The Conference assembled more than sixty-five delegates including PhD. students and other representatives from the universities.
The scientific contributions involved a wide range of research fields:

- Building Information Modeling
- Building Physics and Services
- Construction Economics, Marketing and Management
- Construction Technology, Organization and Management
- Environmental Engineering
- Indoor Environment
- Hydrotechnical Engineering
- Innovations in Construction Design and Technology
- Material Engineering and Recycling
- Statics, Dynamics and Modeling
- Structural Engineering and Bridges
- Sustainable Architecture and Energy Efficiency
- Sustainable Civil and Environmental Engineering
- Sustainable Water Management
- Transport and Geotechnical Engineering
- Urban Engineering

Forty-five contributions have been selected to be published in the volume of MATEC Web of Conferences. The organizing committee appreciates the enormous work of the both the editorial board and reviewers in the preparation of this volume. Many thanks also to all authors for their flexibility and meeting deadlines.

Issue editors
Peter Mésároš
Vinayakam Jothiprakash
Kamila Kotrasová
Martina Zeleňáková

# Statement of Peer review

In submitting conference proceedings to *Web of Conferences*, the editors of the proceedings certify to the Publisher that

1. They adhere to its **Policy on Publishing Integrity** in order to safeguard good scientific practice in publishing.
2. All articles have been subjected to peer review administered by the proceedings editors.
3. Reviews have been conducted by expert referees, who have been requested to provide unbiased and constructive comments aimed, whenever possible, at improving the work.
4. Proceedings editors have taken all reasonable steps to ensure the quality of the materials they publish and their decision to accept or reject a paper for publication has been based only on the merits of the work and the relevance to the journal.

Title, date and place of the conference

Young Scientist 2023, 29th-31th March 2023, High Tatras, Slovakia

Proceedings editor(s):

Peter Mésároš, Vinayakam Jothiprakash, Kamila Kotrasová, Martina Zeleňáková

Date and editor's signature

June 15, 2023

Peter Mésároš, Vinayakam Jothiprakash, Kamila Kotrasová, Martina Zelenáková

17, avenue du Hoggar - PA de Courtabœuf – BP 112 - 91944 Les Ulis Cedex A (France)
Tél. : 33 (0)1 69 18 75 75 - Fax : 33(0)1 69 07 45 17 – www.edpsciences.org

# *SAFEFORM*: Usability analysis of a safety-based knowledge management system

*Vigneshkumar* Chellappa[1*], *Peter* Mesaros[2], and *Grzegorz* Ginda[3]

[1]Amity School of Design, Amity University, 201303 – Noida, India
[2]Faculty of Civil Engineering, Technical University of Kosice, Kosice -042 00, Slovakia
[3]Department of Strategic Management, AGH University of Science and Technology, Krakow- 30-067, Poland

**Abstract.** Globally, the prevention of falls in construction is always challenging to contracting organizations. A web-based safety knowledge management (KM) system, SAFEFORM, was developed to prevent falls during vertical formwork operations in Indian construction projects. The purpose of this study was to conduct a usability analysis for SAFEFORM using a system usability scale (SUS). Fifteen potential end users were targeted to evaluate the SAFEFORM. The results indicated that the respondents considered SAFEFORM excellent and acceptable. It is expected that the SAFEFORM could assist users in preventing falls during vertical formwork operations in Indian CPs and enhance overall safety performance.

## 1 Introduction

Fall from height (FFH) accounts for many accidents and injuries in the construction industry (CI; [1]). In many countries, including India [2], Malaysia [3] and Saudi Arabia [4], FFH fatalities were reported more than half of the overall construction accidents. Such accident rates result in work loss and delays in construction projects (CPs) and cause serious financial damage to the construction environment and individuals [5].

Globally, in CI, enhancing safety performance through achieving zero accidents is the current mantra [6]. Therefore, many researchers (e.g., [7]) proposed different approaches to prevent falls, but the fall rates continue to impact the CI [8]. Sanni-Anibire et al. [9] indicated that these methods were too complicated to implement in practice; hence, a straightforward, comprehensive, and practical approach is still lacking to enhance the safety performance of CI. This could be achieved by the fall risk assessment (RA) approach. RA is the critical process in CI, and all other procedures will likely fail if the RA method fails [10].

Safety planning entails recognizing and evaluating safety risks involved in construction activities and selecting effective control measures to control the impact of risks [11]. According to Celik and Gul [12], RA is a core of safety planning and needs proper planning and monitoring. In any CPs, RA is essential and usually performed by safety managers/heads [13]. An effective RA approach could reduce the number of accidents/injuries [14].

* Corresponding author: cvkumar@amity.edu

Nevertheless, users face significant challenges, including recognizing hazards/causes of accidents, identifying the sequence of accidents, and having insufficient time to conduct the RA process [15].

Incorporating knowledge management (KM) into safety planning could eliminate the challenges related to safety in CPs [16]. According to Kamardeen [17], utilizing and transforming knowledge from different sources for continuous corporate improvement is the strategy of KM. KM is viewed to improve organizations' efficiency and competitiveness [16]. In the same line, Hallowell [18] pointed out that an effective safety KM could enhance the performance of safety in CI. India is a developing country like other regions, and CI contributes to poor safety performance [19]. The CI in India also faces challenges such as insufficient medical facilities, poor accident reporting methods, poor safety culture, and lack of implementation of safety regulation all of which contribute to the industry's poor safety culture [19]. Given the above, a web-based safety KM system, SAFEFORM, was developed for the RA process in Indian CPs to prevent falls during vertical formwork activities.

SAFEFORM is a safety-based KM tool where users can access the safety knowledge for the fall RA process focusing on preventing falls during vertical formwork operations in India. SAFEFORM contains the breakdown of formwork activities into sub-activities, risks involved in each sub-activity, causes of falls/hazards, and individuals at risk, and control measures for each sub-activity to prevent fall risks. The development of SAFEFORM can be found in [20].

One of the essential aspects that SAFEFORM must consider is user experience (UX) while interacting with the system. UX is subjective because it depends on everyone's views and opinions [21]. According to ISO ((International Organization for Standardization; [22]), UX describes the satisfaction of the users and their pleasure with a product or service as they use, see and experience it. The main goal of website creation is to provide users with a positive experience by effectively displaying useful information. This paper presents and discusses the usability analysis carried out for SAFEFORM. To answer the question of what impact the usability of SAFEFORM has from the respondents' point of view, the author proposed using the system usability scale (SUS) to evaluate SAFEFORM's usefulness.

## 2 Materials and methods

The usability of SAFEFORM was measured using the SUS questionnaire in the CPs in southern India.

### 2.1 Research object

In this study, the authors chose a web-based safety KM system, SAFEFORM, that the authors developed to prevent fall accidents/injuries during vertical formwork operations in Indian CPs. The SAFEFORM was developed using the PHP programming language. Fig. 1 shows the front end of the SAFEFORM.

In the front end, the SAFEFORM describes the overview of the system, publications related to SAFEFORM, safety regulations and safety practices for vertical formwork operations, how to use SAFEFORM, and accident statistics of formwork activities. The user can start performing the fall RA by clicking the assess risk tab on the front end. The user should start performing the fall RA by entering the project details and mail IDs (see Fig. 2).

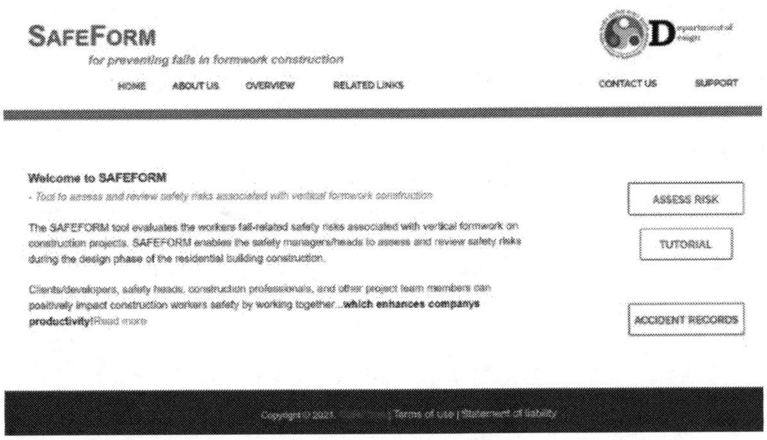


**Fig. 1.** The front end of the SAFEFORM.

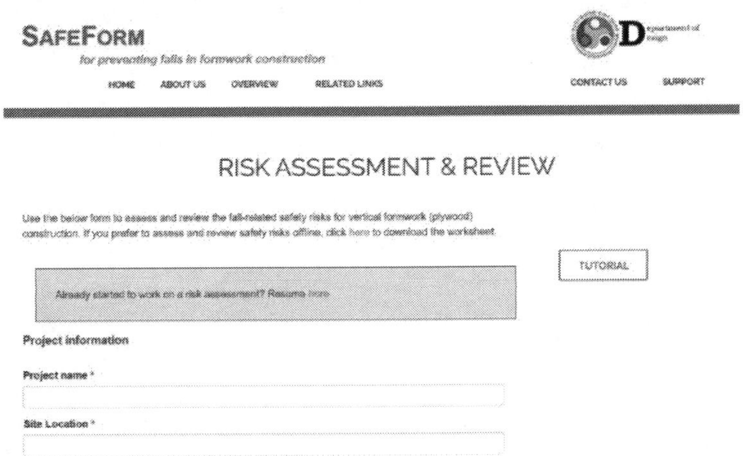


**Fig. 2.** The page display of fall RA.

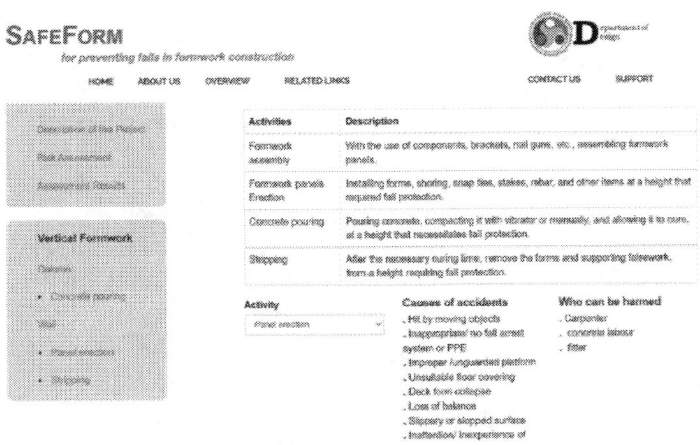


**Fig. 3.** The page display of trade (wall/column) for vertical formwork activities.

Once the users enter the RA page, they can start selecting the trade for which they want to perform fall RA. At this stage, the user can select the activity involved in any trade. For instance, Fig. 3 shows the vertical formwork activity "panel erection" under the trade "column." Once the activity is selected, the causes of falls/hazards involved will be visualized along with the individuals who might be harmed during that activity. The user must choose the height of work to be carried out and select the appropriate control measures from the list that will be already visualized in SAFEFORM. Once the user finishes performing RA, the entire result will be displayed on the final page (see Fig. 4). Here the user can modify the list of activities by adding or rejecting them. The user can finally generate the worksheet and printing option and forward it to the management team for approval.

**Fig. 4.** The page display of RA results.

## 2.2 Questionnaire design

In 1986, John Brooke developed SUS, which has been used to measure usability in many studies [23] due to its effectiveness. SUS is a questionnaire with ten statements on a Likert scale that provides views on the usability of the products or services.

## 2.3 Target users and participants

To measure the usability of SAFEFORM, 20 random potential end-users were contacted. Among these, 15 were accepted to participate in the study, which was an adequate sample as recommended by [17] for the usability study. The users carried out a live demonstration and they were asked to fill out the SUS questionnaire in which they would be able to express their thoughts on various systems aspects. Along with this, users were also asked to indicate their designation, experience in construction safety, and educational background. Safety managers/heads were targeted as they are the ones who perform fall RA in CPs.

## 2.4 Data analysis and processing

Once the data was collected, data were analyzed according to SUS rules as follows [24]:

a. For odd number statements, each question score is calculated by subtracting one from the user score.

b. For even number statements, each question final score is calculated by deducting its user score from 5.
c. To determine the SUS score, the outcome of each question's score addition is multiplied by 2.5.

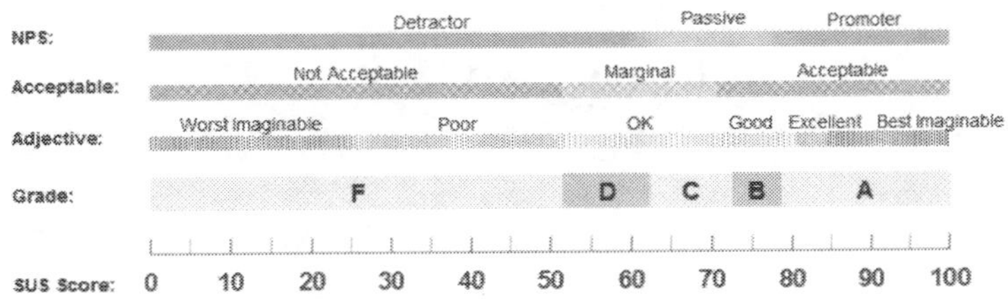


**Fig. 5.** SUS score scale (source: [25]).

The following equation calculated the SUS score:

$$\text{SUS score} = ((Q1\text{-}1) + (5-Q2) + (Q3-1) + (5-Q4) + (Q5-1) + (5-Q6) + (Q7-1) + (5-Q8) + (Q9-1) + (5-Q10)) * 2.5) \quad (1)$$

One respondent was subject to the scoring formula's rules. The mean of the SUS score for each respondent was calculated by summing the full scores and dividing the total number of respondents. The formula used to calculate the SUS score was

$$\bar{X} = \sum x/n) \quad (2)$$

Where $\bar{x}$ = mean score, x = the total score of each respondent, and n = the number of respondents.

The category of the final SUS score was determined in the evaluation process using the SUS score scale (see Fig. 5). Following the final SUS score calculation, the SUS score scale will show whether the system falls inside the acceptable category or not based on percentile ranks, grades, adjectives, acceptability, and promoters and detractors [25]. Overall, the system is acceptable if the mean score of SUS is higher than 70 [26].

## 3 Results and discussion

### 3.1 Respondents' characteristics

Table 3 shows the complete participant's background. Out of 15 respondents, most of them were safety heads (60%), followed by safety managers (27%) and safety officers (13%). Most of the respondents were undergraduate degree holders (47%), and all the respondents' experience in RA ranges from ten to twenty-four. The participant's level of safety competency in Table 1 indicates that they had sufficient knowledge and expertise to evaluate the proposed system.

**Table 1.** User's characteristics.

| **Study variables** | |
|---|---|
| **Highest level of qualification** | |
| Undergraduate degree | 7 (47%) |
| Postgraduate degree | 3 (20%) |
| Diploma | 5 (33%) |
| **Work experience (in years)** | |
| Mean (SD) | 15.2 (4.2) |
| Range | 10-24 |
| Position at work | |
| Safety manager | 4 (27%) |
| Safety officer | 2 (13%) |
| Safety head | 9 (60%) |

## 3.2 SUS score for SAFEFORM

Usability testing was carried out in this study to determine how effective, efficient, and satisfied the users were with SAFEFORM. This test aimed to measure users' satisfaction with the systems' use. Usability testing could assist designers in determining if the designed system or product is in line with users' demands or not [27].

SUS is a comprehensive evaluation of usability factors (effectiveness, efficiency, and satisfaction) as users subjectively feel them. The SUS score could indicate the user's acceptability range. Brook [26] states that a mean SUS score greater than 70 is deemed acceptable. The average SUS score obtained for SAFEFORM was 82.16 after calculating the data according to the rules of calculating SUS score. The SUS score for SAFEORM is indicated in the SUS scale (see Fig. 6).

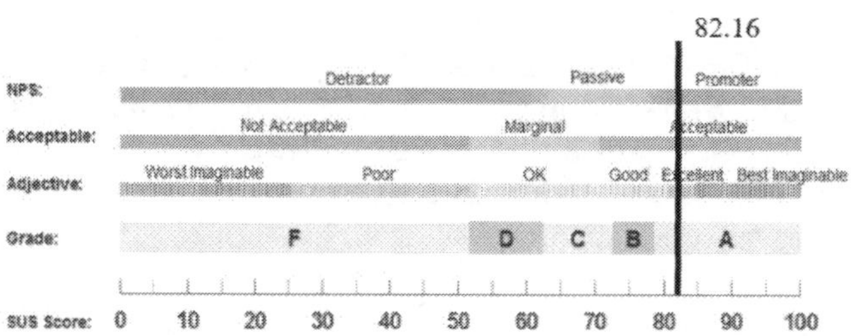


**Fig. 6.** SUS score for SAFEFORM.

Sauro [25] described the SUS score category. The SUS score can be interpreted in five ways: percentile ranks, grade, adjectives, acceptability, and promoters and detractors.

The obtained SUS score was converted into grades and percentile ranks. The grade reflects the usability rate from A+ (superior performance) to F (failed performance). In contrast, percentile ranks visualize the usability rate as a percentage (%) [28]. The SAFEFORMs' SUS score, i.e., 82.16, was compared with the grade scale, and the grade score was in grade A. The SUS score that falls within the percentile range is the percentile score. Accordingly, a SUS score above 68 is considered above average, whereas a score below 68 is considered below normal [25]. According to the SAFEFORMs' SUS score, the percentile range was between 90 to 95. The obtained SUS score was above average in both grade and percentile ranks.

The adjective scale includes best imaginable – excellent – good – ok – poor - worst imaginable. According to Sauro [25], a SUS score above 80.7 is excellent. As SAFEFORMs' SUS score was more than 80.7, the usability of SAFEFORM, according to respondents, was excellent. Acceptability is the fourth way of describing SUS. The SUS score is categorized as acceptable if the SUS score is greater than 71 [29]. The SUS score for SAFEFORM was higher than 71 and categorized as acceptable.

SUS and Net Promoter Score (NPS) are strongly and consistently correlated. The range between 30% and 50% in users' likelihood to suggest the application is described by the average SUS. According to Sauro [25], the NPS classifies recommenders into three categories: promoter (if SUS score is higher than 78.8), passive (if the SUS score is above 62.6), and detractor (if SUS score ranges between 0 and 62.6). The relationship between NPS and SUS is shown in Fig. 6, and it is evident that the SAFEFORMs' SUS score, 82, falls into the promoter category. The detractor classification is less likely to recommend a system or product to a friend and the promoter classification is more likely to do so (Derisma, 2020). As a result, the study's respondents were recognized as some of the first to promote the SAFEFORM.

## 4 Conclusion

A web-based KM system, SAFEFORM, was developed to improve safety performance to prevent falls during vertical formwork operations in Indian CPs. This study conducted the usability analysis using a SUS for SAFEFORM with fifteen potential end-users. The results show that the average SUS score for SAFEFORM was 82.16. The interpretation of the results indicated that the obtained SUS score was in grade A and the percentile range was between 90 and 95. These findings show that the respondents believed the SAFEFORM was excellent and acceptable. Respondents also act as early promoters when seen in the context of the NPS category. Overall, it is expected that the SAFEFORM could assist users in preventing falls during vertical formwork operations and enhance the overall safety performance of construction organizations.

### References

1. C. Vigneshkumar, U.R. Salve, Constr. Econ. Build. **20** (2020)
2. C. Vigneshkumar, Proc. Inst. Civ. Eng.: Forensic Eng. **175** (2022)
3. A. Ayob, A.A. Shaari, M.F.M. Zaki1, M.A.C. Munaaim, M. A. C. IOP Conf. Ser.: Earth Environ. Sci. **140** (2018)
4. E. Abukhashabah, A. Summan, M. Balkhyour, Saudi J. Biol. Sci. **27** (2020)
5. J. Yuan, W. Yi, M. Miao, L. Zhang, Int. J. Environ. Res. Public Health **15** (2018)
6. G.I. Zwetsloot, M. Aaltonen, J.L. Wybo, J. Saari, P. Kines, R.O.D. Beeck, Saf. Sci. **58** (2013)
7. C.M. Zuluaga, A. Albert, P. Arroyo, J. Cons. Eng. Manag. **144** (2018)
8. HSE, http://www.hse.gov.uk/statistics/pdf/fatalinjuries.pdf. (2020)
9. M.O. Sanni-Anibire, A.S. Mahmoud, M.A. Hassanain, B.A. Salami, Saf. Sci. **121** (2020)
10. F.A. Manuele, On the practice of safety (John Wiley & Sons, Inc Press, 2013).
11. C.Vigneshkumar, U.R. Salve (2023) Proc. Inst. Civ. Eng.: Manag. Procurement and Law (to be published)
12. E.Celik, M. Gul, Autom. Const. **127** (2021).
13. V. Chellappa, U.R. Salve, R.Y.M. Li, R. Liias, J. Stat. Manag. Syst. **23** (2020)

14. C. Vigneshkumar, R.Y.M. Li, R. Liias, U.R. Salve, *Developing a database to capture, store and share fall-related safety knowledge to enhance fall prevention in construction industry,* in proceedings of the 19th International Conference on Construction Applications of Virtual Reality, 11-12 November 2019, Bangkok, Thailand (2019).
15. C.Vigneshkumar, U.R. Salve, Lect. Notes Civ. Eng. **231** (2022a)
16. V. Chellappa, U.R. Salve, J. Know. Mang. **21** (2022)
17. I. Kamardeen, Architect. Sci. Rev. **54** (2011)
18. M.R. Hallowell, J. Manag. Eng. **28** (2012)
19. C. Vigneshkumar, K. Saravanamuthu, Des. Sci. Innov. **63** (2021)
20. C.Vigneshkumar, U.R. Salve, Lect. Notes Civ. Eng. **231** (2022b)
21. J. Laurance, E.R. Kaburuan, TURCOMAT **12** (2021)
22. ISO, Ergonomics of human–system interaction - Human-centred design for interactive systems (2010).
23. J.L. Emil, R. Kaburuan, TURCOMAT **12** (2021)
24. D.W. Ramadhan, B. Soedijono, E. Pramono, JIPI **4** (2019)
25. J. Sauro, *5 Ways to Interpret a SUS Score*, https://measuringu.com/interpret-sus-score/ (2018).
26. J. Brooke, *Usability Evaluation in Industry* (Taylor & Francis, 1996)
27. B. Lestari, P.I. Rifiani, A.B. Gati, Eur. J. Bus. Manag. Res. **6** (2021)
28. D. Derisma, Int. J. Interact. Mob. Tech. **14** (2020)
29. A. Bangor, T. Saff, P. Kortum, J. Miller, T. Staff, J. Usability Stud. **4** (2009)

# Energy consumption of smart facade

*Radek* Salajka[1*], *Jiri* Znebejanek[1], and *Ales* Rubina[1]

[1] Institute of Building Services, Faculty of Civil Engineering, Brno University of Technology, 60200 Brno, Czech Republic

**Abstract.** Expanding awareness of multi-layer smart facades and their use and comparison with external thermal insulation composite system (ETICS). Numerical calculations being used to present energy use of given facade in comparison with ETICS on the same building with identical boundary conditions.

## 1 Introduction

Ventilated facades are nowadays already a well-known and commonly used way of realizing the building envelope. The construction of ventilated facades is suitable for their attractive appearance and a wide range of used cladding materials, durability, and many other advantageousness. Above all, they are used mainly for their advantages from the building physics point of view. For example, compared to the external thermal insulation composite system (ETICS), ventilated facades have a milder course in terms of the annual temperature balance inside the structure, and they drain moisture well from the perimeter shell. The course of heat flux is the subject of this article.

A key component of ventilated facades is the ventilated gap. This is the area between the external cladding, which is in direct contact with the exterior, and the perimeter wall, or its thermal insulation located on the outside of the perimeter structure. This ventilated gap allows air to move freely. Air is sucked into the structure at the bottom of the facade through an entrance opening usually equipped with a grid to prevent animal entry, is driven along the length of the facade behind the external cladding and is then released to the exterior through an outlet opening in the upper part of the facade. The air movement itself is ensured by a phenomenon called natural convection. In the vertical gap, warmer air rises upwards due to its lower bulk density – the so-called chimney effect. Air flows naturally in most ventilated facade systems. Forced ventilation is used in the system of double facades [1], which by their own design are more space-intensive and create larger air gaps.

The air is heated up the height of the facade in two ways – the first and most significant is the influence of solar radiation, which heats the outer cladding and then transfers the heat to the air in the ventilated gap. The second way with less influence is the heat loss of the building (from the perspective of the interior), where heat flows from a warmer to a cooler environment, in other words, from the interior to the ventilated gap. This particular phenomenon occurs at the moment when the air temperature in the gap is lower than the temperature on the interior side, mainly in the colder period of the calendar year. In this

* Corresponding author: radek.salajka@seznam.cz

ingenious way, the heat load causing overheating of the building envelope is effectively removed, and consequently the requirements for possible cooling of the interior spaces in the summer season are reduced. In winter, a ventilated facade, very similar to ETICS, causes milder heat flows than a facade without any thermal insulation measures. Compared to ETICS, however, the ventilated facade may not be more effective in winter [2].

The concept of a smart facade developed at the Institute of Building Services, Faculty of Civil Engineering, Brno University of Technology, consists in the ability to regulate the air flow in the gap using a system of forced ventilation and closing flaps on the side of the inlet and outlet opening depending on the changing boundary conditions. Fans with adequate input power and performance located in a smart facade can ensure greater air flow to get rid of a greater heat load. Shut-off flaps ensure the movement or stopping of air flow to achieve the desired air temperature in the ventilated gap. They work in two modes: "open" and "closed". Multiple control steps in the terms of regulation for these flaps seem irrelevant at the moment due to the leaks in the outer casing - the lining through which air can freely escape from the ventilated gap. These leaks are some between 10% to 40% depending on the design of the cladding and the size of the gaps between the individual cassettes.

The FSVM software (acronym that stands for "Façade with Ventilated Gap" in Czech language), also created at our faculty, helps us calculate the heat flows through the ventilated facade and the temperatures in the specified facade. This software allows us to calculate the heat transfer coefficient of the perimeter wall construction with a ventilated facade based on the input parameters of the facade, taking into account the thermal bridges from the anchors supporting the external cladding. This calculation also appears in the freely available version of the program [3]. Other results are the parameters of the facade in hourly steps during the entire calendar year. The FSVM program calculates with corrected interior temperatures and takes into account the influence of climatic influences throughout the year, including the significant influence of solar declination after months and heat gains from solar radiation also throughout the whole year. A very valuable output is the simulated heat flows over the year by month steps. The heat flows are simulated for the specified ventilated facade, for the virtual relevant ETICS system (for comparison with the ventilated facade) and for the smart facade. The software is thus able to display the percentage reduction of heat flows by using a smart facade. The calculations of the FSVM software went through numerous checks and careful development within the project supported by TAČR (abbreviation for Technological Agency of the Czech Republic) FW03010062: Smart facade with optimized energy properties.

Our main goal in the smart facade research is to achieve a possible prediction of the structural physical behavior inside the facade based on measurable boundary conditions. Another goal is to be able to learn to respond to this behavior, to achieve optimal properties inside the perimeter wall structure with a ventilated facade, to achieve the smallest possible heat flows through the structure and thus ensure significant energy savings.

## 2 Methods

By measuring on real ventilated facades, we obtained valuable data regarding the temperatures on the inner and outer surface of the facade cladding, the temperatures inside the ventilated gap and the temperatures in the thermal insulation. In addition, by measuring the mentioned temperatures, we obtained relevant data related to the weather effects acting on the measured facade in real time.

The FSVM software is developed on the basis of data measured from a south-east oriented ventilated facade with orange and gray cladding material surface color on a building located in the Czech Republic in the regional city of Brno. Corrected calculations are used to approximate the calculation results of ventilated facades with other parameters. These corrections will be gradually refined with the gradual development of the software and the evaluation of data from other measurements already carried out on other facades.

Calculations in the software follow the procedures and relationships according to the standard [4], or when calculating the heat transfer coefficient with anchors according to the standard [5]. Our corrections were examined according to the already mentioned evaluation of data from the measured ventilated facade in Brno. The procedure for the appropriateness of using our corrected calculations is through the approximation of the own correction factors used in the normative calculations. After being used in the calculation, these must come close to the results of real measured data with the smallest possible error.

The input data for calculating the heat fluxes in the FSVM software are:

- structure composition and facade area,
- design temperatures of the interior and exterior in winter and summer and the amplitude of temperature fluctuations,
- geographic location and sun orientation of the facade,
- estimated dimensions of the facade,
- parameters of ventilation holes,
- wind map region, orthography, solar constant and atmospheric pollution,
- specific heat capacity and air density,
- position of flaps for the smart facade in individual months (closed by default in winter, open the rest of the year),
- thermal insulation properties of the external cladding.

A detailed breakdown of all input quantities can be seen in the freely available demo version of the software online [3]. The software is currently only available in the Czech language.

For further measurement and testing of the system of flaps and fans of the smart facade, a prototype of the facade located in the research center of the Technical University of Brno AdMaS (abbreviation for Advanced Materials, Structures and Technologies) was built and is being used. Research on this part is still in development.

## 3 Data

By measuring with the data logger AHLBORN ALMEMO 5690-2 and a series of temperature sensors at several control points on the facade and pyranometers, we obtained the values of temperatures, air flow speed, air humidity and the intensity of solar radiation acting on the measured facade. Subsequently, the research team took over the data from the measuring station after a certain time interval and transferred all the measured values to the MS Excel software, in which it was much easier to analyze the data.

The control points on the facade were set up to measure two series of temperatures in the lower part of the facade and two series of temperatures in the upper part of the facade, i.e., 4 control points. The height of the facade is 14 meters. There were 2 control points on the gray facade, one series of sensors in the lower part of the gray facade and one series of sensors in the upper part of the gray facade. In the winter, the measuring center was equipped with a heat-insulating cover against possible freezing.

We recorded all the data every 10 minutes 24 hours a day throughout the year and got a real insight into the physical phenomena present on the facade while processing and analyzing it.

## 4 Results

After processing the measured data, we were able to analyze and get an idea of the course of constructional-physical phenomena present on/in the ventilated facade. In the following graphs (Fig. 1., Fig. 2.) we can see the course of temperatures and the influence of solar radiation and outdoor temperature on air temperatures in the gap of the ventilated facade. The data always represent two successive reference days in one month of the given season. We distinguish two values of air temperatures in the gap for one moment in time, namely the temperature in the upper part of the facade (at a height of about 14 meters from the lower part) and the temperature in the lower part of the facade (height of the facade about 0.5 meters). The 14-meter-high ventilated facade itself was located at a height of 3 meters above the landscaped area around the building. In the graphs (Fig. 1, Fig. 2), these values are represented by a green curve for the upper part of the facade and a blue curve for the lower part of the facade.

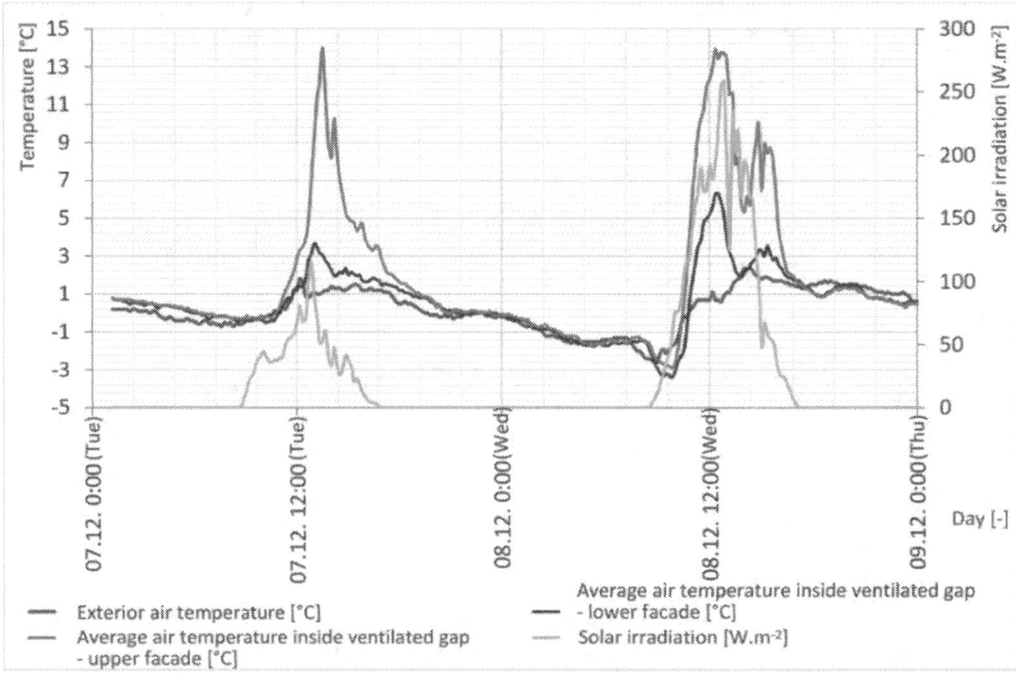


**Fig. 1.** The effect of outdoor temperature and solar irradiation on the temperature in the gap – winter, December 2021.

We can see that solar radiation causes a big spike on air temperature inside ventilated gap. The bigger intensity of solar radiation the bigger effect on both parts of façade – upper and lower. The upper part of façade gets heated more significantly because natural convection from lower part of façade brings the heated air upwards and the air is getting heated continuously more and more during its whole path of travel up to the point where the air exits façade. Meanwhile lower part of façade is supplied by exterior air that enters ventilated gap.

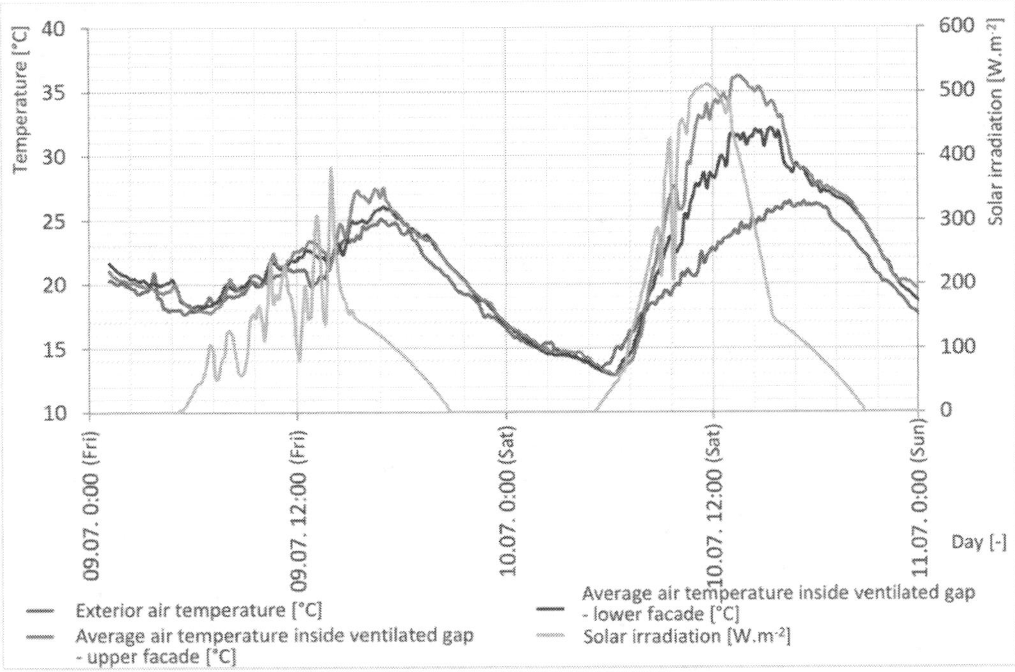


**Fig. 2.** The effect of outdoor temperature and solar irradiation on the temperature in the gap – summer, July 2021.

The graphs (Fig. 1., Fig. 2.) show how much influence solar radiation has on the temperature inside the ventilated gap. Both in winter and in summer part of the year. However, temperatures in graphs are real measured values. Rising temperature of air inside ventilated gap is not affected only by solar radiation but also from heat flow from interior. This phenomenon is not clearly visible in these graphs (Fig. 1, Fig. 2.) for comparison. Adding more information would make these graphs (Fig. 1., Fig. 2.) unclear. More information about more weather effects or more information about heat flow from interior to ventilated gap that affects air temperature inside gap is described in a thesis [2].

As part of determining the correction factors, it was necessary to approximate the values of the absorption coefficient of the cladding material $\varepsilon$ [-] and the heat transfer coefficient $\alpha$ [W.m$^{-2}$.K$^{-1}$] to calculate the surface temperature of the cladding $t_p$ [°C] according to the equation (1):

$$t_p = t_e + \frac{I * \varepsilon}{\alpha} \qquad (1)$$

where:

$t_p$ – surface temperature of the cladding [°C]
$t_e$ – outdoor air temperature [°C]
$I$ – intensity of solar radiation [W.m$^{-2}$]
$\varepsilon$ – coefficient of light absorption of the cladding [-]
$\alpha$ – thermal transmittance coefficient [W.m$^{-2}$.K$^{-1}$]

We have approximated the values in such a way that the resulting temperature of the cladding surface is as close as possible to the external temperature of the cladding surface measured by us on the real facade. The result was one value of $\alpha$ and one value of $\varepsilon$ for each month separately. The determined values of the coefficients in simplicity included the influence of the weather and climate conditions for the given month on the facade with the given parameters [2].

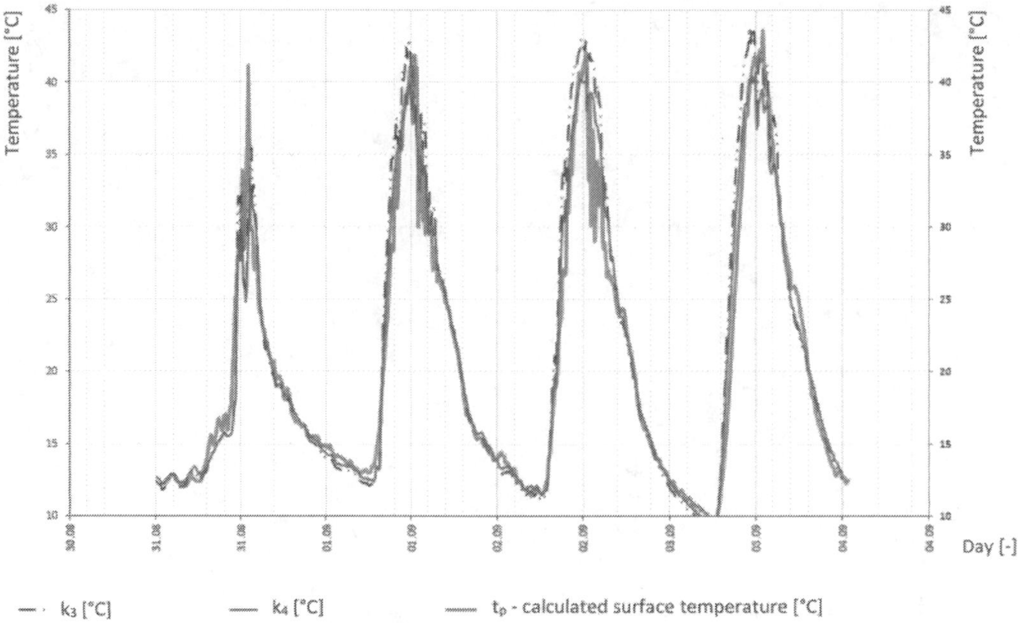


**Fig. 3.** Example of used calculations (1) with already approximated coefficients compared to measured temperatures on real facade. "k3" stands for third control point located on upper left part of the facade and represents temperature of outer cladding surface. Analogically "k4" stands for fourth control point located on the upper right part of the facade and represents temperature of outer cladding surface as well. 4 days of September 2021.

The approximate values of the coefficients $\alpha$ and $\varepsilon$ are used for correction of calculations in the FSVM software. The following results will be presented as results directly from this program.

**Table 1.** Wall layers and their parameters.

| Perimeter wall | | | | |
|---|---|---|---|---|
| Construction No. | No. of layer | material | d [m] | λ [$W.m^{-1}.K^{-1}$] |
| S1 | 1 | Thermal insulation | 0,060 | 0,037 |
| | 2 | Bearing wall structure | 0,380 | 0,111 |
| | 3 | Interior plaster | 0,005 | 0,880 |

where:
$d$ – layer thickness [m]
$\lambda$ – coefficient of thermal conductivity of the material of the layer [$W.m^{-1}.K^{-1}$]

**Table 2.** FSVM input parameters [3].

| Parameter | Value |
|---|---|
| Wall surface | 100 $m^2$ |
| Wall openings | 20 $m^2$ |
| Cladding type | Medium heavy (up to 40 $kg.m^{-2}$) |
| Bearing material | Steel |
| Pad thickness | 10 mm |
| Winter exterior air temperature | -12 °C |
| Winter interior air temperature ($t_{iz}$) | 20 °C |
| Facade location latitude (lat) | 50 ° |

**Table 2.** FSVM input parameters [3]. (Continues from previous page)

| Parameter | Value |
|---|---|
| Facade location longitude (long) | 16 ° |
| Facade location altitude (H) | 300 m AMSL |
| Facade angle with a horizontal plane (α) | 90 ° |
| Facade azimuth (γ) | 120 ° |
| Mean calculated facade width (š) | 1 m |
| Facade height (h) | 15 m |
| Facade height above ground ($h_0$) | 1 m |
| Mean calculated facade height ($h_x$) | 7,5 m (temperatures in the centre of facade) |
| Bottom inlet coefficient | 0,4 |
| Upper outlet coefficient | 0,6 |
| Ventilated gap depth ($d_m$) | 0,12 m |
| Narrowing of the intake vent (Z) | 80 % |
| Narrowing of the exhaust vent ($Z_0$) | 70 % |
| Czech wind map area (O) | 2 |
| Orthography (o) | City |
| Solar constant (I) | 1370 $W.m^{-2}$ |
| Atmospheric pollution (z) | 4 (throughout year) |
| Temperature fluctuations amplitude ($A_m$) | 7 K |
| Summer interior air temperature ($t_{il}$) | 25 °C |
| Air density ($\rho_w$) | 1,2 $kg.m^{-3}$ |
| Specific heat capacity of air ($c_w$) | 1010 $J.kg^{-1}.K^{-1}$ |
| Closing flap settings | Oct-Mar closed, Apr-Sep opened |
| Cladding thickness ($d_p$) | 0,08 m |
| Thermal conductivity of cladding ($\lambda_p$) | 0,115 $W.m^{-1}.K^{-1}$ |
| Relative solar absorptivity ($\varepsilon_p$) | 0,7 |
| Default convection heat transfer coefficient of outer cladding surface ($\alpha_e$) | 20 $W.m^{-2}.K^{-1}$ |
| Air permeability of the cladding ($n_p$) | 20 % |

Possible graphic and numerical outputs from the FSVM software as of 17.3.2023:

- diffuse solar radiation,
- direct solar radiation falling on the facade,
- total solar radiation falling on the facade,
- equivalent solar temperature on the azimuth-oriented facade,
- outdoor temperature according to the season,
- corrected interior temperature according to the season,
- the temperature of the outer surface of the facade cladding,
- heat gain from total solar radiation on the facade,
- air temperature in the ventilated gap for the facade height $h$,
- air temperature in the ventilated gap for the selected height on the facade $h_x$,
- heat flux from the building into the air gap with opened shot-off flaps,
- heat flux from the building to the air gap with smart ventilation control,
- heat flux from the building to the exterior without a ventilated gap (virtual ETICS).

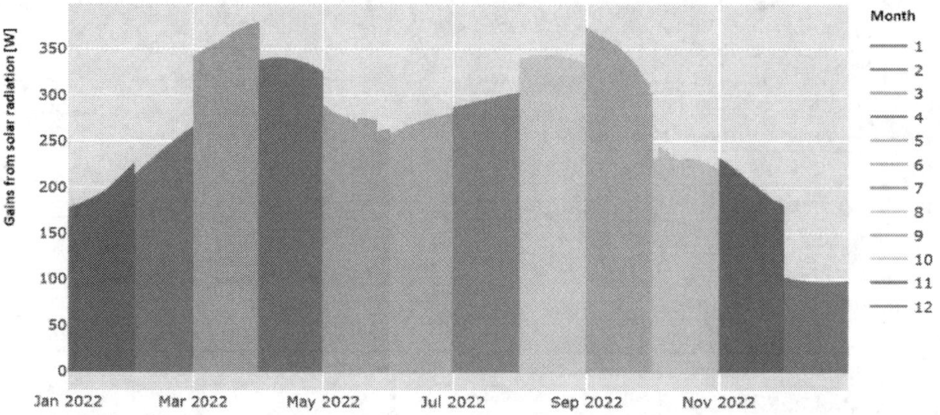


**Fig. 4.** Example of chart of simulated values of facade gains from solar radiations throughout the year – software FSVM.

Sudden and visible differences between months are an example of differences between correction coefficients for solar gains. Corrections differ by month, and they represent (among other weather conditions) simulation of how sun rays act on our specified surface – on 90° wall in our case. At our specified geographical location sun rays hit our surface at the most efficient angle in March and September. Thus, the solar irradiation heats the surface of cladding in an efficient way, and this produces significant solar gains which heats up air in the ventilated gap.

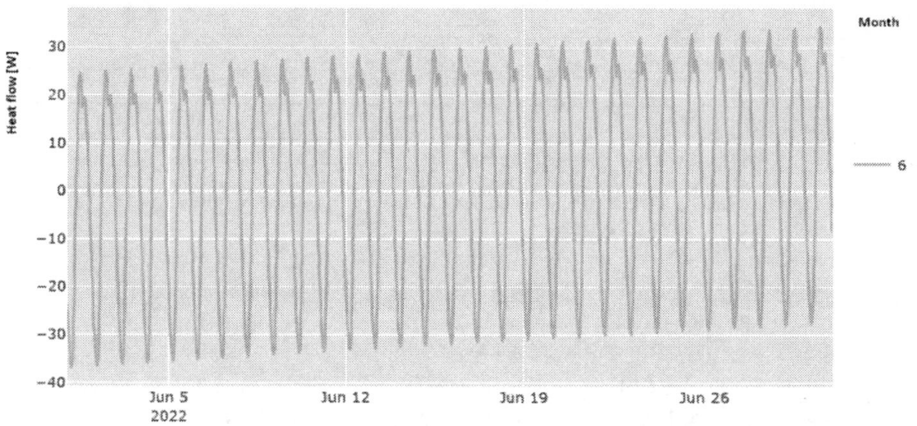


**Fig. 5.** Example of chart of simulated heat flow from interior to ventilated gap – closer look at month of June reveals that the detail of continuous graph is indeed represented by hourly simulated values. Increase in heat flow during month seems constant. This is caused by simplified calculations in our simulations which tells us with an approximation that the month's weather and building usage is considered in a simpler correction. More complex settings during specific months can be set in an upgraded version of software.
– software FSVM.

The table of heat flows passing from the interior to the exterior or to the ventilated gap (Table 3.) will give us a more detailed view of the simulated heat gains and losses during the year on the specified facade. The units used here are facade heat flows in kW recalculated for a period of one month. Therefore, the unit kWh/month, or $x$ kWh per 720 h, where "$x$" value represents the output value from the software, is given here. The percentage reduction of heat flows through a smart facade expresses a relative saving of energy, or a change in the character of their action from negative values (heat losses) to positive values (heat gains) when using the flap system of a smart facade instead of a standard ventilated facade with natural and unrestricted air movement.

**Table 3.** Simulated total heat flow values for considered type of facade.

| | ETICS [kWh/month] | Ventilated facade [kWh/month] | Smart facade [kWh/month] | Percentual reduction of heat flow with smart facade |
|---|---|---|---|---|
| 1 | -41,75 | -57,09 | -48,82 | 14,48% |
| 2 | -24,41 | -36,88 | -29,53 | 19,91% |
| 3 | -9,79 | -21,85 | 2,62 | 111,99% |
| 4 | -1,66 | -12,55 | -12,55 | 0,00% |
| 5 | 4,02 | -7,28 | -7,28 | 0,00% |
| 6 | 8,19 | 1,51 | 1,51 | 0,00% |
| 7 | 13,30 | 6,80 | 6,80 | 0,00% |
| 8 | 17,35 | 11,66 | 11,66 | 0,00% |
| 9 | 11,95 | 6,68 | 6,68 | 0,00% |
| 10 | -3,27 | -10,59 | 3,33 | 131,44% |
| 11 | -17,04 | -26,06 | -16,51 | 36,63% |
| 12 | -37,00 | -48,83 | -44,76 | 8,33% |

Simulated temperatures from the FSVM software in the following graphs (Fig. 6., Fig. 7.) point to the fact that the use of a ventilated facade is more effective in the summer months (Fig. 7.) and contact insulation, on the other hand, shows better efficiency in the winter months (Fig. 6.). According to the data from the graph (Fig. 6.), it can even be said that by using a smart facade, at noon on the simulated day of January 15, we even achieve a small thermal gain from the facade for a relatively short moment in time. In the summer months, the computer simulation did not consider the closing of the flaps and thus the effect of the smart facade on the change in heat flows is identical to the classic ventilated facade.

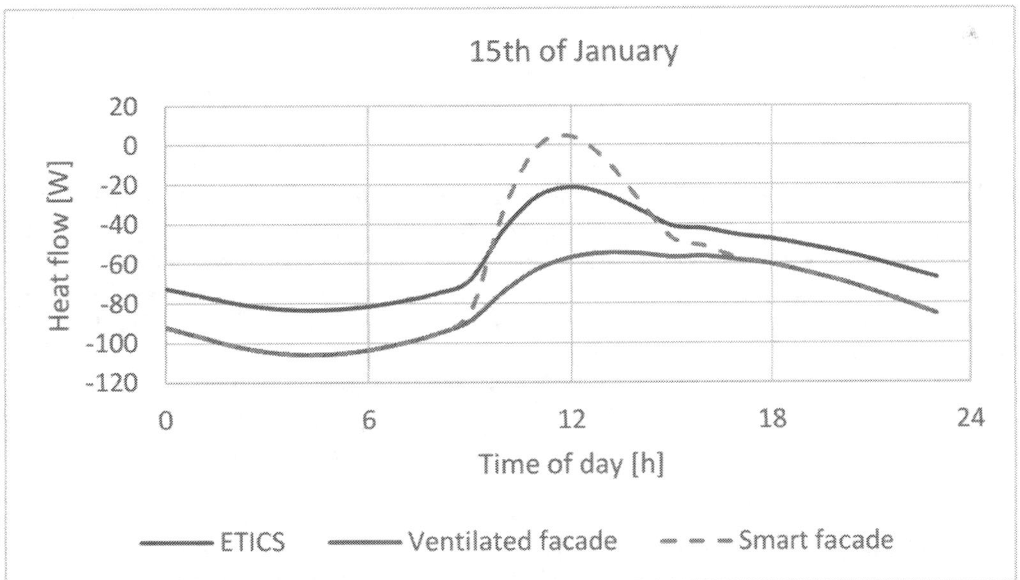


**Fig. 6.** Simulation of heat flow through a different type of facades throughout reference winter day on 15th of January.

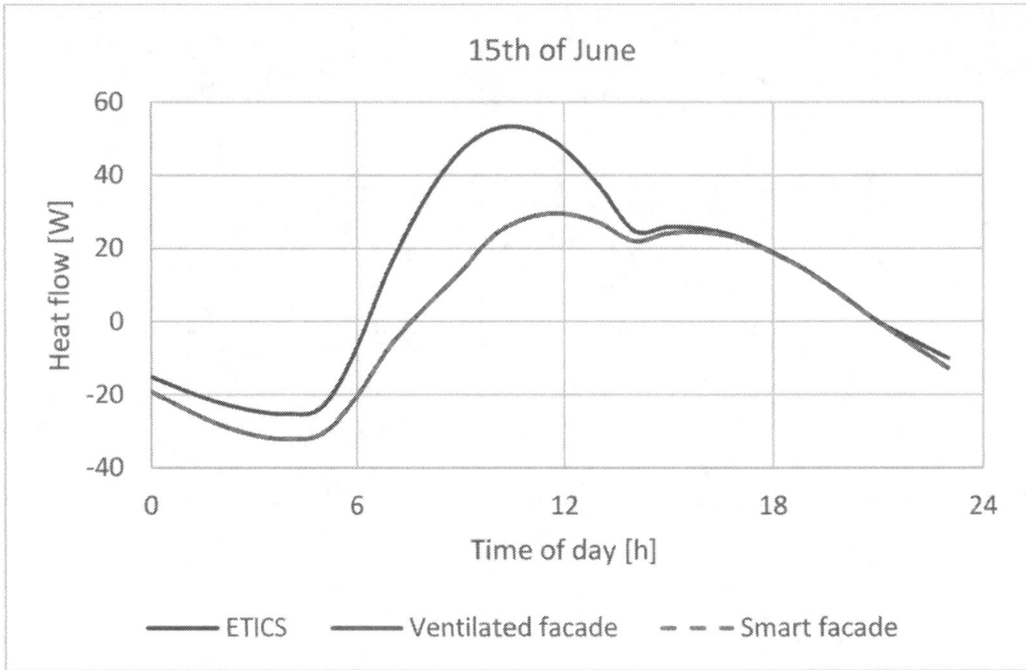


**Fig. 7.** Simulation of heat flow through a different type of facades throughout reference summer day on 15th of June.

## 5 Discussion

The FSVM program has proven to be an effective computational tool. It was created on the basis of our research, which deals with the solution of basic questions connected with the development of a smart facade. The results of the simulations correspond to a large extent to the actual condition that we measured and examined on the facade described in chapter 2 of this publication. However, the program is still being developed in terms of universality and accuracy, so that its calculations can be applied to as many types of classic ventilated facades with different parameters as possible. The outputs from the software are very rich and can serve as a cornerstone in the design and implementation of a possible future smart facade.

The most significant variable factors that will influence the thermal technical parameters of the facade will be the outside air temperature and the intensity of solar radiation. Our goal in research and development (besides improving the FSVM software) will further be the ability to respond to these variable influences acting on the facade in real time and to control the airflow in ways other than closing and opening the shutters at the air inlet and outlet of the air gap. The motive for the development of a smart facade is certainly energy saving, which should ideally be equal to zero consumption. With our abilities, we would like to at least come close to this ideal.

According to the professional publication [6], experts have been working on numerical calculations in the field of ventilated facades for over 40 years. Along with the expansion of the scope and possibilities of computing technologies, a lot of approaches arose, to which the authors of this publication devoted their attention and experimentally verified various methods, possibly their advantages, disadvantages, and their limitations [6].

Smart facade research also includes the investigation of modules with fans for air flow control, the influence of modules with a so-called "green facade" and the effect of plants on the properties of a ventilated facade or modules with air flow regulation together with photovoltaic cells.

# 6 Summary

Based on our results, it can be said that we can already partially predict the structural physical behavior on the surface and inside the ventilated facade according to variable boundary conditions. This result is a key step for further possible development and research of the smart facade system, from which we promise a significant contribution in the field of construction in terms of energy saving and construction sustainability.

This article's creation has been supported by specific research called Experimental research of heat and mass transfer in systems of building services, marked FAST-S-23-8176.

## References and literature

1. Ing. Erik Šagát Větrání obvodových plášťů budov. Brno, 2016. 130 s., 2 s. příl. Dizertační práce. Vysoké učení technické v Brně, Fakulta stavební, Ústav pozemního stavitelství. Vedoucí práce doc. Ing. Libor Matějka, CSc., Ph.D., MBA
2. SALAJKA, Radek. Energetické využití fasády. Brno, 2023. 103 s., 23 s. příloh. Diplomová práce. Vysoké učení technické v Brně, Fakulta stavební, Ústav technických zařízení budov. Vedoucí práce Aleš Rubina.
3. RUBINA, Aleš; UHER, Pavel; UHER, Václav; RUBINOVÁ, Olga; BEČKOVSKÝ, David; HRON, Libor; ILČÍK, Jiří. Výpočet tepelně technických parametrů fasády s větranou mezerou (R 5.2). *PROFESIS: Profesní informační systém ČKAIT* [online]. Praha: ČKAIT, [cit. 2023-03-17]. Dostupné z: https://profesis.ckait.cz/r-5-2/
4. ČSN EN ISO 10211 Tepelné mosty ve stavebních konstrukcích - Tepelné toky a povrchové teploty - Podrobné výpočty. ÚNMZ, Praha, 2020, 60 s
5. ČSN 73 0540-2:2011 Tepelná ochrana budov - Část 2: Požadavky. ÚNMZ, Praha, 2011, 56 s
6. DE GRACIA, Alvaro, Albert CASTELL, Lidia NAVARRO, Eduard ORÓ a Luisa F. CABEZA. Numerical modelling of ventilated facades: A review. *Renewable and Sustainable Energy Reviews* [online]. 2013, **22**, 539-549 [cit. 2023-03-17]. ISSN 13640321. Dostupné z: doi:10.1016/j.rser.2013.02.029
7. RUBINA, Aleš; UHER, Pavel; VRÁNA, Jakub; NOVOTNÝ, Miloslav; NESPĚŠNÝ, Ondřej; SKŘEK, Daniel; ŠUHAJDOVÁ, Eva; VYSTRČIL, Jan; FORMÁNEK, Marian. Heat Flow through a Facede with a Controlled Ventilated Gap, *Buildings* **2023**, *13*, x. https://doi.org/10.3390/xxxxx, www.mdpi.com/journal/buildings

# The relation between apparent and true water colour for pond water

*Tereza* Prokop Kubinova[1*], *Katerina* Brozová[1], and *Miroslav* Kyncl[1]

[1]Department of Environmental Engineering, Faculty of Mining and Geology, VSB-Technical University of Ostrava, Czech Republic

**Abstract.** The colour of the water can be considered as apparent colour, which is determined by the amount of suspended matter such as clay, phytoplankton and colloidal particles, and true colour, which is the proportion of suspended matter passing through a membrane filter of pore size the 0.45 μm. This study aimed to investigate the apparent and true water colour of pond water and evaluate the impact of filtration on water colour. Using a spectrophotometer, measurements and analyses were conducted on pond water samples. The results of the study showed that filtration can significantly affect the water colour, with a clear difference observed between the apparent and true water colour. The research findings provide a deeper understanding of the factors that influence water colour.

## 1 Introduction

The colour of the water may be of natural or anthropogenic origin. In the case of natural waters, the colour is mainly due to the presence of humic substances, such as fulvic acids, which make the water looks like yellow to yellow-brown. In addition to dissolved substances, water may also be coloured by suspended substances such as clay or phytoplankton. For this reason, a distinction is made between the true colour of the water, colour due only dissolved substances passing through the 0.45 μm pore size filter, and the apparent colour, which is expressed by the colour due both dissolved and undissolved substances, usually of a colloidal nature. For example, the green to green-blue colour of heavily eutrophicated waters is apparent because it is caused by the presence of cyanobacteria and algae, which can be eliminated by filtration. The colour of the water and its intensity is usually dependent on the pH value and must, in any case, be related to the value [1-4]. The studies on the determination of water colour discuss the objective evaluation, the relation of various water colour determinations and actual legislative standards [3-7]. The filtration process is needed for true water determination and the effect of various pore sizes and filter effects is studied [8-10].

* Corresponding author: tereza.prokop.kubinova@vsb.cz

# 2 Experiment methodology

## 2.1 Determination of water colour

The Standard Methodology for the Determination of Water Colour specifies various methods for the determination of water colour. These include both visual colour determination and determination of the true colour of the water by optical instruments. The determination of water colour is carried out according to the Czech technical standard ČSN EN ISO 7887 of July 2012. This standard list four different methods for determining the colour of water [2].

The first method involves the determination of the apparent colour by visual observation of the sample in the bottle. This provides only preliminary information useful, for example, in field work. Only the apparent colour data is reported. The second method involves the determination of the true colour of a water sample measured with an optical instrument. The method can be used for raw and drinking water and for less coloured industrial water. The third method involves determining the true colour of a water sample using an optical instrument by comparison with the concentration of hexachloroplatinate at a wavelength of $\lambda = 410$ nm. The fourth method involves the determination of colour by visual comparison with standard solutions of hexachloroplatinate, which is applicable to raw and drinking water [2].

In this research, the method chosen was to determine the colour of a water sample using an optical instrument by comparison with the concentration of hexachloroplatinate at a wavelength of $\lambda = 410$ nm. This method was used to measure both the true colour of the water and the apparent colour of the water in order to compare the results as accurately as possible [2].

## 2.2 Characteristics of the measured data

The area of interest was a pond with a size of 30 $m^2$ and a volume of 10 $m^3$ located in a private garden in the village of Polanka nad Odrou. The pond is provided with aeration due to Koi carp breeding. Data were collected from June 2021 to March 2023. There were 142 pond water determinations made as part of the analysis. 71 measurements of apparent water colour and 71 measurements of true water colour. Water colour was determined according to EN ISO 7887 spectrophotometrically. Another measured value was the pH value.

# 3 Experiment evaluation

## 3.1 Basic description of measured values of apparent and true colour

The graph in Fig. 1 shows the colour values obtained by spectrophotometry in a series of measurements of the waters of the area of interest. The apparent colour value is shown by the full height of the bar, the true colour is shown in blue. The pH value has also been added to the graph on the minor axis. The pH values ranged from 7.1 to 10.4. When examining the values of the true colour, we can see lower values in the case of lower values of the apparent colour.

In addition to the graphical representation, it iss possible to summarize the measured values using descriptive statistics:

- The measured value of the apparent water colour ranged from 19 to 111 $mg \cdot l^{-1}$ Pt. The mean value was 49 $mg \cdot l^{-1}$ Pt and the standard deviation was 29 $mg \cdot l^{-1}$ Pt. For half of the determinations, the apparent water colour did not exceed 29 $mg \cdot l^{-1}$ Pt. In half of the measurements, the apparent colour of the water ranged from 25 to 78 mg·l$^{-1}$ Pt.

- The measured value of the true water colour ranged from 5 to 31 mg·l$^{-1}$ Pt. The mean value was 13 mg·l$^{-1}$ Pt and the standard deviation was 6 mg·l$^{-1}$ Pt. For half of the determinations, the true water colour did not exceed 10 mg·l$^{-1}$ Pt. In half of the measurements, the true water colour ranged from 8 to 18 mg·l$^{-1}$ Pt.

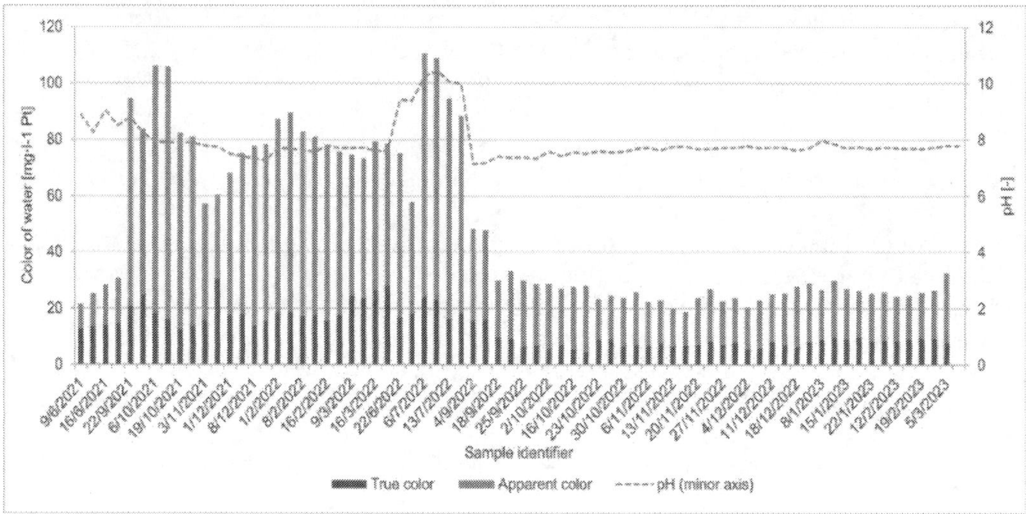


**Fig. 1**: Overview of apparent and true colour.

## 3.2 Evaluating the filtering effect when tracking apparent and true colour

In the data, the ratio of the difference between the apparent and true colour relative to the apparent colour was calculated. This ratio is used to monitor the effect of filtration in the determination of water colour.

Due to the wide range of apparent colour values, the investigated dataset was divided into 2 parts according to the water colour value of 40 mg·l$^{-1}$ Pt, which was chosen when considering the histogram of apparent colour values in Fig. 2.

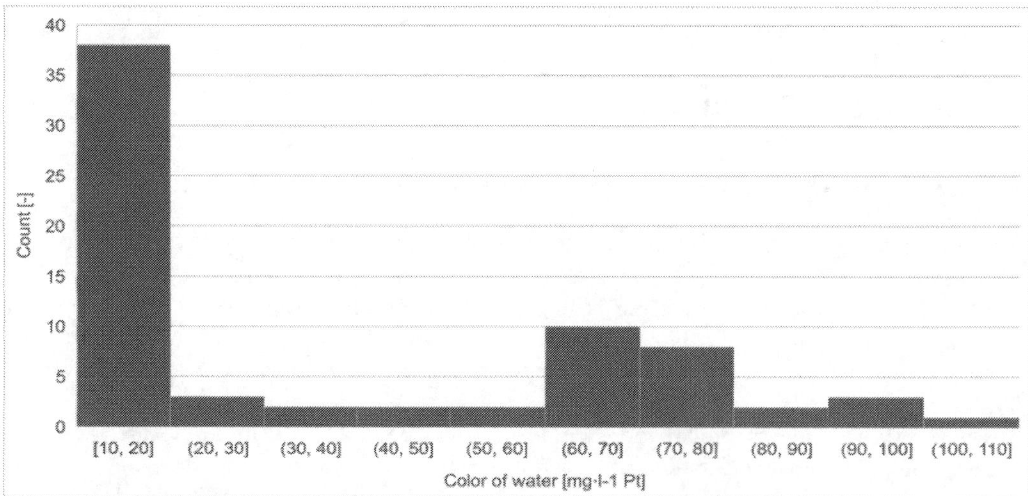


**Fig. 2**: Histogram of apparent water colour values.

Dividing the measured data according to the value of the apparent colour resulted in 2 groups of 40 measurements in the colour range 10-40 mg·l$^{-1}$ Pt and 30 measurements in the range 41-110 mg·l$^{-1}$ Pt.

The filtering effect ratio is shown in Fig. 3 using a box plot showing the range of the filtering effect. The first observed group of apparent colour value from the range 10-40 mg·l$^{-1}$ Pt shows a percentage reduction in apparent colour of 40-83 %. Typically, the reduction in apparent colour was in the values of 63-72 %. The second observed group achieved a reduction in apparent colour in the range of 49-84 %, with fifty percent of the cases showing a 68-80 % colour reduction effect. The fact of higher colour reduction and therefore the effect of filtering on the colour is higher for the group with higher values of apparent colour.

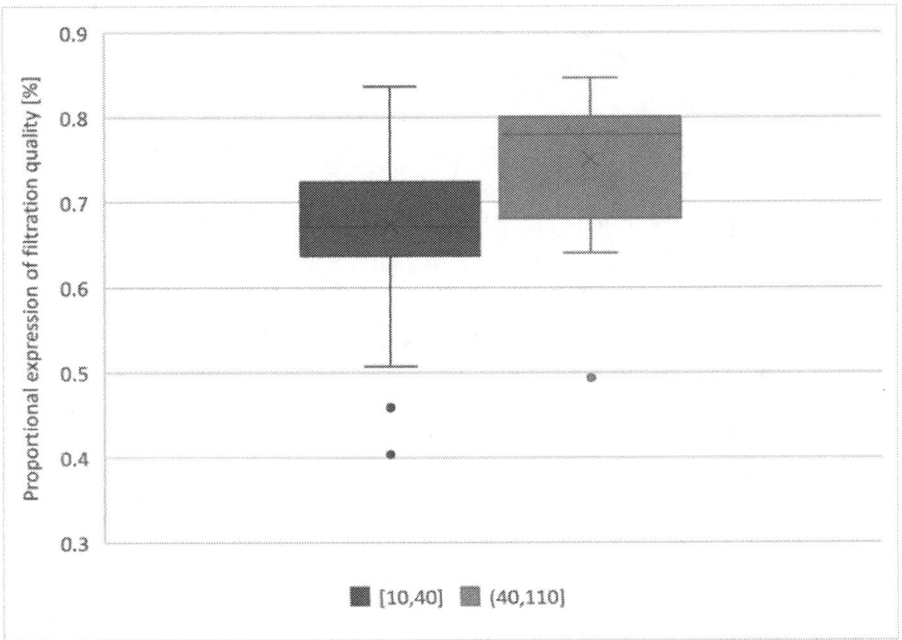


**Fig. 3**: Range of proportional reduction of water colour by filtration.

Pearson's correlation coefficient was calculated to evaluate the relationship between apparent and true colour. The correlation value of 0.81 indicates a strong linear dependence between the colour values. However, this dependence is not 100%, as confirmed by the dispersed values of the proportional colour reduction described above.

## Conclusion

Measurements and analyses of pond water samples were carried out using a spectrophotometer. The results of the study showed that filtration can significantly affect water colour, and a clear difference between apparent and true water colour was observed.

In this study, a ratio was used to express the relationship between apparent and true colour. Due to the observation of the ratio value and correlation coefficient, it was concluded that it was not possible to find a mathematical conversion to express the true colour using the apparent colour. Therefore, it is necessary to make the determination of the true colour according to the standard and it is not possible to omit the filtering step and use only the apparent colour and its conversion.

## References

1. P. Pitter. Hydrochemistry. (2015)
2. ČSN EN ISO 7887. International standard: Water quality – Examination and determination of colour (2011)
3. L. Bennett, M. Drikas. The evaluation of colour in natural waters. Water Research - WATER RES. **27**. 1209-1218. (1993)
4. G. Mitchell, A. Mcdonald. Relationship between Different Methods of Colour Measurement in Potable Water. Measurement and Control. **24**. 308-309 (1991).
5. T. Kubínová, M. Kyncl. Comparison of standard methods for determining the colour of water in several countries. IOP Conference Series: Earth and Environmental Science. **900.** (2021).
6. APHA. Standard Methods for the Examination of Water and Wastewater. **23.** (2017)
7. E. C. Wert, F. L. Rosario-Ortiz, S. A. Snyder. Using ultraviolet absorbance and colour to assess pharmaceutical oxidation during ozonation of wastewater. Environmental science & technology. **43.** 4858-4863. (2009)
8. K. Sutherland, G. Chase. Filters and filtration handbook. **5.** (2011)
9. D. Hongve, G. Åkesson. Spectrophotometric determination of water colour in Hazen units. Water Research. **30.** (1996)
10. S. X. Campos. Colour removal by coagulation, flocculation and sedimentation from water containing humic substances with different apparent molecular sizes. Journal of Water Supply: Research and Technology. **56.** 327 – 333. (2007)

# Reliability of the grouting of tendons under consideration of uncertainties under on-site conditions

*Vlastimil* Muron[1*], and *Ladislav* Klusacek[1]

[1]University of technology Brno, Faculty of civil engineering, Institute of Concrete and Masonry Structures, Veveri 331/95, 602 00 Brno, Czech Republic

**Abstract.** This paper contributes to the body of knowledge from German research work Deutscher Ausschuss für Stahlbeton and from author's experiences, to improve the reliability of cable ducts grouting. The complete and reliable filling of post-tensioned tendons with grout is necessary to protect the prestressing steel against corrosion and for bond between prestressing steel and the concrete structure, which is in both limit states applied. Diagnostics of bridges built in the last century showed the cases of faulty grouting of cable ducts. These faults are cause of corrosion prestressing steel often. Based on these diagnostics, bridges are after 35 to 70 years of service life demolished, although service life of these bridges was to 100 years declared. In the studied research work were investigated the effects of individual components, procedures and conditions of the grout and results of grouting investigated. Tests have been performed both on the experiment cable ducts and the real structures. Based on this paper, certain principles have been recommended that aim to prevent grouting faults and to improve the reliability of the grouting. In these recommendations are suggestions of controls of grouting included.

## 1 Introduction

Injection of cable ducts with grout is very important process when constructing construction from post-tensioned concrete. Any faults during grouting are unacceptable, as the load-bearing capacity, functionality and consequently the lifetime of the structure would not be ensured. This paper responds to increasing of faults during the grouting of wire prestressing cables, which were used in beams KA-61, KA-67, KA-73 and I-62, I-67, I-73, IS-73 from 1950 to 1993 in Czechoslovakia. These faults are cause of corrosion prestressing steel often. This corrosion is closely monitored and then these constructions are demolished after 35 to 70 years of service life, although service life of these constructions was to 100 years declared. Principles of grouting have not changed, and risk of faults persists. It is important to know these risks and do everything to avoid these faults. The purpose of this report is to draw attention to possible risks and consequences bad grouting cable ducts, repeat basic knowledge

* Corresponding author: muron.v@fce.vutbr.cz

of grouting from design of cable duct diameter, mixing mortar to injection, and indicate possible directions for increasing the reliability of injection works, including possible control.

## 2 Risk of bad grouting

Grout ensures permanent anti-corrosion protection of prestressing steel and bond between the prestressing steel and concrete structure, which is in both limit states applied. Although the knowledge of bad grouting increased, diagnostics of existing construction repeatedly show faults of grouting. Badly grouted cable ducts are often a source of corrosion of the prestressing steel because the prestressing steel is not protected in any way against corrosive substances. These substances entering to prestressing steel mainly from the area of anchors, through bad expansion joints, see Fig. 1. Reactions of bridge managers on bad expansion joints are often delayed. Bridge managers or bridge owners can act after recommendations of bridge inspections according to standard ČSN 73 6221. These inspections do not provoke an appropriate reaction in stage of construction condition from 2 to 5. Constructions are often operated for several years to decades. In this time rainwater or water saturated with chlorides (from road salt) penetrates to anchor.

**Fig. 1.** Demonstration of corrosion of the prestressing steel in the anchors, due to a bad bridge joint.

Bad grouting can be caused by e.g., inappropriate properties of the grout, impermeability of the cable duct, inappropriate arrangement of drain and vent, insufficient knowledge of the workers, unfavourable conditions during the grouting, insufficient grouting technology, etc. To this day, there are no sensors for performing quality of grouting and status of grouting during lifetime of the structure. Checking whether the cable duct is filled with grout is crucial for a reliable statement about intactness prestressing steel. Such a statement is only possible using a semi-destructive method. The procedure for this implementation of the method is as follows:

1. Diagnostics of the position of the prestressing steel - localization of the prestressing steel in the longitudinal and transverse direction, from the project documentation and on site using non-destructive methods, see Fig. 3.

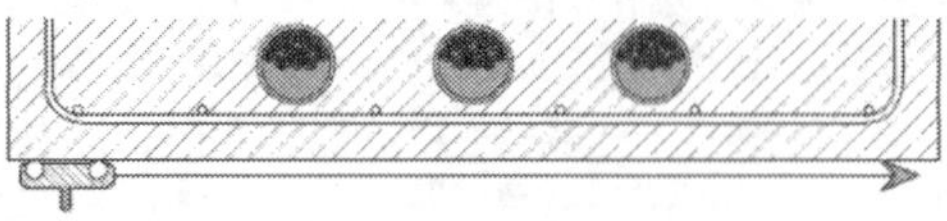

**Fig. 2.** Diagnostics of the position of the prestressing steel.

2. Probe – drilling of the cable duct, first with a drill of smaller diameter (stabilizing hole) and then drilling with a larger diameter, e.g., with a drill with a diameter of 40 mm, see Fig. 3.

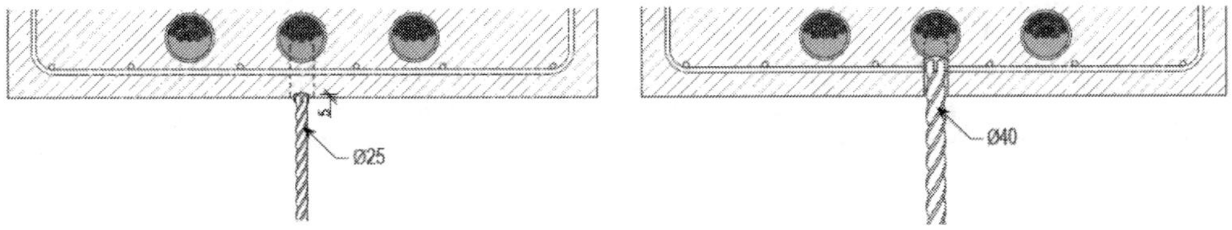


**Fig. 3.** Drilling the probe.

3. Inspection of the grout – visual inspection of the condition of the grouting of the cable duct and the condition of the grout, see Fig. 4.

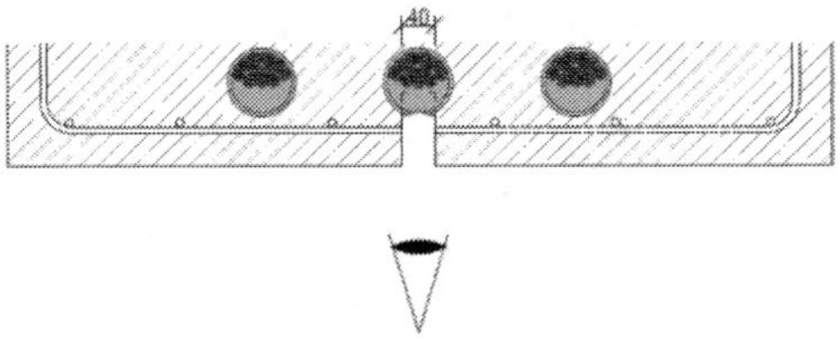


**Fig. 4.** Inspection of the grout.

4. Extending the probe – drilling the probe through the grout up to the prestressing steel, possible mechanical loosening of the grouting, see Fig. 5.

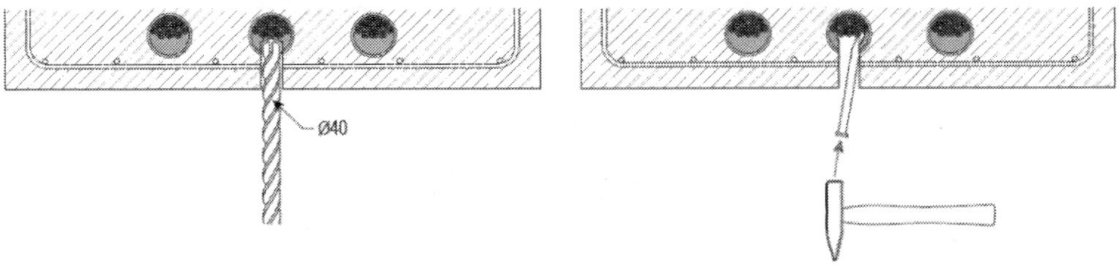


**Fig. 5.** Extending the probe.

5. Inspection of prestressing steel – visual inspection of the state of the prestressing steel and mechanical inspection of the state of tension (LFT = lever force test), see Fig. 6.

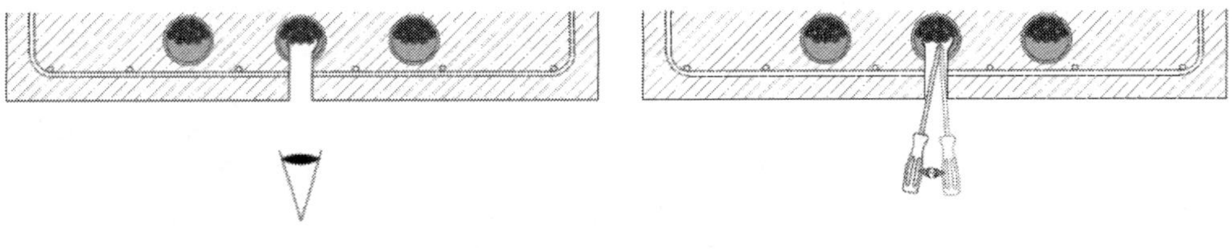

**Fig. 6.** Inspection of prestressing steel.

6. Filling - cleaning the walls of the borehole, filling the drill hole, see Fig. 7.

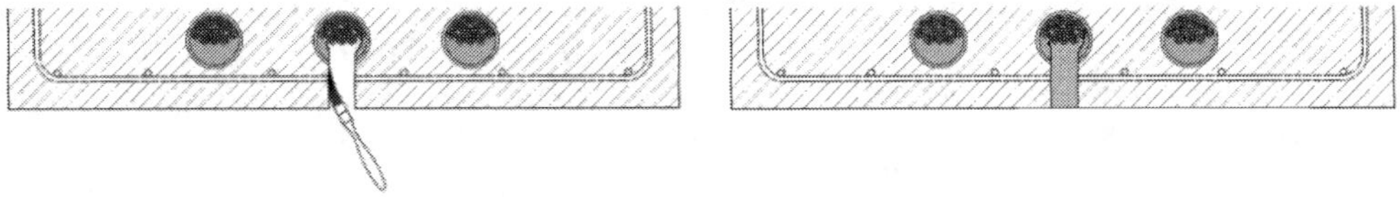

**Fig. 7.** Filling the drill hole.

The possible results of the semi-destructive method are shown in the Table 1. The difference between correctly and incorrectly injected cable channels is shown in Fig. 8. Other examples of poor grouting are shown in Fig. 9 and Fig. 10.

**Table 1** Possible results of the semi-destructive method.

| | *Filling cable duct with grout* | *Grout condition* | *Composition of water* | *Condition of prestressing steel* | *Condition of construction* | *Suggested method of repair* |
|---|---|---|---|---|---|---|
| **Cable duct** | Fully grouted | Dry grout | Without water | Prestressing steel without corrosion | It fulfills its function | No measures |
| | | Wet grout | Water without chlorides | Prestressing steel without corrosion | It fulfills its function | Waterproofing repair |
| | | | | Prestressing steel with corrosion | According to the degree of corrosion | According to the degree of corrosion |
| | | | Water with chlorides | Prestressing steel with corrosion | Critical condition | According to the overall condition of the structure |
| | Partly grouted | Dry grout | Without water | Prestressing steel without corrosion | It fulfills its function | Cable duct reinjection |
| | | | Without water | Prestressing steel with corrosion | According to the degree of corrosion | According to the degree of corrosion |
| | | Wet grout | Water without chlorides | Prestressing steel without corrosion | It fulfills its function | Cable duct reinjection + waterproofing repair |
| | | | | Prestressing steel with corrosion | According to the degree of corrosion | According to the degree of corrosion |
| | | | Water with chlorides | Prestressing steel with corrosion | Critical condition | According to the overall condition of the structure |
| | Nongrouted | Dry grout | Without water | Prestressing steel without corrosion | It fulfills its function | Cable duct reinjection |
| | | | Without water | Prestressing steel with corrosion | According to the degree of corrosion | According to the degree of corrosion |
| | | Wet grout | Water without chlorides | Prestressing steel without corrosion | It fulfills its function | Cable duct reinjection + waterproofing repair |
| | | | | Prestressing steel with corrosion | According to the degree of corrosion | According to the degree of corrosion |
| | | | Water with chlorides | Prestressing steel with corrosion | Critical condition | According to the overall condition of the structure |

In the Fig. 8 (left) the grouting was performed incorrectly. Individual wires are not covered with grout and corrosion is obvious here. Also using LFT (lever force test), any tension stress was not found in the individual wires. The wires were loose in the cable duct and could be moved freely. The parallelism of individual wires is not evident here. Fig. 8 (right) is an example of a properly grouted cable duct. The individual wires are completely covered with grout and the corrosion of the prestressing steel is not visible here.

In Fig. 9 (left) a partially grouted cable duct can be seen. The individual ropes are partially covered with grout. Partial corrosion due to poor grouting can be seen here. Fig. 9 (right) is almost a deterrent case. The cable duct is without grout and the individual ropes lie in the water. One of the causes is poorly made hydro isolation, and therefore water can flow up to the prestressing steel.

Lumpy grout can be seen in Fig. 10 (left), this can be caused by e.g., a bad mixing technique or a wrongly chosen plasticizer. In Fig. 10 (right) completely wet grout can be seen. In this case, it leaks somewhere into the structure and the amount of water in the cable duct is as large as in Fig. 9 (right) but here the cable duct is filled with grout.

**Fig. 8**. Poorly (left) and well (right) grouted cable duct.

**Fig. 9.** Partially grouted cable duct (left) and water-filled cable duct (right).

**Fig. 10.** Lumpy grout (left) and wet grout (right) in the cable duct.

Current knowledge about the grouting is based on extensive experiments of grout, grouting of cable ducts and various experiences. Grout represents a significant factor of uncertainty in the grouting of cable duct. One of the reasons is the fact that the grout must be often prepared under unfavourable on-site conditions. The properties of the grout strongly depend on the chemical-physical properties of cement and the environmental conditions during the grouting that affect them. This paper contributes knowledge, a selection of knowledge, experiments, and experiences both from German literature [1] and from the authors' own experience and knowledge and recommendations of Czech standards [2], [3] and [4] to improve the future reliability of cable duct grouting.

## 3 Description of the current state of problematic

Characteristics of grout depend on the used materials (cement, water coefficient w/c and used admixtures or additions), mixing techniques and injections techniques and mainly conditions during the grouting (temperature, processing time), see Fig. 11. As the length of the cable and the amount of prestressing steel increases, so does the volume of grout and possible of any failures. For example, water can be drained away between individual strands (bleeding), in this case cable duct can be blocked. Furthermore, if there is a too many prestressing strands in the cable duct, this makes it difficult for the grout to flow transversely between the individual strands, so that the individual strands do not have to be coated on all sides with grout. Therefore, requirements must be placed on the grout for its fluidity, prevention of segregation and limitation of bleeding [4]. Grouting is complex process into which a lot of unknowns enter.

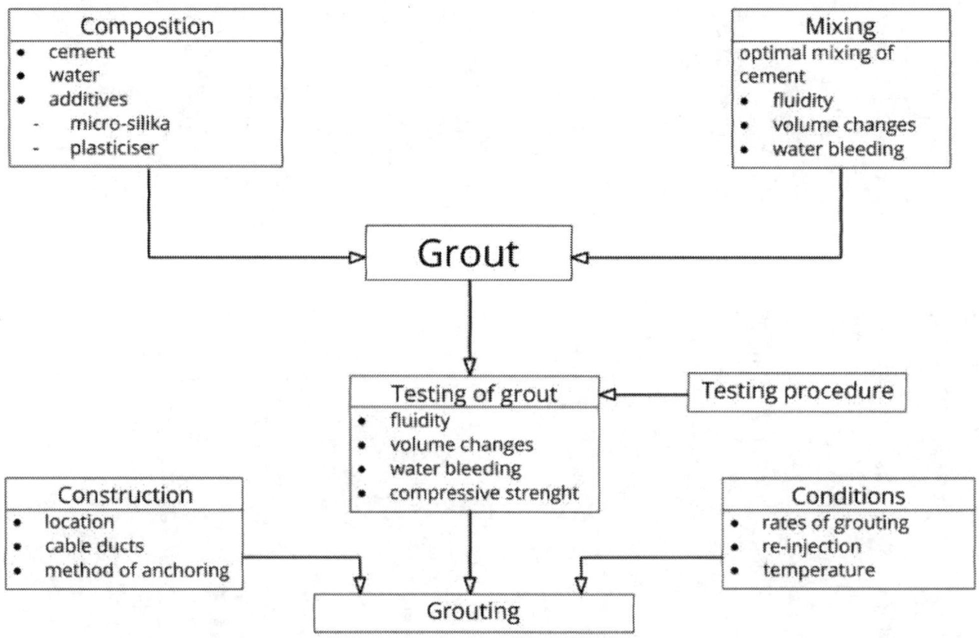


**Fig. 11.** Elements of grouting of cable duct and their interrelationship [1].

Grout mainly consists of cement and water. According to the standard [4], the water coefficient (w/c) must be maximum of 0.40. Water coefficient is decisive for the properties of fresh and hardened grout. It must be confined at the top to minimize bleeding. This also ensures certain strength requirements. If the w/c ratio is too high, water can accumulate in the top of cable duct and then result in spacing in the grout. However, the w/c ratio cannot be reduced arbitrarily, as fluidity is limited. Additional additives or admixtures can be added to grout to improve the properties of grout, especially workability. These are mainly plasticisers or micro-silika.

A mixer and injection pump are required for the grouting itself. The mixing process is decisive for the properties of the grout (homogeneity, fluidity). The intensity of mixing is a basic prerequisite to produce homogeneous mixture. The correct grouting procedure must be selected (injection pressure, injection rate, injection procedure in the structure), the location of vents and drains, and the workers must be properly trained.

The quality of the grouting itself can also be influenced by the structural engineer when designing the structure. The ratio of the areas of prestressing steel and the cable duct is very important. If this ratio were too large, there would be no space for the grouting itself and thus proper grouting of the cable duct is not guaranteed. The probability of blockage the cable duct is much bigger. In addition, there can be a bleeding, where water is drained away through the individual strands. On the other side, with a decreasing ratio, the grouted cross-section increases, and this promotes sedimentation, which can create cavities from rising air bubbles and separated water. The optimal ratio resulted from many years of experience in the range of 0.20 to 0.45 [5]. Another important aspect in the design is the location of the cable duct in the structure. If the cable duct were too close to the surface and the volume changes were large, the structure could be damaged.

A very important factor is the conditions during the grouting. We must monitor the temperatures in the structure, the grout and in the air during the grouting. According to the standard [3], the temperature must be between +5 and +35 °C, during the grouting and in the following 48 hours. The reasons will be explained later.

Another monitored variable is the injection pressure, which must not exceed 20 bars, primarily to ensure the safety of the workers. It is important to monitor the injection speed. The maximum speed is usually given by the injection device, but speeds above 30m/min lead to segregation of the grout. Speeds of 5 to 15 m/min are recommended.

# 4 Literature review

As already mentioned, some experiments, procedures and results presented in this chapter are taken from German literature [1] and the aim is to pass this information on.

## 4.1 Range of experiments

A total of 14 types of CEM I cement from different manufacturers were used for the experiments according to the literature. This type of cement is preferred in practice for grouting. All cements were subjected to chemical analysis. First were produce and tested grout with a standard formula (w/c = 0.38, T = 20 °C). Then had some selected grouts their formula modified, for example by adding water, micro-silica, or plasticiser. In another series of experiments, the properties of the grout were tested at different temperatures or using different mixing equipment. Grout with ultrafine cements were tested also.

The grout tests were carried out according to the then valid standards. Fluidity, volume changes, compressive strength and bleeding were mainly determined. Fluidity was verified using a dip test as well as a funnel flow test. Volume changes and bleeding were determined on a measuring cylinder with a volume of 1000 ml and a height of 950 mm. Compressive strengths were tested on typical beams. One of the results of the tests was the suitability of the test methods under on-site conditions, this topic is not presented in any way.

Injection experiments were also performed on a model cable without prestressing steel and with prestressing steel. The model was formed in a top arch with a plan length of 15.0 m and a top height of 1.40 m. Vents were located at the top and 1.5 m to the left and right of the top. Various types of material for cable ducts (steel, PVC, PE) were also tested. The shape of the grout during grouting was studied here.

### 4.2 Results of experiments

First, the connection between ingested cement and the properties of fresh grout was sought. The cements used were therefore investigated about their chemical and physical properties. From the results, it could be said that the fluidity decreased both with increasing $C_2S$ (belite), see Fig. 12 (left), and alkali content. Also, with increasing heat of hydration, the fluidity of the grout decreased. When choosing cement, it should be considered that the mentioned properties interact with each other, and the individual effects could add up or interfere with each other. Cements with good fluidity are suitable for grouting, and bleeding can be minimized by a suitable proportion of fine particles. Slow-reacting Portland cements should be preferred. Therefore, the content of $C_3A$ (characterized by very rapid hydration immediately after mixing) in the cement should be reduced, or the reactivity of the cement should be reduced.

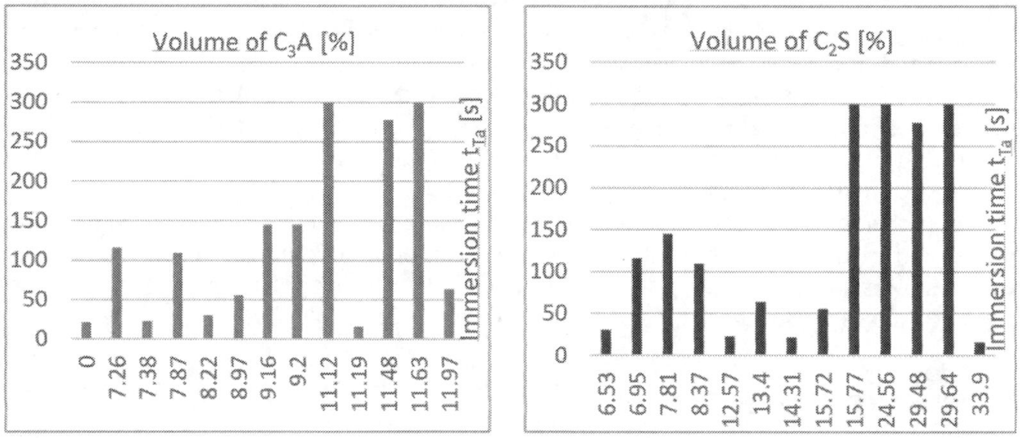


**Fig. 12.** Liquidity depending on the proportion of $C_3A$ (left) and Liquidity depending on the proportion of $C_2S$ (right) [1].

Furthermore, tests were carried out on grouts with the addition of additives and admixtures. The results showed that the plasticiser can caused a loss of compressive strength of the grout (the strength values still met the minimum permissible values), but the processing time of the grout was significantly increased, see Fig. 13 (left), as the consistency of the grout was more fluid. Although the figure shows fluidity as immersion time and this test is not used here, a shorter immersion time indicates a more fluid mixture. A more liquid mixture, due to the addition of a plasticiser, enabled a reduction in the water coefficient and thus a better increase in the strength of the grout. With the addition of micro-silica to the mixture, the fluidity of the mixture increased, and the bleeding decreased, see Fig. 13 (right).

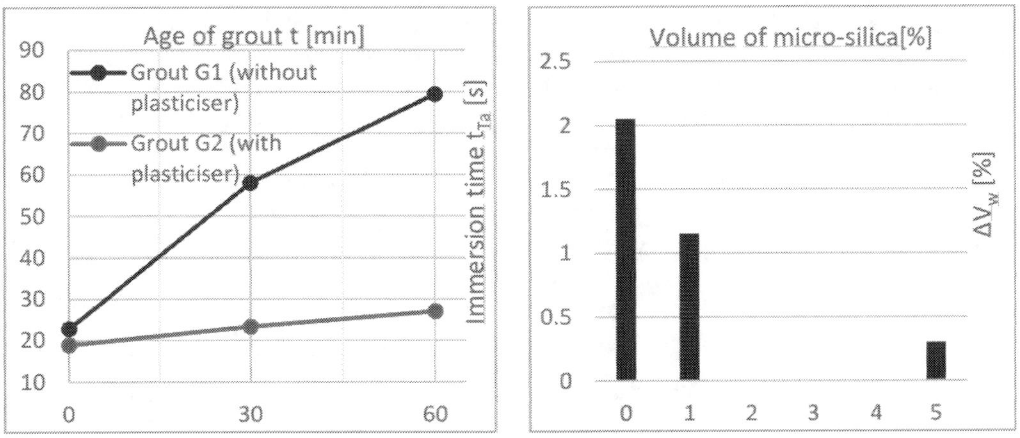


**Fig. 13.** Comparison of flowability of grout with and without plasticiser (left) and bleeding depending on the volume of micro-silica (right) [1].

Ultrafine cement was also considered, but the grout was usually too liquid, and the compressive strength was below the minimum permissible value. Positive features were small volume changes and almost zero bleeding. Ultrafine cement is therefore unsuitable for grouting cable ducts but has found application in remediation or re-injection of cable ducts.

From other test results, it was evident that the volume changes did not depend on the cement but were to some extent determined by the swelling effect of the additive. In addition, the mixing technique and the ambient temperature of the grout influenced the volume change in the first hours of setting. As the temperature increased, so did the volume change. Volume changes also increased with high mixing energy. Furthermore, it follows from the conducted tests that bleeding depended on the cement, the mixing principle, and the temperature of the surrounding environment during the setting of the grout. Bleeding was also investigated under elevated pressure, but the results were not significantly different from conventional tests. The test results show that the volume change of the grout was completed about 3 hours after mixing, see Fig. 14. And as can be seen from this figure, bleeding took place at approximately the same time. Bleeding was completed between 3 and 4 hours after mixing, with only a small amount of water being separated from the third to the fourth hour. After 24 hours, the water was reabsorbed by the grout, which is the cause of the voids in the grout.

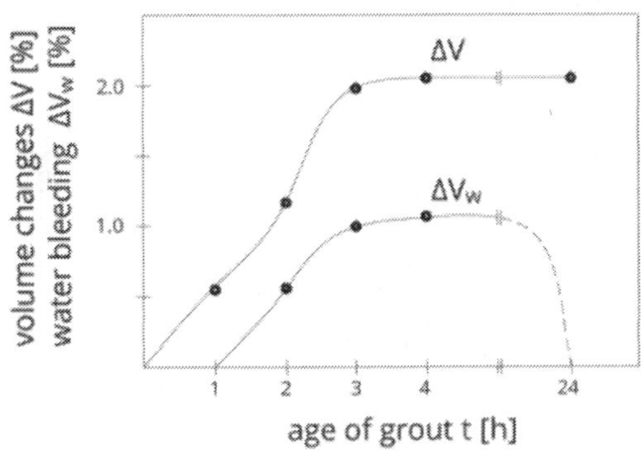


**Fig. 14.** Volume changes and bleeding of grout [1].

Investigation of the effect of low and high temperatures was the subject of further tests, see Fig. 15. The results showed that the fluidity of the mortar decreased with increasing temperature. Furthermore, it was found that volume changes increased with increasing temperature for all investigated grouts. On the other hand, bleeding increased with decreasing temperature, with solid grouts bleeding was significantly lower than with liquid grouts. A possible reason for bigger bleeding at low temperatures is that the reaction between cement and water is slower. Thanks to this, water does not bind to the cement grains as quickly and excess water can settle.

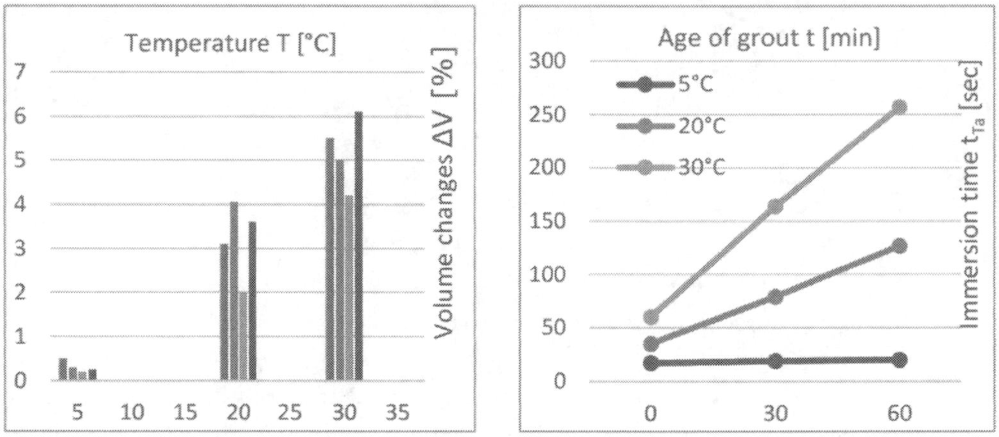


**Fig. 15.** Volume changes depending on the temperature (left) and Fluidity depending on temperature (right) [1].

Furthermore, the shape of the grout during the grouting was studied on the model cable, see Fig. 16. The grouting took place from the bottom upwards. Grout moved smoothly upwards. In the case of grouts with low viscosity, the level was almost horizontal, if the viscosity was higher, the face of the grout was slightly convex. Once the top was passed, the thin grout flowed downwards and the section behind the top was gradually filled upwards and the air escaped through the vent in the top and behind the top. This can cause air voids in the grout. As the stiffness of the grout increased, the level of the grout in the descending part was no longer horizontal. The face of the grout kept its shape, and the protector was filled with grout from top to bottom. In this case, the vent at and near the top was filled with grout and air could only escape at the end of the cable duct. This ensured the correct grouting of the cable duct with grout without the formation of air cavities. The advantage was also the smaller area on the face of the grout, where residual water could mix with the grout. The grout thus remained more homogeneous.

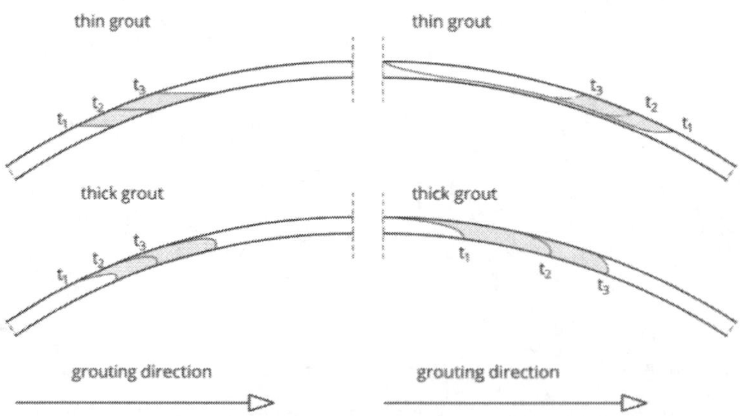


**Fig. 16.** Flow conditions of grout in the rising and falling cable duct depending on the consistency of the grout [1].

Another thing that was investigated in this test was bleeding and volume changes, see Fig. 17. It was noticeable that the separation of water was usually about 4 hours after grouting. Water collected in the ridge of the cable duct, in the peaks and specific points prone to the formation of cavities (expansions, areas behind the anchor, etc.) from where it could not escape anywhere. After 24 hours, the accumulated water completely disappeared or was absorbed by the grout and thus a cavity was created in the cable duct. Additional vents should therefore be placed at these points. Volume changes must also be paid attention to, because when combined with an inappropriate location of the cable duct in the structure (too close to the surface), tearing of the structure may occur.

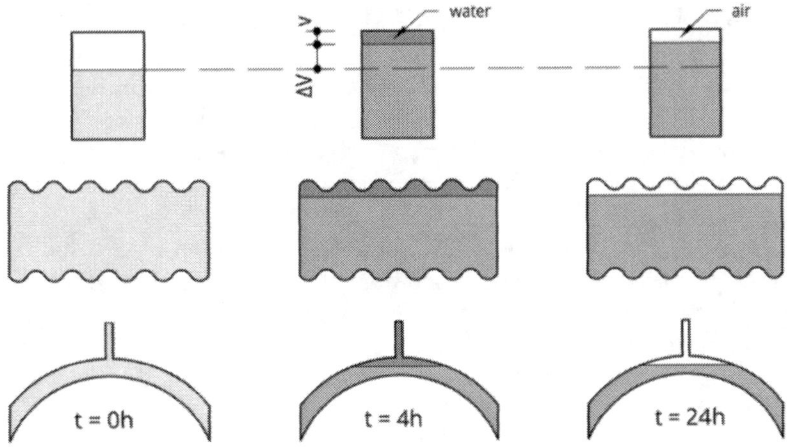


**Fig. 17.** The formation of voids due to the water bleeding during solidification [1].

## 5 Conclusion

The reliable grouting of prestressing steel in cable ducts with grout depends on many factors, some of which interact with each other. The presented results show that the properties of the mortar strongly depend on the cement and its properties. Based on experiments, it was possible to determine to a certain extent the connection between the physic-chemical properties of cement and the properties of fresh grout. The properties of the grout depended not only on the starting materials, but also on the mixing process and the temperature.

Grouting tests have shown that in grout with a strong tendency to sedimentation, voids are formed due to the bleeding, therefore it is important to pay attention to the separation of water in grout. It was also found that the bleeding takes place approximately 3 hours after mixing the mixture and the volume change is almost at its maximum at this time. During the next 24 hours, water is absorbed by the grout, creating voids.

It is not so easy to say what the optimal water factor is. The final properties pretty much depend on a lot of things. According to [4], the value of the water coefficient must be less than 0.4 and it must also meet the requirements according to [2]. From the authors' own experience, the given requirements are met when using a water coefficient of 0.32 and the corresponding amount of plasticizer.

Most of the results were verified and carried out according to the then valid standards. It is certainly necessary to carry out similar tests again and adapt the conclusions to today's requirements and trends. Special attention should also be paid to the still undeveloped diagnostic methods of grouting reliability.

## 6 Recommendation

From the above, the following principles can be recommended:

- When designing the structure, the designer should ensure that the ratio of the areas of the prestressing steel and the cable duct ($A_p/A_k$) is in the range of 0.2 to 0.45, otherwise there is a risk of defects during the grouting of the cable duct.
- According to [4], the value of the water coefficient (water/cement ratio) must not be less than 0.4. It is not recommended to choose a water coefficient that is too low, as the fluidity of the grout would not be guaranteed, and the defects mentioned above could arise. The optimal water coefficient is in the range of 0.3 to 0.4.
- Use a previously proven cement (CEM I) to ensure compliance with the grouting tests according to [2]. It is recommended to choose cement with a low content of $C_2S$, $C_3A$ and alkali to ensure fluidity and good workability of the grout.
- It is possible to add a plasticiser to the grout formula to reduce the viscosity of the grout and delay setting, which leads to a longer workability of the grout. Care must be taken to ensure that the volume changes and strength of the grout meet the conditions in [2].
- Micro-silica can be added to the formula, as it helps reduce bleeding. This is particularly suitable in colder conditions when bleeding is higher due to the low temperature. The recommended volume of micro-silica is 3 to 4% of the grout.
- It is necessary to consider the correct execution of the grouting. The grouting itself must meet the requirements of standards [3] and [4], and the grout must meet the requirements of standard [2].

To verify the correctness of the performed grouting, it is desirable to develop and implement certain diagnostic elements, such as:

- Introduction of sensors into critical cross-sections, to verify the filling of cable ducts with grout. This is especially the area behind the top of the arch, the area in front of the anchor, extensions, etc.
- Placement of grout moisture sensors in the cable channel. These sensors can warn of excessive moisture in the grout, which could indicate that it is leaking into the structure.

The authors gratefully acknowledge the financial support from the state budget through TA ČR as a part of the project No. CK01000042 "Determining the residual load capacity of prestressed bridges".

## References

1. F.S. Rostásy, A.W. Gutsch, Reliability of the frouting of tensons under consideration of uncertainties under on-site conditions (1997)
2. EN 445:2007. Grout for prestressing tendons - Test methods (2007)
3. EN 446:2007. Grout for prestressing tendons - Grouting procedures (2008)
4. EN 447:2007. Grout for prestressing tendons - Basic requirements (2008)
5. T. Telford, Grouting of tendons in prestressed concrete (1990)

# Post- realization phase of an interior green wall: a case study

*Ivana* Halaszova[1*], *Renata* Baskova[1]

[1]Institute of Construction Technology, Economic and Management, Faculty of Civil Engineering, Technical University of Kosice, Vysokoskolska 4, 042 00 Kosice, Slovakia

**Abstract.** One of the popular modern ways to bring greenery back into interiors are interior green walls, which are also known as vertical gardens. Along with the increase in awareness of green walls, their technical solutions are also increasing and their innovation is progressing rapidly. In recent years, green walls have rapidly moved from the simplest walls with the need for mechanical service to maintenance-free walls connected to the digital environment using various sensors. The contribution is focused on the inputs of monitoring in the post-implementation phase of the interior green wall located on the premises of the Technical University in Košice. The goal of the monitoring is to define the inputs for the transition to digitized maintenance of the green wall. The article analyses of the inputs of 3-month monitoring of a green wall with simple manual maintenance. At the end of the paper, the requirements for setting up a digital device ensuring suitable conditions for the sustainability of green construction are specified.

## 1 Introduction

Green walls, also known as vertical gardens [1,2], or vertical greenery [3] are important for several reasons. One of the benefits of green walls is that they improve air quality, as plants absorb carbon dioxide as well as other pollutants and release oxygen into the air. This reduces the amount of toxins inside and outside the building.

Green walls also have a positive effect on psychological health and well-being. Since plants have a calming effect, they can help reduce stress and improve mood and overall quality of life. A lot of research is devoted to the impact of green walls on improving the quality of the environment, which focuses on green walls from different points of view. For example, from the point of view of reducing the concentration of $CO^2$ in the space [4], from the point of view of reducing the cooling load in the building [5,6], from the sociological point of view of the perception of these elements by the users of the building [7,8]. In the case of exterior green components, we also come across the impact on the urban structure in the publications.

* Corresponding author: ivana.halaszova@tuke.sk

Green walls represent a living system that needs to be taken care of during the entire life cycle, from the construction design, through implementation, the phase of use and the end of the service life. The main phase of the life cycle is the use phase, which in the long run represents the longest part. For the longest possible use of the green wall, it is necessary to maintain it as effectively as possible. Green wall technology offers different options for the level of maintenance, from manual systems to fully self-maintenance systems. Regardless of the system by which the green wall is procured, the element should be tracked, maintained, requirements optimized and addressed in a timely manner.

When transitioning from a manual maintenance method to a mechanical one, it is necessary to know how to collect data correctly and forward it to the equipment for evaluating the requirements of the green wall. This article, which is dedicated to a specific green wall in the interior, is focused on data collection during 12 weeks of operation after the start of the use phase. It also records errors arising during the adaptation of the wall in the post-realization phase.

## 2 Green wall construction

The green wall, located in the building of the Faculty of Architecture of the Technical University in Košice, is one of the outputs of the project "Improving the parameters of the internal environment by adapting a green wall to the classroom space", financed from the funds of a research grant for young scientists at TUKE in 2021. The first part of the project was aimed at determining the state of selected air quality parameters of the indoor environment in the addressed room (temperature in the room, relative humidity, radiation temperature, operational temperature, speed of air movement, concentration of solid particles PM2.5 and PM10, volatile organic substances, CO2 concentration, temperature and emotional well-being, noise intensity) before the installation of the green wall. After installing the green wall in the room, the same measurements should be taken in the future. After the measurements have been completed, both detected conditions should be evaluated and compared. The goal of the second part of the project was the actual production and application of a green wall prototype with a total size of 2.4m x 2.4m. to the assessed room no. 227 at the TUKE Faculty of Civil Engineering, Vysokoškolská 4, Košice.

The construction of the green wall, together with the substrate and plants, was designed and implemented by the company Lusaro s. r. about. The wall is assembled from prefabricated plastic panels with flower pots for placing the substrate and greenery. The company also developed a proposal for artificial wall lighting, as the daylight in the room was not sufficient in terms of light requirements for the growth of planted plants. **Figure 1** shows the design of the green wall in question.

**Fig. 1** The construction of the solved green wall (source:Authors).

The construction system of the green wall prototype was designed from 16 pieces of prefabricated plastic wall panels with a design structure. Individual panels measuring 600x600 mm connect to each other (both horizontally and vertically). Each panel is anchored to the wall with a pair of anchors in the upper part of the panel. Plastic flower pots with a width of 600 mm are anchored to the wall separately using anchors and are arranged in 4 horizontal rows. The wall includes 12 plastic flower pots with a drainage system for water overflow into the lower rows and 4 flower pots in the lower row with a full bottom. Artificial lighting of the wall 2 pcs of lamps, attached to the ceiling at a distance of 2 m from each other, approx. 1 m from the green wall, is set for 8 hours a day and is controlled by a time switch.

For planting plants in interior green walls, fewer types of plants with different creeping and climbing properties are recommended, which would contribute to the integrity of the wall in the long term. For the prototype of the green wall, living plants suitable for the interior with the assumption of climbing were selected. The bottom row is made up of fern species, the remaining rows are made up of three types of plants, namely Golden Potosovec, Syngonuium peduncle and Asparagus Sprengerov.

## 2.1 Implementation of a green wall

The implementation of the green wall took place in November 2022. Before the installation of the wall, renovation work on the floor and walls was completed in the room, and preliminary preparations for future lighting were made. For financial reasons, a new door is not installed in the room at the moment. The implementation procedure is documented in **Fig. 2.**

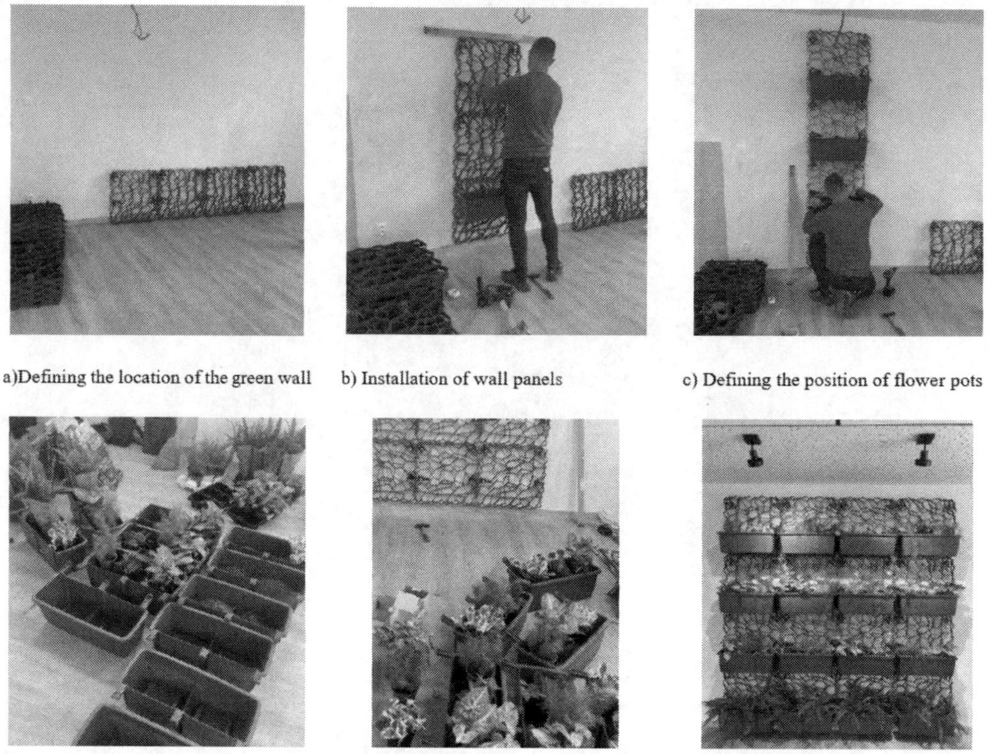

a)Defining the location of the green wall b) Installation of wall panels c) Defining the position of flower pots

d) Plant and substrate preparation e) Instalation of flower pots f)Setting the lighting conditions and finishing works

**Fig. 2** Procedure for the implementation of the green wall (source:Authors).

The process of implementing the green wall consisted of the following steps:

- ***Defining the location of the green wall -*** Before starting the implementation of the green wall, it was necessary to visually inspect the underlying structure. Subsequently, the exact position of the future wall structure was determined in relation to the location of the cabling for the artificial green lighting.
- ***Installation of wall panels*** - When installing the plastic wall panels, proceed vertically from the left. Each panel had to be carefully checked to see if it was in a horizontal position. Subsequently, the individual panels were anchored to the wall, creating a solid and integral surface with a design structure.
- ***Defining the position of flower pots*** - In the next step, the height level of all rows of flowerpots was defined. The first reference column of flower pots was established, according to which other lines were established. In this step, the flower pots were not yet firmly anchored to the construction of the wall surface.
- ***Plant and substrate preparation -*** The substrate was placed in the pots and the plants were planted. The bottom row of pots was planted with only one type of greenery, which required the least amount of lighting. The other three rows had a combination of plants planted to form a decorative part of the green wall.

- ***Instalation of flower pots*** - The pre-prepared planted pots were anchored to the wall panel in this step. For the joint, anchors were used with anchoring at two points of the flower pot, which firmly connected them to the wall panel system.
- ***Setting the lighting conditions and finishing works*** - The last step was the installation of artificial lighting with the setting of optimal lighting conditions for the growth of planted greenery. Finally, the floor of the room was cleaned and the wall construction and the horizontality of the entire system were checked.

The total implementation time was 2 days, while the green wall construction was installed on the first day, the wall lighting was installed on the second day and the system was checked.

## 3 Maintenance and monitoring of the green wall

Since the green wall does not have a self-maintenance system installed, the greenery of the wall requires manual maintenance on a weekly basis. The main part of maintenance is watering the substrate with plants. The weekly care of the green wall also includes simple tasks, such as collecting dead plant parts and checking the lighting.

The post contains partial conclusions that emerged from the 12-week maintenance and mapping of the state of the green wall

**The goal of the monitoring** is to map the selected parameters of the green wall as one of the inputs necessary for the analysis and subsequent design of the parameters of the digital solution for the self-maintenance system of the green wall. An equally important goal is to gain personal experience in the requirements and costs of maintaining a green structure.

During the maintenance of the green wall in its post-realization phase, a measuring device was installed and the following **monitoring methodology** was chosen:

1. At the beginning of the post-implementation phase, the expected amount of irrigation was set to 15-20 l applied manually once a week. The amount of water for one flower pot was approx. 0.9-1.25 l per week.
2. Monitoring was carried out at the time of watering/maintenance with a weekly frequency, the monitored parameters included:
   - maintenance time
   - amount of water used for irrigation
   - air temperature in the place of construction of the green wall (before and after watering)
   - relative humidity at the site of the green wall construction (before and after watering)
   - notes (maintenance performed, error and deficiencies found,...)
3. Photo documentation of the state of planted plants

An instrument for measuring the relative humidity of the air and a thermometer installed directly on the construction of the green wall (with sufficient distance from the axis of the living component) were used to monitor the parameters. In addition to the recording of monitored parameters, a photo of the prosperity of the live components was taken every week, detected errors and deficiencies were recorded and documented.

**Partial results from the 12-week monitoring** of the green wall are processed in an overview table, in **Fig. 3**. For detected errors, documented Fig. 4, measures for their elimination were continuously proposed and implemented.

| Year | Month | Date | Time | Amount of water [l] | Air Temperature $T_0$ [°C] | Air Temperature $T_1$ [°C] | Relative humidity $\phi_0$ [%] | Relative humidity $\phi_1$ [%] | Note |
|---|---|---|---|---|---|---|---|---|---|
| 2022 | Nov. | 14.11 | Completion of the green wall realization | | | | | | |
| 2022 | Nov. | 21.11 | 10:05 | 19 | 19.5 | 21 | 99 | 96 | |
| 2022 | Nov. | 28.11 | 10:00 | 19 | 19.8 | 20.1 | 98 | 98.9 | |
| 2022 | December | 5.12 | 10:00 | 13,5 | 19.1 | 20 | 94 | 96 | The fern dries up |
| 2022 | December | 12.12 | 9:45 | 12 | 19 | 19.1 | 92 | 96 | |
| 2022 | December | 19.12 | 9:59 | 15 | 18 | 18.5 | 91.5 | 92 | |
| 2022 | December | 26.12 | 9:10 | 15 | 18.5 | 19 | 98 | 99 | *Issue 1 |
| 2023 | January | 2.1 | 8:30 | 16,5 | 18 | 18 | 99 | 99.5 | |
| 2023 | January | 9.1 | 9:30 | 6,5 | 19.1 | 19.8 | 99 | 99.5 | |
| 2023 | January | 16.1 | 12:10 | 5 | 19 | 20 | 5 | 99.5 | |
| 2023 | January | 23.1 | 10:00 | 5 | 18 | 19 | 98 | 99 | |
| 2023 | January | 30.1 | 9:50 | 5 | 18 | 19 | 90 | 91 | *Issue 2 |
| 2023 | Feb | 6.2 | 10:00 | 5 | 19 | 18 | 90 | 91 | Fertilizer |

**Fig. 3** Analysis of the results from the 12-week monitoring of the green wall in the post-implementation phase (Source: Authors).

During the monitoring of the green wall, several errors and deficiencies were discovered that occurred during use in the post-implementation phase of the green wall, e.g. as a result of the overflow, there was a poor prosperity of the fern in the bottom row of pots, which appeared around the third week. Error 1 detected at week 6 was an overflow of the pot and overflowing the system, or error 2 at week 11 was the onset of an increased number of small insects. Photo documentation of the identified deficiencies is in **Fig. 4**.

a) Overflow of the bottom row of the flower pots

b) Worse adaptation of some types of plants

c) Effect of spraying on the underlying construction

d) Increased occurrence of insects in the room

**Fig. 4** Deficiencies detected during the monitoring of the green wall in the post-implementation phase (Source : Authors).

As a result of poorly set irrigation, the system overflowed during the first weeks (Fig. 4 a). The consequence of this spill was, in addition to the worse condition of the fern in the bottom row of flower pots, also the leakage of excess water from the bottom row of flower pots. With the bottom row secured with plugs, the plugs fell out (or overflowed). The set expected amount of water for irrigation had to be adjusted after 7 weeks to an irrigation amount of approx. 5 l per week.

Another error found was the poor adaptation of some types of plants. The prosperity of individual plant species did not proceed equally, some plant species required more frequent maintenance (collection of dead plant parts). The deteriorated condition of the plants, which can be seen in Fig. 4 b), was probably due to insufficient intensity of artificial lighting in the given part of the wall. Due to the larger dimensions of the wall, there was a different intensity of lighting in individual parts. While the plants thrived in the part with proper lighting, the plants thrived worse in the upper row of pots, where the range of lighting was lower.

When fertilizing and spraying the plants, the underlying structure of the green wall got wet (Fig. 4 c). Even if its damage was not obvious at the time of monitoring, in the long term it could mean damage to the surface of the wall or some part of it.

During the monitoring of the green wall, an increased occurrence of small insects was detected (Fig. 4 d), not only in the immediate vicinity of the wall, but in the entire room. The insect death did not damage the structures or the vegetation, but it represented the need for additional work (harvesting, removal, cleaning) during the weekly maintenance. The problem was also that the exact source of the insect infestation of the room had not yet been identified.

## 4 Conclusion

The implementation of the green wall in the premises of the Technical University in Košice represented a short period from the point of view of its life cycle, the main phase is the onset of the post-implementation phase - the use phase. Since it is a living system with living plants, in order to maintain the efficiency of the system, it is important in the long term to focus on setting the correct maintenance of greenery and setting optimal conditions to support the sustainability of the green wall. The investigated green wall currently does not have a self-irrigation system, nor a connection to sensors for a self-maintenance system, so all maintenance must be performed manually.

As it turned out during the monitoring, manual maintenance of the green wall does not exclude the risk of bad maintenance settings.

During the maintenance of the wall in the post-realization phase, several shortcomings were found: worse adaptation of some types of plants, higher incidence of small insects in the room, influence of irrigation on the surface of the underlying wall structure, overflow of the system due to poor setting of the irrigation of the green wall. These errors could be prevented by choosing a different level of maintenance, by adapting sensors for soil moisture, sensors for monitoring the quality of the environment in the room (humidity, temperature, air quality, ...), or even by introducing an irrigation system into the existing green wall.

The green wall has the prerequisite of connection to digital monitoring and control of lighting and irrigation, which would help to better and more efficiently monitor the needs of the living system that is the green wall. Such a connection would not only make the maintenance of the wall more efficient, but it would also bring better adaptation and more stable conditions for living components.

The ongoing monitoring of the green wall can be the basis for further research. The research could be focused on the transition of the manual maintenance system to a self-maintenance system, it could also be interesting to analyze the results of long-term data collection and monitor the impact of the type of maintenance on the green wall. In the future, it is expected that the conditions in the room will change (teaching in the room, completion of the room modernization project and associated new furniture, new filling structures, etc.), which may also cause a change in the requirements for the maintenance of the green wall so that it fulfills its ecological and aesthetic function .

This work was supported by project under the VEGA 1/0336/22 Research on the effects of Lean Production/Lean Construction methods on increasing the efficiency of on-site and off-site construction technologies.

## References

1. Ö.B. Timur and E. Karaca, *Vertical gardens.*, in Advances in Landscape Architecture. In techOpen (2013)
2. E. Ekren, *Advantages and risks of vertical gardens*, in Bartın Orman Fakültesi Dergisi 2017 , pp.51-57 (2017)
3. G. Pérez, J. Coma, I. Martorell, and L.F. Cabeza, *Vertical Greenery Systems (VGS) for energy saving in buildings: A* review, in Renewable and sustainable energy reviews 39, pp.139-165 (2014)
4. F.R. Torpy, M. Zavattaro, and P.J. Irga, *Green wall technology for the phytoremediation of indoor air: a system for the reduction of high CO2 conentrations*, in Air Quality, Atmosphere & Health 10, pp.575-585 (2017)
5. K.C. Dahanayake, and C.L. Chow, *Comparing reduction of building cooling load through green roofs and green walls by EnergyPlus simulations*, in Building Simulation vol.11, pp. 421-434 (2018)
6. J. Li, B. Zheng, W. Shen, Y. Xiang, X. Chen, and Z. Qi, *Cooling and energy-saving performance of different green wall design: A simulation study of a block*, in Energies, pp.2912 (2019)
7. A. Eihady, A. Elhalafawy, and R. Moussa, *Green wall benefits perception according to the users versus experts* views, in International Journal of Engineering Research and Technology, pp.975-3154 (2019)

8. T. Liberalesso, R. Mutevuie Júnior, C. Oliveira Cruz Matos Silva, C. and M. Manso, *Users perceptions of green foors and green walls: An analysis of youth hostels in Lisbon, Portugal*, in Sustainability (2020)

MATEC Web of Conferences **385**, 01006 (2023) https://doi.org/10.1051/matecconf/202338501006
*Young Scientist 2023*

# Seasonal decomposition of different types of rivers in Slovakia: Implications for water management and agriculture

*Almikaeel Wael*[1*] and *Cubanova* Lea[1]

[1]STU, Department of Hydraulic Engineering, Radlinskeho 2766/11 Bratislava, Slovakia

**Abstract.** This study conducts an analysis of the seasonal decomposition of various river types in Slovakia in order to enhance the understanding of hydrological time series analysis and its potential implications for water management and agriculture. A comprehensive statistical methodology including decomposition of time series and correlation analysis was applied to the data collected from multiple river gauging stations across Slovakia. The results of the study indicate distinct seasonal variations in the hydrological characteristics of different river types, with a clear differentiation between lowland and mountain rivers. These findings have crucial implications for water management and agriculture as they suggest that a deeper understanding of seasonal patterns in river flow can lead to more efficient and effective utilization of water resources. Overall, this study highlights the significance of hydrological time series analysis in gaining a better comprehension of water resources and their management in Slovakia.

## 1 Introduction

In Slovakia, the surface runoff regime is characterized by increased spring runoff. Because snow melts slower in mountainous places at higher altitudes, enhanced spring runoff occurs later in the season than in lowland streams [1, 2]. The summer-autumn period in most Slovak streams marks a period of low water content (particularly the months of August to October, which is a crucial period of the growing season). In mountainous areas, the winter low water period is particularly significant (especially the months of December to February) [3]. Snowfall is connected with the low-water winter season, which does not contribute to immediate runoff during periods of low temperatures (below freezing) produced by partial or complete freezing of the stream [4].

Understanding river flow seasonality is critical for sustainable water management and agriculture. Seasonal river flow patterns can have a substantial impact on both the quantity and quality of accessible water resources, with significant impacts on water supply, agriculture, hydropower generation, and other water-dependent industries. Hydrologists and agriculture can better plan [5] and optimize their use of water resources by understanding seasonal variations in river flows, improving productivity, and reducing loss. This is especially essential in locations with limited water resources, where the ability to catch and

* Corresponding author : wael.almikaeel@stuba.sk

store water during high-flow periods can be critical for meeting demand during low-flow periods. Additionally, a deeper understanding of the seasonality component of river flows can help to inform the development of climate adaptation strategies, as changing patterns of precipitation and temperature are likely to have significant impacts on the availability and variability of water resources over time [6].

In many parts of the world, seasonality can play a significant role in escalating drought conditions [7]. Since precipitation patterns are often correlated with water availability, the presence of seasonal cycles can amplify the effects of low rainfall or prolonged drought. The availability of water during times of the year when demand is the greatest, such as during the growing season for agricultural crops, may frequently be reduced due to seasonal differences in rainfall and evapotranspiration [8, 9].

Additionally, seasonality can have a significant role in exacerbated flood disasters. Precipitation patterns are crucial in determining the likelihood and severity of floods, just like they are with droughts. Prominent seasonal cycles can result in times of greater runoff and high flow conditions, which may raise the risk of flooding. The timing and size of floods are frequently closely correlated with seasonal patterns of precipitation and snowmelt in many places, which can cause a concentration of high-flow occurrences at particular times of the year [6, 10].

Effective water management and planning require a comprehensive understanding of the correlation between seasonality and water scarcity. This information can help with the development of strategies to reduce the effects of water scarcity and low rainfall. Similarly, understanding how seasonality and floods interact is essential for efficient flood management and planning. It can also help with the creation of mitigation plans for high-flow events. Hence, for effective management of water resources, a comprehensive understanding of the seasonality of water availability and its effect on water-related problems is required [10, 11].

In the field of water management, multiple hydrological time series decomposition approaches have been widely utilized. These methods can assist in identifying and detecting the various components of a hydrological time series, such as trend, seasonality, and noise. The classical decomposition method, the empirical mode decomposition (EMD) method and the wavelet analysis method are some of the most commonly used methods. The classical decomposition method is based on separating the original time series into trend, seasonal, and irregular components using moving averages and seasonal adjustment techniques. The EMD method is a data-driven methodology that decomposes a time series into intrinsic mode functions (IMFs) based on the data's local properties. The wavelet analysis method decomposes a time series into various time-frequency components using a frequency-based approach. These methods have been utilized in a variety of water-related investigations, including the study of precipitation and streamflow data to identify trends and seasonality patterns [12, 13]. For instance, the EMD method was used to decompose streamflow time series and identified the long-term trend and seasonality patterns [14]. Another study used wavelet analysis to decompose precipitation time series and identify the different frequency components [15]. These studies demonstrate the importance of time series decomposition methods in water management and their potential to provide valuable insights for decision-making.

## 2 Study area

The Topľa River is located in the eastern part of Slovakia (Figure 1., 2.). and is a major tributary of the Bodrog River. The river basin covers an area of approximately 654 $km^2$, which includes both agricultural land and forested areas. The hydrological regime of the Topľa River is characterized by seasonal variations in discharge, with peak flows typically occurring during the spring snowmelt period and summer rainfall events. In contrast, the river experiences low flows and occasional ice cover during the winter months [3, 16].

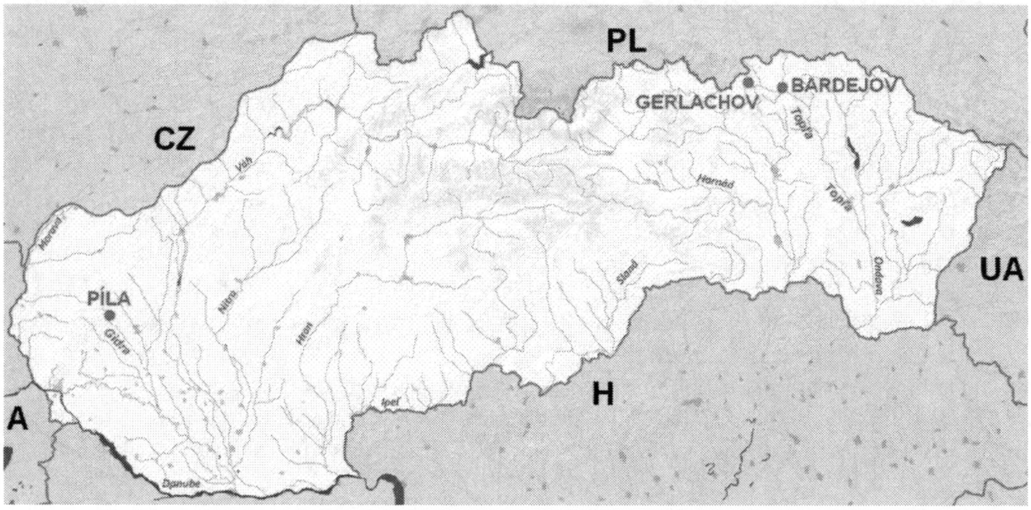


**Fig. 1.** Location of analysed gauging stations.

The long-term mean annual discharge for the years 1961 – 2000 is $Q_{a\ 1961\text{-}2000} = 2.978$ $m^3 \cdot s^{-1}$, according to Slovak Hydrometeorological Institute (SHMI) [17]. The Gerlachov gauging station, with a catchment area of 139.4 $km^2$, is located at 118.6 rkm in Topľa and has an elevation of 358.39 m above sea level. The second water gauging station is situated in the town of Bardejov, has a catchment area of 325.8 $km^2$, and is located at 103.5 rkm with an elevation of 265.05 m above sea level [3].

The river's discharge exhibited high variability, with both floods and droughts occurring throughout the historical record. Seasonal distribution of discharge also varied, with the highest discharge occurring in the spring and summer months, and the lowest discharge occurring during the winter months [3]. Understanding the hydrological characteristics of the Topľa River is crucial for effective water management and the mitigation of extreme events such as floods and droughts.

**Fig. 2.** Topľa River – upstream section, near Gerlachov village and gauging station in Bardejov.

The Gidra River, a typical foothill stream, is a right-hand tributary of the Dudváh River with a length of 38.5 km. It originates below Baďurka in the Little Carpathians at an altitude of approximately 470 m above sea level. In 2011, the Gidra River gained media attention due to the catastrophic flood it caused. The only water gauging station in the area is located above the village of Píla at rkm 33.30 (Figure 1., 3.). and an altitude of 270.04 m a. s. l. The catchment area of the Gidra River at the water gauging station is 32.95 $km^2$, and the available measured data on water stages have been recorded since 1961. The analysis of mean daily discharges for the 59-year period from 1961 to 2020 was conducted. The long-term mean annual discharge reported by SHMI for the period 1961–2000 is $Q_{a\ 1961–2000} = 0.298\ m^3 \cdot s^{-1}$ [17].

**Fig. 3.** Gidra River – gauging station above Píla village.

# 3 Methodology and Data

## 3.1 The seasonal decomposition method

The seasonal decomposition method is a time series analysis technique that separates a time series into its trend, seasonal, and residual components using moving averages. Mathematically, the decomposition can be expressed as:

$$y_t = T_t + S_t + R_t \tag{1}$$

where $y_t$ is the observed value at time $t$; $T_t$ is the trend component at time $t$; $S_t$ is the seasonal component at time $t$, and $R_t$ is the residual component at time $t$.

The seasonal component $S_t$ can be modeled using one of two methods: additive or multiplicative. The additive method is used when the magnitude of the seasonal fluctuations is constant over time, while the multiplicative method is used when the magnitude of the seasonal fluctuations varies with the level of the time series. Mathematically, the seasonal component can be expressed as:

Additive:
$$S_t = \sum_{i=1}^{m} s_i \tag{2}$$

Multiplicative:
$$S_t = \prod_{i=1}^{m} s_i \tag{3}$$

where $m$ is the number of seasonal periods (e.g., 12 for monthly data), and $s_i$ is the seasonal factor for period $i$.

To estimate the trend and seasonal components, the seasonal decomposition method uses a moving average and a period-adjusted average, respectively. The moving average is applied to the time series to estimate the trend component, while the period-adjusted average is used to estimate the seasonal component. Mathematically, the trend component $T_t$ can be expressed as:

$$T_t = \frac{1}{2q+1} \sum_{i=-q}^{q} y_{t+i} \tag{4}$$

where $q$ is the order of the moving average.

The seasonal component $S_t$ can be estimated by first calculating the average value for each seasonal period, then adjusting these averages to have a mean of zero. Mathematically, the seasonal component can be expressed as:

$$S_t = \frac{1}{m} \sum_{i=1}^{n} \left( y_t - \frac{1}{n} \sum_{j=1}^{n} y_{i,m+j} \right) \tag{5}$$

where $n$ is the number of seasonal periods in the time series, and $y_{i,m+j}$ is the observation for the $j$th seasonal period in the $i$th year [18, 19].

In this study, seasonal decomposition was performed using the Statsmodels library [20] in Python programming language. The Statsmodels library provides a range of tools for time series analysis, including methods for time series decomposition. Specifically, it was used the seasonal decomposition function from the Statsmodels library to decompose given time series data into its seasonal, trend, and residual components. This approach enabled to gain a better understanding of the underlying patterns and trends in analysed data.

### 3.2 Data

This research utilizes data on the daily mean discharge for Gidra and Topľa rivers from January 1, 1981, to December 31, 2018. The data was provided by the SHMI. Discharge data is a critical component in hydrological analysis and modeling, as it provides information about the amount and timing of water flow in a particular river. The long-term nature of this dataset makes it ideal for studying trends and patterns in river discharge, as well as for identifying potential changes in hydrological regimes over time. The accuracy and reliability of the data provided by the SHMI ensures the robustness of the results obtained from this study.

The seasonality component of the data is particularly important because it provides insights into the natural variations in river flow that occur within a year. This allows us to understand how river discharge changes throughout the year and how it might be impacted by changes in climate, land use, or other factors. The long-term nature of this dataset provides a comprehensive view of the seasonal patterns in river discharge, making it an ideal resource for studying trends and identifying potential changes in hydrological regimes over time. The reliability and accuracy of the data provided by the SHMI ensure the validity of the conclusions drawn from this study.

## 4 Analysis

To extract the seasonality, trend, and residuals from the discharge data of the Gidra and Topľa rivers, the seasonal decomposition method was applied [20]. This method is a powerful technique for identifying the underlying patterns in time series data, which can then be used to better understand and model the data. The seasonal decomposition method separates the data into three components: the seasonal component, which represents the periodic fluctuations that occur within a year; the trend component, which shows the long-term changes in the data; and the residual component, which captures the random fluctuations in the data that cannot be explained by the other two components. This information is essential for understanding how changes in climate, land use, or other factors might impact the hydrological regimes of these rivers over time.

The present study aims to investigate the seasonal variability of Topľa River discharge during the period of 1981 – 2018, by analysing the seasonal component of the river's discharge. The obtained results are presented in Figure 4., which displays the seasonal component of the discharge over time. The analysis reveals that the seasonal component of the discharge is relatively low during the first two months of the year, followed by a steady increase that reaches its peak in mid-April. This peak is most likely due to the onset of the spring snowmelt in the catchment area. From mid-April to mid-May, the seasonal discharge component decreases slowly, followed by two peaks in May and June where a significant increase in the seasonal discharge component is observed.

Due to summer storms in June, it is uncertain whether the increase can still be attributed to ice melting, a typical phenomenon during this time of the year. After mid-June, the seasonal component of the discharge decreases slightly, reaching another significant peak in the last week of July and the first week of August, which dominates the discharge component during this period. Following this peak, the seasonal discharge component decreases until it reaches its minimum in October, with a slight increase observed in November and December.

These findings provide a comprehensive understanding of the seasonal pattern of the Topľa River discharge, which is crucial for water management and planning purposes. Further studies are recommended to examine the factors that affect the discharge variability and its potential implications for the local ecosystem and water resources management.

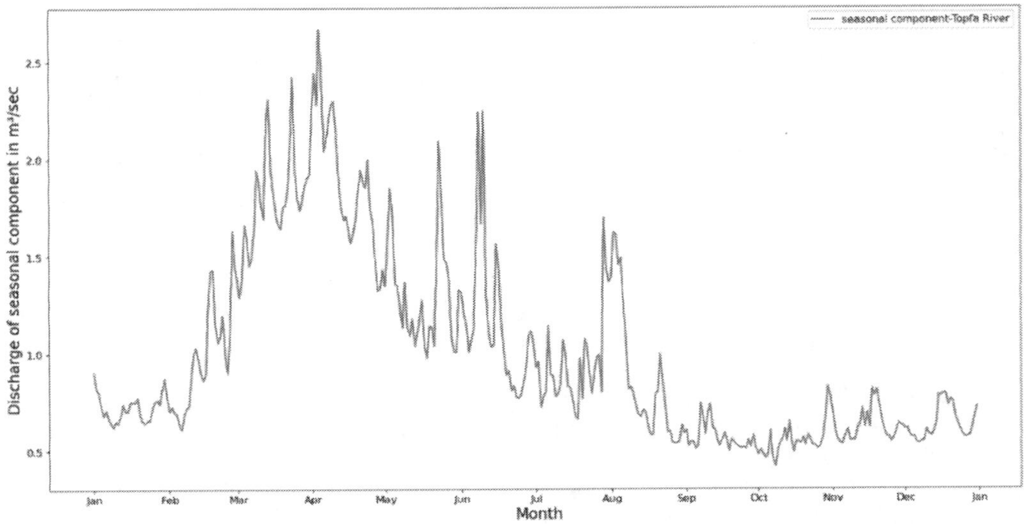


**Fig. 4.** The seasonal component of discharge in Topľa River.

The Gidra River is a typical representative of the foothill river with its seasonal discharge exhibiting distinct patterns throughout the year Figure 5. The river's discharge is relatively high in January compared to the summer season, and this trend continues until February, where it reaches one of its peaks. During March and April, the discharge reaches its highest point, with a steady increase from March until the first week of April. Following this peak, the seasonal discharge declines steadily until it reaches its minimum in late August or early September. This phenomenon is attributed to the reduced rainfall and warmer temperatures during the summer season. However, there are significant increases in the seasonal discharge in mid-May, mid-June, and mid-July, which could be attributed to ice melt or summer storms in the mountains.

The factors influencing the seasonal discharge of Gidra River are complex and dynamic, with fluctuations in response to various environmental factors throughout the year. It is noteworthy that the river's discharge during March and April is higher than any other season, which highlights the importance of early spring precipitation in the river's hydrological regime. Moreover, the increase in discharge during mid-May, mid-June, and mid-July coincides with the period of ice melt or summer storms in the mountainous regions. This finding emphasizes the crucial role of mountainous areas in providing water resources for downstream areas.

From mid-August until mid-October, the seasonal discharge of Gidra river follows a constant rate at its minimum. Finally, the discharge begins to increase from mid-October until the end of December, with a relatively small slope. This increase could be attributed to the onset of the winter season, which brings in more precipitation and lower temperatures, leading to a higher rate of snow accumulation in the mountains. In summary, the seasonal discharge patterns of Gidra river are indicative of its hydrological regime, and understanding these patterns is critical for the sustainable management of water resources in the region.

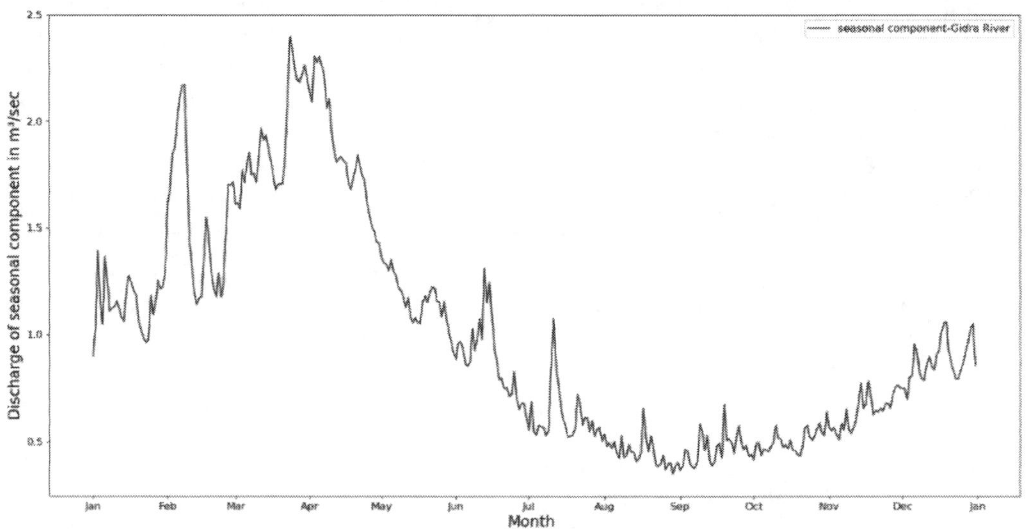


**Fig. 5.** The seasonal component of discharge in Gidra River.

As was shown on the analysed seasonal component, the discharge patterns of both rivers have significant observed trend, which can be included in the planning of water management, agriculture, and ecological systems.

It is worth mentioning that the model deployed in this study is a multiplicative model. This type of model is particularly useful for time series data that exhibit non-constant variance over time. Additionally, in this study, the period chosen for the model to analyse seasonality is 365 days, which corresponds to the yearly cycle and is expected to capture any yearly patterns that may exist in the time series.

# 5 Results and conclusion

## 5.1 Results

The discharge seasonality of rivers is a complex phenomenon, and the comparison between Gidra and Topľa rivers highlights several similarities and differences. In particular, the higher discharge rates observed in Gidra River during January and February are typical for under-mountainous regions, where the terrain favours a faster runoff response time, leading to more rapid discharge rates. The steeper slopes and higher elevation of mountainous areas result in faster snowmelt and higher seasonal discharges during the winter months. In contrast, the lower discharge rates observed in Topľa River during the same period are associated with the flatter terrain and lower elevation of lowland areas, which results in lower discharge rates. During March and April, the onset of snow melting in the catchment area leads to increased discharge rates in both Gidra and Topľa rivers. The snowmelt effect is more significant in lowland rivers, such as Topľa River, where the melting of the snow cover occurs later in the season, resulting in higher discharge rates in June and July. The similarities in the seasonal patterns of both rivers during this period reflect the influence of snowmelt on the river discharge.

The summer season reveals a significant difference in discharge rates between the two rivers, with Topľa River exhibiting higher seasonal discharges. This is due to the melting of the snow cover in the mountains and the higher rainfall rates in the lowland areas. These factors result in a more pronounced hydrological response in lowland rivers, which are more sensitive to precipitation events than under-mountainous rivers. The higher discharge rates of Topľa River during the summer season reflect this sensitivity, whereas Gidra River experiences a smaller increase in seasonal discharge.

During September and November, both rivers exhibit low discharge rates due to the absence of snowmelt and lower precipitation rates. However, the faster progression of Gidra River discharge rates from mid-November to December is linked to the onset of winter conditions and the accumulation of snow cover in the upper catchment area. This difference in seasonal patterns reflects the influence of topography on river discharge and highlights the importance of considering local hydrological conditions when analysing river systems.

Overall, the discharge seasonality of Gidra and Topľa rivers (Figure 6.) reflects the complex interplay of various factors, such as precipitation, temperature, snowmelt, vegetation cover, and topography. The differences and similarities observed in the seasonal patterns of both rivers provide insights into the hydrological functioning of river systems and the effects of climate change on these systems. By understanding these factors and their interactions, researchers and policymakers can develop more effective strategies for managing and protecting river ecosystems.

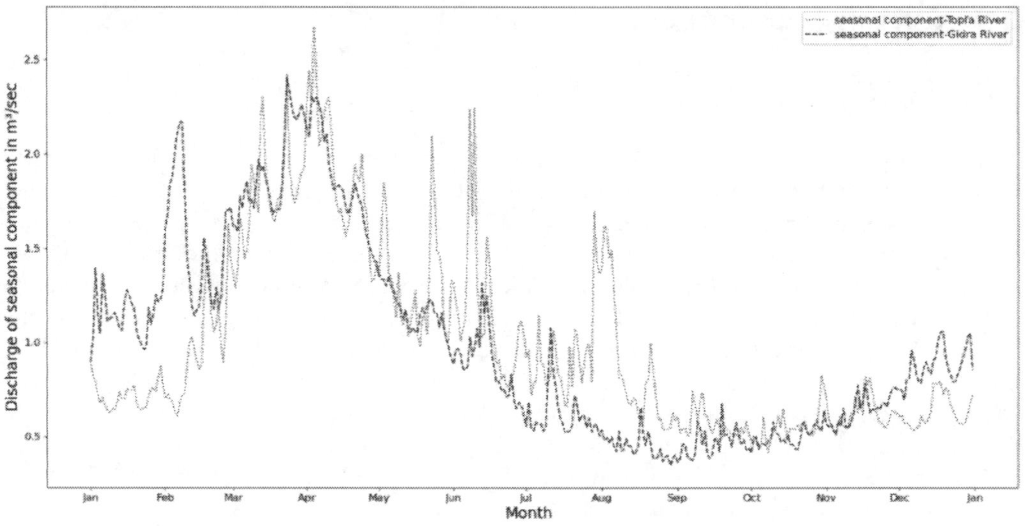


**Fig. 6.** Discharge seasonality comparison between Gidra and Topľa rivers.

## 5.2 Conclusion

The study focused on analysing the seasonal patterns in two different types of rivers across Slovakia. The study used a comprehensive statistical methodology, including the decomposition of time series and correlation analysis, to analyse the discharge data collected from multiple river gauging stations.

The study's results showed that there were distinct seasonal variations in the hydrological characteristics of different river types, with a clear differentiation between lowland and mountain rivers. These seasonal variations were found to be significant and could potentially have important implications for water management and agriculture (e.g. Gidra River is utilized for water supplying of several ponds, reservoir Budmerice and also for Ronava reservoir, placed in neighbouring catchment).

The study highlighted the significance of hydrological time series analysis in gaining a better comprehension of water resources and their management. It is suggested that a deeper understanding of seasonal patterns in river flow can lead to more efficient and effective utilization of water resources.

It is important to note that all of the results presented in this study were obtained solely from the analysis of discharge data. While this provides valuable insights into the seasonality of the studied river, it is important to acknowledge that a more comprehensive understanding of the underlying processes would require the inclusion of additional data types such as precipitation, snow cover, and temperature. By including such data, it would be possible to gain a more complete understanding of the interplay between these variables and the observed patterns in discharge. Moreover, in order to better understand the reasons for peaks in seasonal component, additional data analysis would be necessary. This could involve examining the relationship between the peaks and factors such as snow melting or storms, by looking at different types of data, including precipitation, snow cover, and temperature, to help identify the underlying mechanisms driving the observed patterns. Therefore, while the present study provides valuable insights, it is important to acknowledge the limitations of the dataset and the need for further research to provide a more comprehensive understanding of the seasonality and its effects.

Overall, the study contributes to enhancing the understanding of hydrological time series analysis and its potential implications for water management and agriculture in Slovakia. The findings of this study can be used to inform water management policies and strategies, with the aim of ensuring the sustainable use of water resources in the country.

This work was supported by The Slovak Research and Development Agency under the contract No. APVV-19-0383, together with APVV-20-0023.

## References

1. M. Fendeková, O. Horvát, L. Blaškovičová, Z. Danáčová, M. Fendek & O. Bochníček, Prognosis of climate change driven drought in the Poprad, Torysa and Topľa river basins. Acta Hydrologica Slovaca, **19**(2), 234-243. (2018).
2. Y. Velísková, R. Dulovičová, R. Schügerl, Impact of vegetation on flow in a lowland stream during the growing season. Biologia, Vol. **72**, No. 8, 840–846. (2017).
3. W. Almikaeel, L. Čubanová, A. Šoltész, Comparison of mean daily discharge data for under-mountain and highland-lowland types of rivers. Acta Hydrologica Slovaca, 23(1), 73-81. (2022)
4. S. Wang, Freezing Temperature Controls Winter Water Discharge for Cold Region Watershed. Water Resources Research, **55**, 10479-10493. (2019).
5. A. Dráb, J. Říha, An approach to the implementation of European Directive 2007/60/EC on flood risk management in the Czech Republic. Natural Hazards and Earth System Science, Volume **10**, Issue 9, 1977 – 1987. (2010)
6. L. O. Serrano, R. B. Ribeiro, A. C. Borges, F. F. Pruski, Low-Flow Seasonality and Effects on Water Availability throughout the River Network. Water Resources Management, **34**, 1289-1304. (2020).
7. V. Sýs, P. Fošumpaur, T. Kašpar, The impact of climate change on the reliability of Water Resources. Climate, **9**(11), 153, 1-15. (2021)
8. A. F. van Loon, E. Tijdeman, N. Wanders, H. A. J. Van Lanen, A. J. Teuling, R. Uijlenhoet, How climate seasonality modifies drought duration and deficit. Journal of Geophysical Research: Atmospheres, **119**, 4640-4656. (2014).
9. B. Kandra, A. Tall, M. Gomboš, D. Pavelková, Quantification of Evapotranspiration by Calculations and Measurements Using a Lysimeter. Water **15**(2), 373, 1-18. (2023)
10. M. Sivapalan, G. Blöschl, R. Merz, D. Gutknecht, Linking flood frequency to long-term water balance: Incorporating effects of seasonality. Water Resources Research, 41. (2005).
11. G. Y. Abawi, S. Dutta, T. R. Harris, J. W. Ritchie, D. Rattray, A. J. Crane, The Use of Seasonal Climate Forecasts in Water Resources Management. In Proceedings of the 3rd International Conference on Water Resources and Environment Research, Part I (pp. 129-136). (2000).
12. Y. F. Sang, A review on the applications of wavelet transform in hydrology time series analysis. Atmospheric Research, 122, 8-15. (2013).
13. S. Lv, Multi-scale analysis of hydrological series using ensemble empirical mode decomposition. Proceedings of the 2nd International Conference on Mechanic Automation and Control Engineering, 974-977. (2011).
14. D. Zhihong, EMD Method for Multiple Time-scale Analysis on Fluctuation Characteristic of Natural Annual Runoff Time Series of Fen River. Water Resources and Power. (2008).

15. P. Pathak, A. Kalra, S. Ahmad, M. Bernardez, Wavelet-Aided Analysis to Estimate Seasonal Variability and Dominant Periodicities in Temperature, Precipitation, and Streamflow in the Midwestern United States. Water Resources Management, **30**, 4649-4665. (2016).
16. B. V. Mitková, Effect of the data length and seasonality on the accuracy of T-year discharges estimation: Case study on the Topľa River. Proceedings of the IAHR World Congress, **28**, 1692-1699. (2020).
17. L. Čubanová, W. Almikaeel, Drought assessment and prediction for Gidra river, Slovakia. IOP Conf. Ser.: Mater. Sci. Eng., 1209. (2021).
18. A. Dokumentov, R. J. Hyndman, STR: A seasonal-trend decomposition procedure based on regression. arXiv: Methodology, (1506.03805). (2015).
19. M. Theodosiou, Forecasting monthly and quarterly time series using STL decomposition. International Journal of Forecasting, **27**, 1178-1195. (2011).
20. Statsmodels. Statsmodels: Time Series Analysis in Python. Retrieved September 10, 2021, from **[https://www.statsmodels.org/stable/tsa.html](https://www.statsmodels.org/stable/tsa.html)**

# FEM analysis of the influence of reinforcing mesh on concrete pavement performance

*Jakub* Vesely[1*], *Petr* Panek[1], and *Ludvik* Vebr[1]

[1] Faculty of Civil Engineering CTU in Prague, Department of road structures, Thakurova 7/2077, 166 29 Prague, Czech Republic

**Abstract.** Reinforcement of concrete pavement slabs with steel reinforcing mesh (KARI mesh) has been regularly used in the Czech Republic. Reinforcing mesh is used for atypical shapes and non-standard (larger) dimensions of pavement slabs (intersections, runways, transhipment areas, etc.). The placement of mesh across the thickness of the slab varies in designs (top, bottom or both surfaces, eventually in centre). The mesh has a positive effect on cracking occurrence during concrete hardening. It is assumed that the reinforcing meshes also contribute to the reduction of tensile stresses caused by cyclic loading from traffic and temperature. The amount of the reduction in concrete depends on the mesh positioning, but also on the specific loading conditions. In this paper, the FEM analysis of concrete pavements with steel reinforcing mesh is carried out. The reduction of maximal value of tensile stress induced by overweight vehicles, temperature and their combination is inspected while using different mesh positioning. The slab dimensions used in the simulation are larger than the usual highway slabs, the dimensions are based on the real transhipment area pavement. The contribution of the reinforcement to limitation of the maximum tensile stress in the pavement due to different reinforcing mesh positioning was determined.

## 1 Introduction

For roads with extreme loads such as airports, motorways or container transhipment points, concrete pavements are proposed instead of asphalt alternatives. These pavements are usually designed as unreinforced jointed plain concrete pavements (JPCP), although in some cases, the joints are reinforced using welded mesh or other reinforcement to ensure the durability of the construction or resistance against cracking. The proposed numerical model uses linear elastic theory, the material properties are retrieved from TP170 technical norm [1], the analysis was performed using Abaqus software.

The objective of this research was to model the reinforcing welded wire mesh in the concrete pavement slab so that a time-saving stress analysis could be performed in the concrete pavement, taking into account the specific geometry of the slab, material properties of the subsoil system, the specific loading, the specific location of the reinforcement within the cross section, and also the geometry of these reinforcement elements. A well-known

* Corresponding author: jakub.vesely.2@fsv.cvut.cz

benefit of reinforcement elements is the reduction of cracking in the phase of concrete hardening [2]. However, this research was focused on the contribution of reinforcement mesh to reduction of the maximum tensile stresses generated during pavement serviceability for selected loading conditions and specific reinforcement locations along the concrete slab thickness. Thus, a series of calculations were performed for wheel loading, temperature loading (positive temperature gradient) and a combination of these scenarios. For the analysis, a concrete slab with a slightly atypical shape designed for a specific transhipment area was selected, for which reinforcement with welded mesh (WWF) was considered. The specific conditions for the pavement subsoil system as well as the loading are taken from the design of the site. For the purpose of the research, the position of the reinforcement nets along the thickness of the slab was chosen in several variations. These include locations at the top surface, bottom surface, in the middle and at both surfaces of the concrete slab. Three different reinforcement elements were modelled with the same mesh size (100 x 100 mm) and different steel bar diameters (8, 10 and 12 mm). In this analysis, the differences in stresses comparing different reinforcement to a plain unreinforced concrete JPCP pavement and the percentage contribution to reducing the maximum tensile stresses for specific load conditions are calculated. The fatigue benefits of the reinforcing elements to the concrete slab during service life lead to a more complex analysis [3] and will be the subject of follow-up research.

## 2 3D FEM model of concrete pavement

As already mentioned in the introduction, the 3D model of the concrete pavement (see Figure 1) was created based on a specific transshipment area and its concrete pavement area.

- The dimensions of the concrete slab (according to the design) are : 8.06 x 4.96 x 0.32 m
- The linear elastic material parameters of the proposed pavement are as follows:

CB 320 mm, E= 35000 MPa, $\mu$=0.2
MCB 150 mm, E = 6000 MPa, $\mu$=0.2
ND 250 mm, E = 400 MPa, $\mu$=0.3
Subsoil 1350 mm, E= 70 MPa, $\mu$=0.35

- Load from Kalmar reach stacker (for container transport) : radius of load area (single wheel) = 0.2559 m, contact pressure 1.0 MPa, axial distance between the load areas of two wheels 0.62 m (Figure 1).

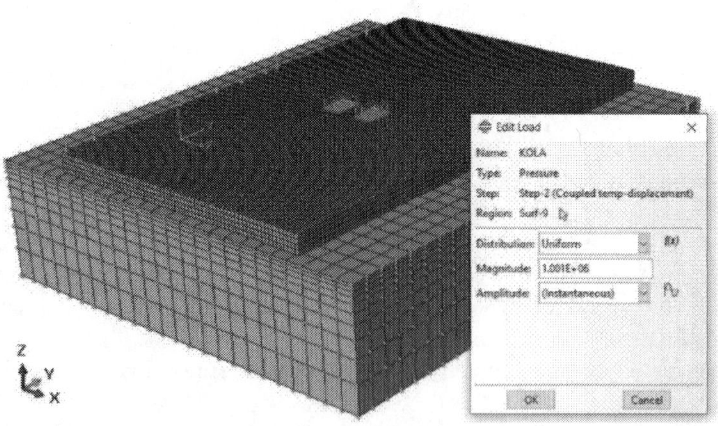


**Fig. 1.** 3D model – wheel loading.

- Temperature loading : a linear positive temperature gradient was considered along the thickness of the plate. The temperature difference between the top and bottom surface of the CB plate was set to 10°C (Figure 2).

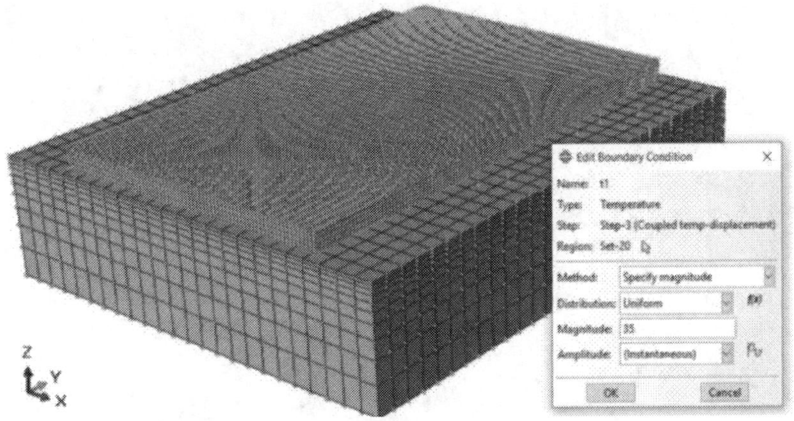


**Fig. 2.** 3D model – temperature loading.

## 3 Modelling of reinforcement elements in ABAQUS

In Abaqus [4], several approaches have been presented to model reinforced regions in concrete. From the point of view of modelling and lower computational time requirements, the most suitable approach was chosen to "embed structural elements in solid elements" [5]. In this technique, reinforcement elements are inserted into host elements, with the proviso that the element networks do not have to coincide. The program searches for elements close to the embedded elements. Then the nodes of the embedded elements are constrained by the response of the host elements. In this work, by first using the element "wire" and then assigning a specific profile, a reinforcement steel network was created (see Figure 3).

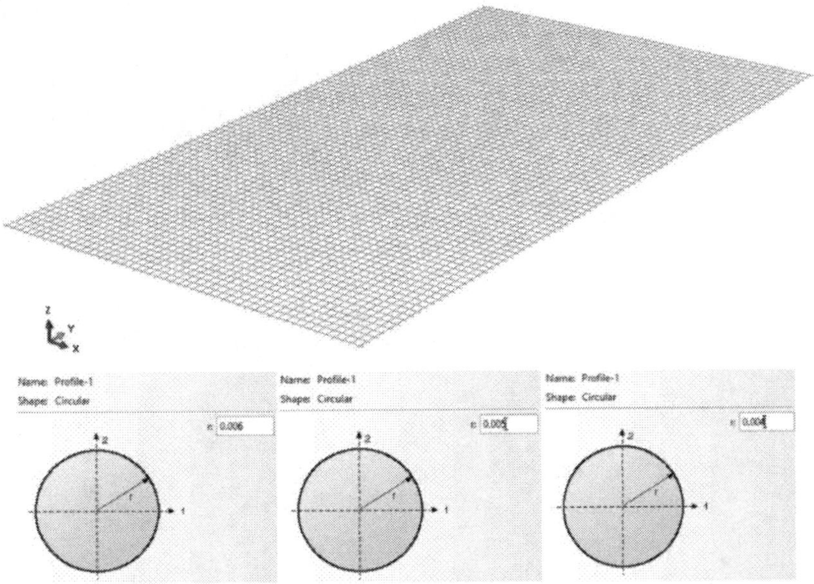

**Fig. 3.** Reinforcing welded mesh. Adjusting the radius of the steel bar of the reinforcing mesh.

This mesh was then placed in the concrete plate model lying on the subsoil system (see Figure 4). When placing the net to the plate surfaces, the prescribed coverage was always maintained.

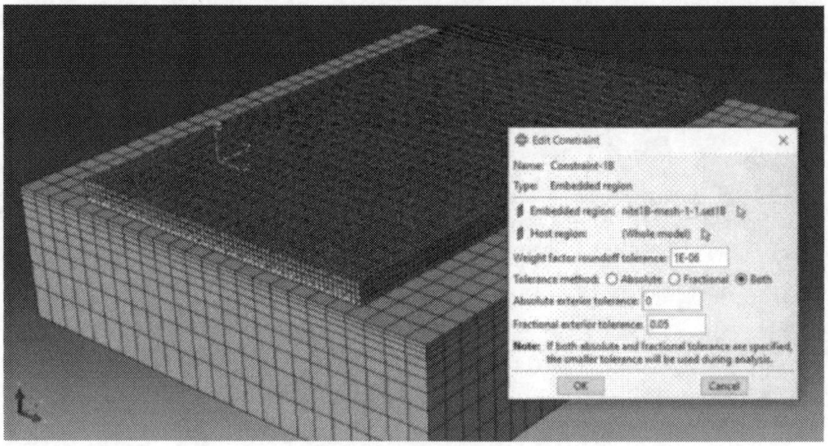

**Fig. 4.** Contact area between pavement and subsoil.

## 4 Calculated stress in the concrete pavement slab

The tensile stresses for the selected loading conditions, i.e. from the reach stacker wheels, from the temperature and from the combination of these loads were determined by means of a computational analysis in ABAQUS (see Figures 5, 6 and 7).

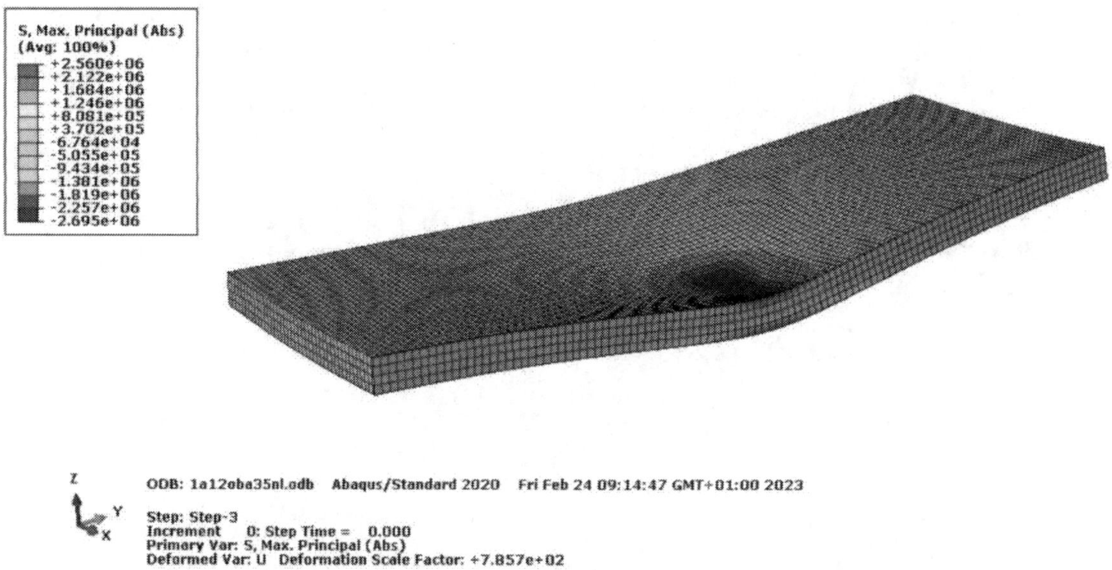


**Fig. 5.** Maximal tensile stress – wheel load.

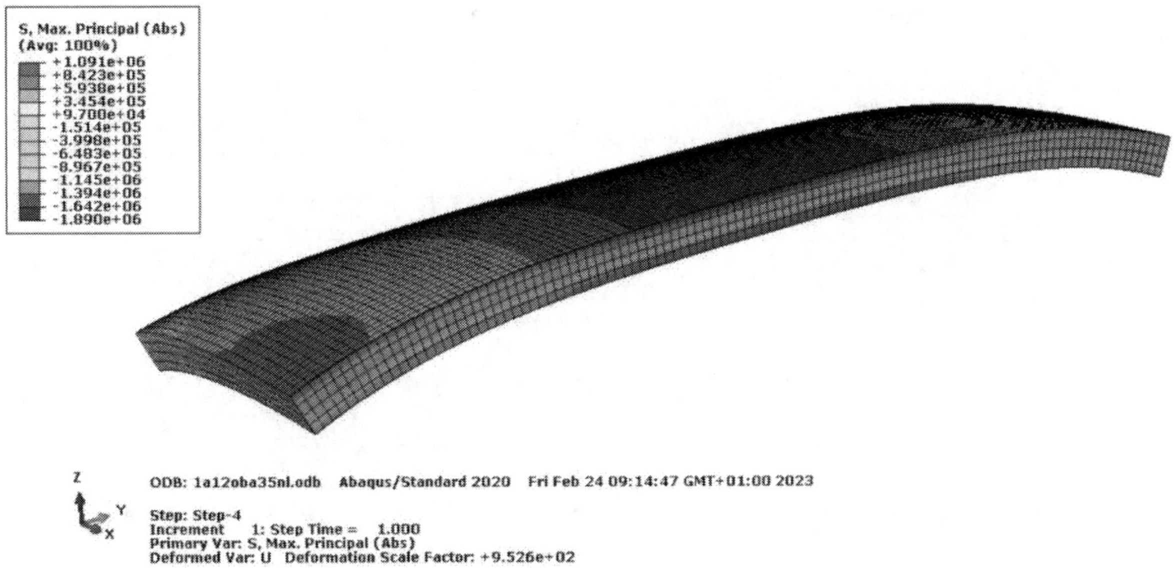


**Fig. 6.** Maximal tensile stress – temperature load.

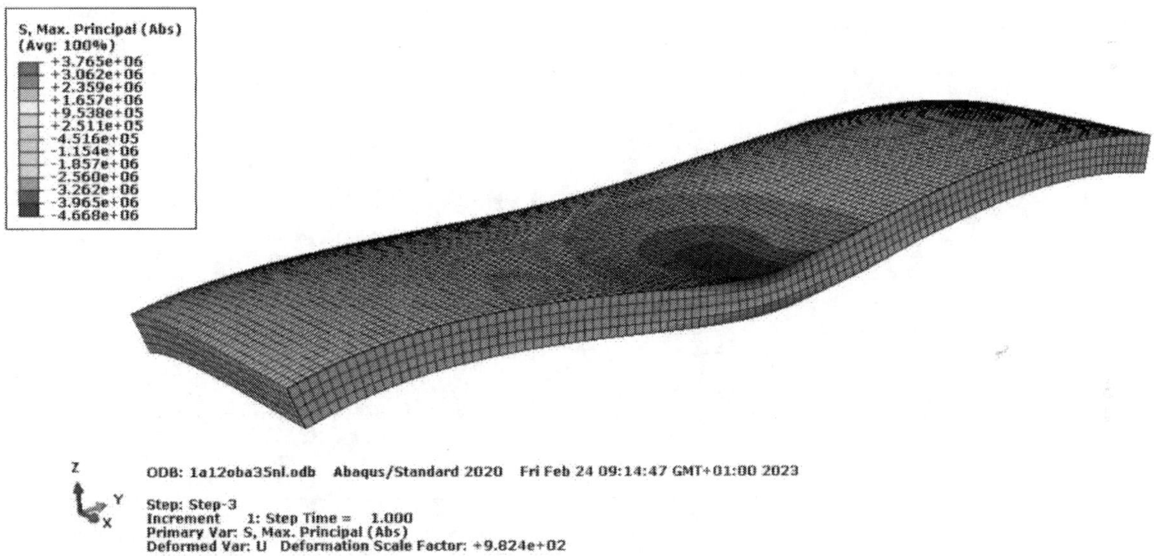


**Fig. 7.** Maximal tensile stress – wheel and temperature load combination.

Table 1 summarizes the values of the maximum tensile stresses in the concrete slab for different positions of the reinforcement welded mesh or without any reinforcement. Table 1 also shows the effect of the diameter of the reinforcing mesh member on the resulting stress magnitude. The differences in stresses (in MPa) compared to the unreinforced slab are shown in Table 2 and the percentage comparison is shown in Table 3.

**Table 1.** Calculated maximal tensile stress for different reinforcement (WWF) positions and diameters (MPa).

| **Ø 8mm/100x100** | both surfaces | bottom surface | top surface | center | without WWF |
|---|---|---|---|---|---|
| wheel | 2.580 | 2.581 | 2.598 | 2.599 | 2.599 |
| temperature | 1.227 | 1.271 | 1.296 | 1.281 | 1.341 |
| wheel + temperature | **3.935** | **4.000** | **4.011** | **4.016** | **4.076** |

| **Ø 10mm/100x100** | both surfaces | bottom surface | top surface | center | without WWF |
|---|---|---|---|---|---|
| wheel | 2.571 | 2.572 | 2.597 | 2.599 | 2.599 |
| temperature | 1.178 | 1.246 | 1.272 | 1.249 | 1.341 |
| wheel + temperature | **3.858** | **3.958** | **3.975** | **3.982** | **4.076** |

| **Ø 12mm/100x100** | both surfaces | bottom surface | top surface | center | without WWF |
|---|---|---|---|---|---|
| wheel | 2.560 | 2.562 | 2.597 | 2.599 | 2.599 |
| temperature | 1.091 | 1.189 | 1.243 | 1.222 | 1.341 |
| wheel + temperature | **3.765** | **3.907** | **3.932** | **3.942** | **4.076** |

**Table 2.** Tensile stress difference between reinforced and not reinforced slab (MPa).

| **Δσ (MPa) Ø 8mm/100x100** | both surfaces | bottom surface | top surface | center |
|---|---|---|---|---|
| wheel | 0.019 | 0.018 | 0.001 | 0.000 |
| temperature | 0.114 | 0.070 | 0.045 | 0.060 |
| wheel + temperature | **0.141** | **0.076** | **0.065** | **0.060** |

| **Δσ (MPa) Ø 10mm/100x100** | both surfaces | bottom surface | top surface | center |
|---|---|---|---|---|
| wheel | 0.028 | 0.027 | 0.002 | 0.000 |
| temperature | 0.163 | 0.095 | 0.069 | 0.092 |
| wheel + temperature | **0.218** | **0.118** | **0.101** | **0.094** |

| Δσ (MPa) Ø 12mm/100x100 | both surfaces | bottom surface | top surface | center |
|---|---|---|---|---|
| wheel | 0.039 | 0.037 | 0.002 | 0.000 |
| temperature | 0.250 | 0.152 | 0.098 | 0.119 |
| wheel + temperature | **0.311** | **0.169** | **0.144** | **0.134** |

**Table 3.** Tensile stress difference between reinforced and not reinforced slab (%).

| Δσ (%) Ø 8mm/100x100 | both surfaces | bottom surface | top surface | center |
|---|---|---|---|---|
| wheel | 0.73% | 0.69% | 0.04% | 0.00% |
| temperature | 8.50% | 5.22% | 3.36% | 4.47% |
| wheel + temperature | **3.46%** | **1.86%** | **1.59%** | **1.47%** |

| Δσ (%) Ø 10mm/100x100 | both surfaces | bottom surface | top surface | center |
|---|---|---|---|---|
| wheel | 1.08% | 1.04% | 0.08% | 0.00% |
| temperature | 12.16% | 7.08% | 5.15% | 6.86% |
| wheel + temperature | **5.35%** | **2.89%** | **2.48%** | **2.31%** |

| Δσ (%) Ø 12mm/100x100 | both surfaces | bottom surface | top surface | center |
|---|---|---|---|---|
| wheel | 1.50% | 1.42% | 0.08% | 0.00% |
| temperature | 18.64% | 11.33% | 7.31% | 8.87% |
| wheel + temperature | **7.63%** | **4.15%** | **3.53%** | **3.29%** |

The above results fulfill the logical assumption that a larger wire diameter of the reinforcing welded mesh means a smaller tensile stress in the concrete pavement and therefore a smaller fatigue stress. For more severe loading conditions with higher values of tensile stresses (with simultaneous loading with wheel and temperature), the percentage savings in tensile stresses for the different reinforcement diameters and their different positions within the cross section are as follows:

- Welded wire mesh Ø 8 mm – 1.47 - 3.46 %
- Welded wire mesh Ø 10 mm – 2.31 - 5.35 %
- Welded wire mesh Ø 12 mm – 3.29 - 7.63 %

The most beneficial in terms of minimizing tensile stresses is the placement of reinforcement networks at both surfaces of the concrete slab. Conversely, the position of the net in the centre of the slab is the least beneficial in terms of minimising tensile stresses. From the point of view of stress reduction in the pavement, the use of larger wire diameters of the reinforcement mesh is suggested. However, due to the significant price differences between different wire diameter nets at the same mesh size (these are easily two to three times the price), it may be more efficient to use smaller profiles.

## 5 Conclusion

Therefore, in addition to the benefit of limiting the induction of cracks during hardening of the concrete, reinforcing welded mesh can also be beneficial in terms of reducing tensile stresses and improving fatigue behaviour against wheel and temperature loading, if properly installed and all technological principles are followed. However, it is important to note that the use of the reinforcing element in the concrete pavement slab and therefore the magnitude of the tensile stress reduction in the concrete is dependent on the specific loading condition and the position of the element within the cross section. This analysis shows that the stress reduction can be in the range of a few percent, but even close to twenty percent for some loading scenarios, which could have a significant positive effect in terms of fatigue behaviour. Thus, the service life of the pavement could be significantly extended. However, only the stress values at the more demanding loading conditions, which are closer to the tensile strength of the concrete (i.e. wheel + temperature combination) should be assessed, as these have a greater influence on the cumulative fatigue damage of the pavement. In the case of the reinforcement for both surfaces, the computational analysis showed a saving of tensile stresses between about 1.5 to 7.6%. A more detailed fatigue analysis and possible cost comparison of mesh reinforced and unreinforced concrete pavements will be the subject of further research.

Authors gratefully acknowledge financial support from Czech technical university, grant number SGS23/043/OHK1/1T/11.

## References

1. J. Kudrna, M. Varaus, L. Vébr et al., Design guide for road pavements TP170 (Ministry of Transport ČR, 2004)
2. Huang, Y. H., Pavement Analysis and Design, Pearson Education, Inc. and Dorling Kindersley Publishing, Inc., India (2003)
3. L. Vébr, P. Pánek, B. Novotný, *On fatigue resistance of pavement concrete slabs*, in Proceedings of the 7th European Congress on Concrete Engng, Balatonfüred, , 413–416, 22-23 September 2011, Hungary (2011)
4. ABAQUS, Dassault Systems Simulia Corp, (2022)
5. CAE assistant - Modeling Reinforcement in Abaqus | Abaqus rebar & concrete simulation, (2020)

# Manganese and cobalt metals interrelations with ammonium performance in soil aquifer treatment

*Abdalkarim S.* Gharbia[1*], *Balazs* Zakanyi[1], and *Marton* Toth[1]

[1] Faculty of Earth Science and Engineering, Institute of Environmental Management, University of Miskolc, Miskolc, Hungary

**Abstract.** This study aimed to discover and explore the impacts and interactions of applying $Mn^{2+}$ and $Co^{2+}$ metals in $NH_4$ removal performance under nitrification conditions through soil aquifer treatment. A laboratory scale column was used to simulate the soil aquifer treatment system using quartz sand as packed media to infiltrate the synthetic wastewater mixture from 35 mg/L of $NH_4$, 60 mg/L $Mn^{2+}$ and 40 mg/L of $Co^{2+}$ concentrations. The experimental and linear regression model results demonstrate a significant relationship between $NH_3$-N concentration with $Mn^{2+}$ and $Co^{2+}$ concentration with a p-value $< 0.05$, indicating a substantial influence of $Mn^{2+}$ and $Co^{2+}$ presence on ammonium bioremediation. $Co^{2+}$ has a negative correlation interaction with $NH_3$-N, which means $Co^{2+}$ has increased ammonium oxidation by stimulating the degradation bacteria and cell growth to supplement and improve the activity of ammonium degradation bacteria. However, applying a high concentration of $Mn^{2+}$ works as an inhibitor for $NH_3$-N bioremediation. Using a high concentration of $Mn^{2+}$ negatively impacts bacteria, causing a toxic effect and decreasing the bacteria's degradability for ammonium.

## 1 Introduction

Ammonium ($NH_4$) flowing freely from agricultural, industrial, and household pollution sources into water streams is currently a worldwide concern. The existence of $NH_4$ in unacceptable high concentration levels in water bodies causes growth in oxygen demand. It influences aquatic biodiversity by obstructing and decreasing the capability to remain alive, develop, and propagate [1]. Various ammonium ($NH_4$) removal methods are based on biological, physical, and chemical removal processes for nitrogen species from water and wastewater effluents. The most common ammonium ($NH_4$) removal method is carried out by biological degradation, which is known as nitrification and denitrification reactions to oxidize the nitrogen species that oxidize $NH_4$ to $NO_2^-$ and $NO_3^-$ at nitrifying response followed by denitrifying $NO_3^-$ to $N_2$ gaseous [2].

* Corresponding author: a.kareem92@hotmail.com

Ammonium biological degradation is considered the most environmentally friendly removal process, compared with physical and chemical removal methods. Nitrification and denitrification are naturally oxidation processes with low energy and operation requirements. Furthermore, the nitrogen species in wastewater effluent can be eliminated using aerobic denitrification bacteria [3]. Aerobic denitrification is a biological process that occurs in the presence of oxygen, where certain microorganisms convert nitrate ($NO_3^-$) into nitrogen gas ($N_2$) [4]. Unlike traditional denitrification, which occurs in oxygen-depleted environments, aerobic denitrification takes place under aerobic (oxygen-rich) conditions. It is a natural mechanism for reducing excess nitrate levels in environments such as oxygenated regions of aquatic systems or well-aerated soils. Aerobic denitrification is carried out by specific bacteria and fungi that possess the necessary enzymes to facilitate the conversion of nitrate into nitrogen gas, contributing to the removal of nitrogen compounds from ecosystems [5]. However, the biodegradation bacteria are sensitive to several parameters to achieve a high nitrogen species removal performance in the wastewater system, like as dissolved oxygen level, carbon resources components, temperature, pH and macro and micronutrients. These factors must be in sufficient amounts and suitable species to avoid any metabolic limitation that may harm the degradation of bacteria efficiency [6]. Nitrogen biodegradation bacteria need a sufficient amount of micronutrients like Manganese (Mn), Cobalt (Co), Zinc (Zn), Copper (Cu), Molybdate (Mo), Iodine (I), Boron (B) and Iron (Fe) to secure the bacterial cells metabolism [7].

Cobalt ($Co^{2+}$) is usually detected in polishing and painting wastewater from the porcelain and steel industries [8]. $Co^{2+}$ has been demonstrated as the minimal poisonous metal between $Ni^{2+}$, $Zn^{2+}$, and $Cu^{2+}$ in nitrifying conditions [9]. Gikas [10] abbreviated the impacts of $Co^{2+}$ on activated sludge and discovered that $Co^{2+}$ is performed as a cell development incentive instead of a controller or inhibitor. Also, he found that low concentrations of $Co^{2+}$, less than 2 mg/L, have never turned toxic to microorganisms. Likewise, manganese ($Mn^{2+}$) is regularly initiated in enriched wastewater involving wastewater with high nitrogen species concentrations. However, the nitrogen species purification in nitrifying anaerobic ammonia-oxidizing capacity is developed by adding $Mn^{2+}$ in long time reactions [11]. Huang, Gao, PengTao [12] investigated nitrogen purification capacity in activated sludge, which will be multiplied when adding an $Mn^{2+}$ concentration of 2.8 mg/L. Many studies have discussed and discovered the effects of heavy metals in ammonium oxidation on aerobic denitrification and activated sludge. However, rare studies deal with heavy metals' impact on ammonium removal under aerobic nitrification conditions. Therefore, this study will present the implications and interrelations of infiltrating two types of heavy metals ($Mn^{2+}$ and $Co^{2+}$) with ammonium ($NH_4$) through soil aquifer treatment columns under aerobic nitrification.

## 2 Materials and methods

A plexiglass laboratory scale column using to simulate the soil aquifer treatment system with a 3 cm internal diameter packed with quartz sand at a 30 cm depth applied to explore the impact and relations of adding $Mn^{2+}$ and $Co^{2+}$ metals with the $NH_4$ in the same synthetic wastewater solution and infiltrating through sand media at room temperature of 22 °C. Quartz sand is used as filtrate-packed media with particle sizes range 0.50 – 0.125 mm, 0.42 porosity and 1.48 dry bulk density. A synthetic wastewater solution was designed by mixing $NH_4CL$, $MnSO_4.4H_2O$ and $CoCL_2.6H_2O$ in sufficient amounts to prepare concentrations of 35 mg/L $NH_4$, 60 mg/L $Mn^{2+}$ and 40 mg/L $Co^{2+}$ and pH 6.5. $NH_4$ was measured as $NH_3$-N using the direct reading spectrophotometer device according to the Nessler method, while $Mn^{2+}$ and $Co^{2+}$ were measured using a microwave plasma–atomic emission spectrometry device (MP-AES). A linear regression model applies to explore the interaction relationship and the impact behavior of $Mn^{2+}$ and $Co^{2+}$ presence on $NH_4$ performance through filtrating in the quartz sand column under the nitrification conditions.

## 3 Results and discussions

Dissolved oxygen (DO), pH and oxidation-reduction potential (ORP) are the critical functions for nitrification oxidizing efficiency. An average DO > 2 mg/L, pH 6.5 and ORP range of 180 – 220 mV were performed to save a nitrifying behaviour through the sand column. Fig. 1 presents the breakthrough curves for $NH_3$-N, $Mn^{2+}$ and $Co^{2+}$ through quartz sand column during six days of synthetic wastewater infiltration.

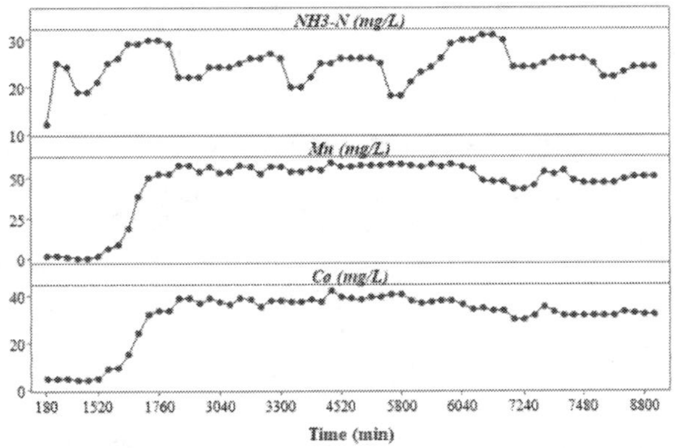


**Fig. 1.** $NH_3$-N, $Mn^{2+}$ and $Co^{2+}$ breakthrough curves with time profile through quartz sand column.

Fig. 2 presents the interaction curves between $NH_3$-N with $Mn^{2+}$ and $Co^{2+}$ and the concentration fluctuation of $Mn^{2+}$ and $Co^{2+}$ over the $NH_3$-N mean. Figure 2 shows a substantial change in $NH_3$ concentration to the $Mn^{2+}$ and $Co^{2+}$ concentrations, indicating a strong interrelationship and impaction between $NH_3$ with $Mn^{2+}$ and $Co^{2+}$.

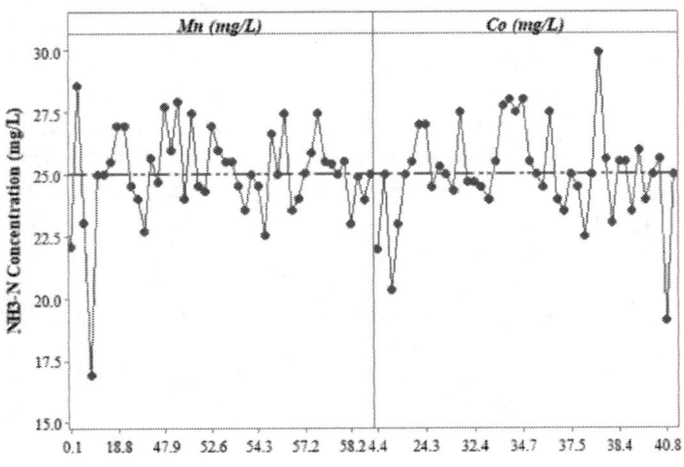


**Fig. 2.** Interaction behavior between $NH_3$-N with $Mn^{2+}$ and $Co^{2+}$.

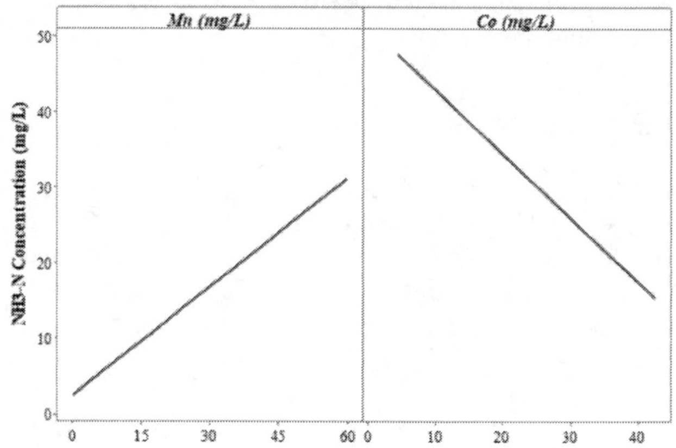


**Fig. 3.** $NH_3$-N factorial regression relationship with $Mn^{2+}$ and $Co^{2+}$.

Perform a linear regression model to show and model the relationship behaviour between $NH_3$-N and $Mn^{2+}$, and $Co^{2+}$ based on P-value and variance inflation factor (VIF) as statistically significant parameters. The interrelation model between $NH_3$-N and $Mn^{2+}$ as linear regression gives a 0.020 p-value and 66.30 VIF value. P-value $< 0.05$ indicates a substantial relationship between $NH_3$-N and $Mn^{2+}$ with a high correlation behaviour between $NH_3$-N and $Mn^{2+}$ according to a high VIF value. Moreover, the interrelation model between $NH_3$-N and $Co^{2+}$ as linear regression gives a 0.037 p-value and 69.40 VIF value. P-value $<$ 0.05 indicates a substantial relationship between $NH_3$-N and $Co^{2+}$ with a high correlation behaviour between $NH_3$-N and $Co^{2+}$ according to a high VIF value. Fig. 3 presents the linear factorial regression model relation between $NH_3$-N with $Mn^{2+}$ and $Co^{2+}$.

As shown in fig. 3, $Co^{2+}$ is negatively correlated to the $NH_3$-N concentration. The biodegradation of $NH_3$ through quartz sand is increased when the $Co^{2+}$ concentration is increased. This means the appearance of $Co^{2+}$ metals are developed the nitrification oxidation as a result of $Co^{2+}$ stimulating the degradation of bacteria and motivating the bacteria growth cells. The positive effect of cobalt ($Co^{2+}$) on ammonium ($NH_4$) degradation in soil column could be due to its role in stimulating the activity of nitrifying bacteria. $Co^{2+}$ is an essential component of vitamin B12, which is an important cofactor in many enzymatic reactions in the microbial community, including nitrification. Therefore, $Co^{2+}$ can enhance the growth and activity of nitrifying bacteria, which in turn can increase the rate of $NH_4$ oxidation to nitrate ($NO_3^-$) in soil. Many researchers have demonstrated this, like Kim, Park, Cho, Nam, ParkBajpai [13] and He, Xie, Ni, LiLi [3]. Despite that, $Mn^{2+}$ is positively correlated to the $NH_3$-N concentration. This means the $Mn^{2+}$ appearance did not stimulate the $NH_3$ oxidation and works as an inhibitor for $NH_3$ degradation that is related to the high applied $Mn^{2+}$ concentration, which may cause toxic bacteria that reduce the $NH_3$-N removal capacity. $Mn^{2+}$ can act as an inhibitor of the microorganisms that perform nitrification, One possible mechanism for this inhibition is the adsorption of $Mn^{2+}$ onto the nitrifying bacteria cells, which can interfere with their metabolism and reduce their activity. Additionally, $Mn^{2+}$ can compete with the nitrifying bacteria for the oxygen and other nutrients that they require for their growth and activity. Overall, the presence of $Mn^{2+}$ in high concentrations can slow down the nitrification process, leading to the accumulation of $NH_4$ in the system. This concluded relation between the $NH_3$-N and $Mn^{2+}$ was demonstrated by He, Xie, Ni, LiLi [3]. However, many studies indicate that $Mn^{2+}$ metals significantly improve anaerobic ammonium oxidation bacteria by applying a low $Mn^{2+}$ concentration. Huang, Gao, PengTao [12] found a double ammonium removal capacity when adding $Mn^{2+}$. Li et al. (2018) also discovered that nitrogen species removal developed when adding $Mn^{2+}$ to the nitrifying anaerobic ammonium oxidizing process. These results indicate that $Mn^{2+}$ is essential for nitrogen species bioremediation.

## 4 Conclusion

These results demonstrated the strong relationship between $NH_4$, $Mn^{2+}$ and $Co^{2+}$. There is a negative correlation between $NH_3$-N concentration with $Co^{2+}$ concentration, and the appearance of $Co^{2+}$ stimulates the bacteria degradation activities and the growth of cells. These factors increase the $NH_4$ oxidation and removal capacity. While $Mn^{2+}$ has positively correlated with the $NH_3$-N concentration, which means $Mn^{2+}$ metals have worked as an inhibitor for $NH_4$ degradation capacity according to the high $Mn^{2+}$ concentration under aerobic nitrification conditions. In general, the interaction between ammonium $NH_4$ and metals can stimulate or inhibitor for $NH_4$ degradation process according to the concentration level and the degradation environment conditions.

## References

1. D. Swathi, P. SabumonS. M. Maliyekkal, International Biodeterioration & Biodegradation, *Microbial mediated anoxic nitrification-denitrification in the presence of nanoscale oxides of manganese.* **119**: p. 499-510 (2017)
2. J. A. CamargoÁ. Alonso, Environment international, *Ecological and toxicological effects of inorganic nitrogen pollution in aquatic ecosystems: a global assessment.* **32**(6): p. 831-849 (2006)
3. T. He, D. Xie, J. Ni, Z. LiZ. Li, Water, *Effect of cobalt, cadmium and manganese on nitrogen removal capacity of Arthrobacter arilaitensis Y-10.* **12**(6): p. 1701 (2020)
4. R. Jin, T. Liu, G. Liu, J. Zhou, J. HuangA. Wang, Applied biochemistry and biotechnology, *Simultaneous heterotrophic nitrification and aerobic denitrification by the marine origin bacterium Pseudomonas sp. ADN-42.* **175**: p. 2000-2011 (2015)
5. A. Abada, M. Sperfeld, R. Carmieli, S. Ben-Dor, I. H. Zhang, A. R. BabbinE. Segev, bioRxiv, *Aerobic bacteria produce nitric oxide via denitrification and trigger algal population collapse.* p. 2021.11. 14.468512 (2021)
6. M. Capodici, S. F. Corsino, D. Di TrapaniG. Viviani, Biochemical Engineering Journal, *Achievement of partial nitrification under different carbon-to-nitrogen ratio and ammonia loading rate for the co-treatment of landfill leachate with municipal wastewater.* **149**: p. 107229 (2019)
7.M. C. van Loosdrecht, P. H. Nielsen, C. M. Lopez-VazquezD. Brdjanovic, *Experimental methods in wastewater treatment.* 2016: IWA publishing.
8. T. DriscollJ. Kartik, WEF Manual of Practice No. FD-3rd; Mcgraw-Hill: New York, NY, USA, *Industrial wastewater management, treatment and disposal.* p. 474-489 (2008).DOI.
9. Z. Hu, K. Chandran, D. GrassoB. F. Smets, Environmental science & technology, *Impact of metal sorption and internalization on nitrification inhibition.* **37**(4): p. 728-734 (2003)
10. P. Gikas, Journal of hazardous materials, *Single and combined effects of nickel (Ni (II)) and cobalt (Co (II)) ions on activated sludge and on other aerobic microorganisms: a review.* **159**(2-3): p. 187-203 (2008)
11. H. Li, H. Yao, D. Zhang, L. Zuo, J. Ren, J. Ma, J. Pei, Y. XuC. Yang, Chemosphere, *Short-and long-term effects of manganese, zinc and copper ions on nitrogen removal in nitritation-anammox process.* **193**: p. 479-488 (2018)

12. X. Huang, D. Gao, S. PengY. Tao, Journal of Environmental Sciences, *Effects of ferrous and manganese ions on anammox process in sequencing batch biofilm reactors.* **26**(5): p. 1034-1039 (2014)
13. J. K. Kim, K. J. Park, K. S. Cho, S.-W. Nam, T.-J. ParkR. Bajpai, Bioresource Technology, *Aerobic nitrification–denitrification by heterotrophic Bacillus strains.* **96**(17): p. 1897-1906 (2005)

# Interface parameters between strengthened flat slab and concrete overlay

*Daniel* Ceres[1*], and *Katarina* Gajdosova[**]

[1]Faculty of Civil Engineering, Slovak University of Technology in Bratislava, Radlinskeho 11, 810 05 Bratislava, Slovakia

**Abstract.** This article aims to the comparison of different parameters for defining the interface characteristics between flat slab without transverse reinforcement and reinforced concrete overlay in non-linear analysis. The input parameters for the interface are stiffness, tensile strength, friction coefficient and cohesion. These coefficients represent uncertainties in non-linear modelling. Non-linear model was developed in ATENA 3D software and simulates punching shear of flat slab without shear reinforcement strengthened by concrete overlay. Two types of interfaces are considered: without reinforcement crossing the interface and with shear reinforcement crossing the interface. Parametric study shows, how different values of interface parameters influence the final punching shear capacity of representative model. Numerical models will be used as a prediction for experimental results of specimens.

## 1 Introduction

In order to overcome the effect of punching shear at connection zone between flat slab and column, the specific type of strengthening is able to enhance the punching shear resistance. The increase in shear resistance of flat slab could be achieved by several strengthening methods. Presented research is focused on enhancing punching shear capacity of flat slab by concrete overlay. The non-linear analytical model presented in this study was created in Atena 3D software and it is based on real experimental fragment of flat slab.

The results from the experimental investigation will be used to calibrate a non-linear model, specifically the input parameters.

## 2 Non-linear analytical model

Atena 3D software provides good corelation between results of analytical model and experimental results, when parameters are appropriately chosen. The major uncertainties of a model represent the number of brick or tetra elements through the depth of the flat slab element, concrete material definition, and interface parameters.

* Corresponding author: daniel.ceres@stuba.sk
** katarina.gajdosova@stuba.sk

## 2.1 Meshing

Parametric study was carried out by Kadlec and Cervenka (2015) [1], according to a number of brick elements through the slab depth. Non-linear models from this parametric study were calibrated with results obtained from experimental tests. Similar parametric study was carried out by Kormošová (2020) [2], where number of elements and type of loading were compared. Reliable results were achieved by using minimum four, better five elements through the slab depth.

Non-linear model presented in this study was created in respect with findings described above. According to flat slab and overlay thickness, the mesh type was applied as brick with absolute size of 30 mm.

## 2.2 Concrete properties

The concrete material is based on fracture-plastic constitutive model. To define interface parameters, the calculation of interface fracture energy is needed. Fracture energy according to Model Code 1990 [3] can be calculated through compressive strength as shown in Equation (1). Dominant value of compressive strength represents the weakest concrete at concrete-concrete interface.

$$G_f = G_{f0}\,(f_c/10)^{0,7} = 30 \cdot (25/10)^{0,7} = 57\ \text{N/m} \tag{1}$$

where $G_{f0}$ is a tabled value of fracture energy of concrete; 30 N/m for $d_{g,max}$=16 mm,
$f_c$ compressive strength of the weakest concrete at interface.

## 2.3 Interface parameters

Interface between the flat slab and concrete overlay is represented by an element with zero thickness, defined only by the contact area between the two layers of specimens. Important interface parameters are normal and tangential stiffness, tensile strength, cohesion, and friction coefficient. Cohesion was calculated considering the Mohr-Coulomb failure criterion in the interface (Figure 1). 3D interface is defined as an open or close one. The open 3D interface allows the realistic behaviour of the interface with full separation of surfaces, while close 3D interface allows mobilization of stresses.

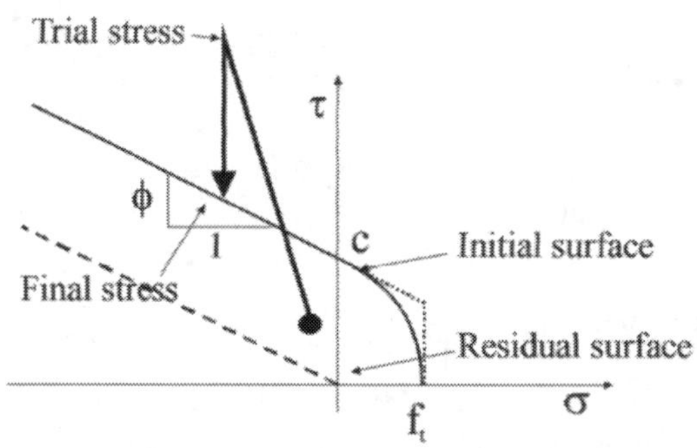


**Fig. 1.** Failure surface according to Mohr-Coulomb [4].

where: $c$ is the cohesion of the interface concrete-concrete,
$f_t$ interface tensile stress,
$\phi$ angle of friction,
$\tau$ interface tangential stress,
$\sigma$ interface normal stress.

The failure criterion is replaced by an ellipsoid in tension. Ellipsoid intersects the axis of normal stress at value of $f_t$ and the axis of tangent stress at a value of $c$ (cohesion) [4]. The cohesion coefficient is not provided in known design procedures, but the cohesion bond strength can be calculated for rough surface as 1.5-2.5 N/mm$^2$ and for a very rough one as 2.5-3.5 N/mm$^2$ [5]. Considering a parametric study of M.E. Mohamad (2014) [5], the theoretical cohesion coefficient and friction coefficient can be empirically determined through peak height of the roughness of interface. Friction coefficient was considered as 0.9 to 1.5 according to Model Code 2010 provisions [6].

For simplification, both normal and tangential stiffnesses ($K_{NN}$, $K_{TT}$, respectively) are considered by the exact values as it is also recommended by the software developers. As Hugo Fernandes stated in his dissertation thesis [7], acceptable results were obtained, when stiffness was calculated according to Model Code 1990 provisions. Equations (2) and (3) represent calculation values used in this study.

$$K_{\mathrm{NN}} = K_{\mathrm{TT}} = \frac{f_{\mathrm{t,i}}^2}{2G_{\mathrm{F,i}}} \tag{2}$$

$$G_{\mathrm{F,i}} = \frac{f_{\mathrm{t,i}}}{f_{\mathrm{t,Conc}}} \cdot G_{\mathrm{F,WeakConc}} \tag{3}$$

where: $f_{\mathrm{t,i}}$ is an interface tensile capacity,
$G_{\mathrm{F,i}}$ interface fracture energy,
$f_{\mathrm{t,Conc}}$ tensile strength of the weakest concrete in interface,
$G_{\mathrm{F,WeakConc}}$ fracture energy of the weakest concrete in interface.

## 2.4 Description of non-linear model

Analysed specimens represent the 1/4 of the experimental flat slab of size of 2.5 x 2.5 m. The original non-strengthened flat slab (marked as $M_0$) was 180 mm thick and its effective depth was 134 mm. Cylindrical compressive strength of concrete element was assumed 25 MPa, the slab was reinforced by longitudinal steel bars of diameter of 16 mm, with spacing of 100 mm in two perpendicular directions $x$ and $y$. The original flat slab element is supported by circular support, which represents column of diameter of 250 mm. Square elements of 50 x 50 mm in width are simulating supports and their position was set on lines of contra-flexure, in which zero bending moments are assumed. Supports serve also as monitoring points, in which the maximum normal force will be monitored.

First set of analysed specimens (marked as $M_1$) imitate strengthening by concrete overlay without shear connectors. Element of concrete overlay also represents the 1/4 of experimental strengthened flat slab strengthened by concrete overlay of size of 2 x 2 m and of 60 mm in thickness. Cylindrical compressive strength of this concrete element representing the concrete overlay was of 30 MPa. The element of concrete overlay was reinforced by the main longitudinal steel bars of diameter of 10 mm, with spacing of 100 mm in two perpendicular directions $x$ and $y$. To achieve the minimum concrete cover to reinforcement with higher reinforcement ratio, steel bars were doubled. To create non-linear fragment marked as $M_{1f}$ Atena 3D default settings [4] of interface concrete-concrete were used and according to low friction coefficient, delamination in interface in early iteration was expected.

Second analysed specimen (marked as $M_{2b}$) imitates strengthening by concrete overlay with shear connectors, modelled as reinforcement with shear heads. Characteristics of elements were the same as described above. Shear connectors were represented by reinforced steel bars intersecting concrete-concrete interface, with diameters of 10 mm. Definition of interface was the same as for $M_{1b}$, therefore similar values were achieved in experimental studies abroad.

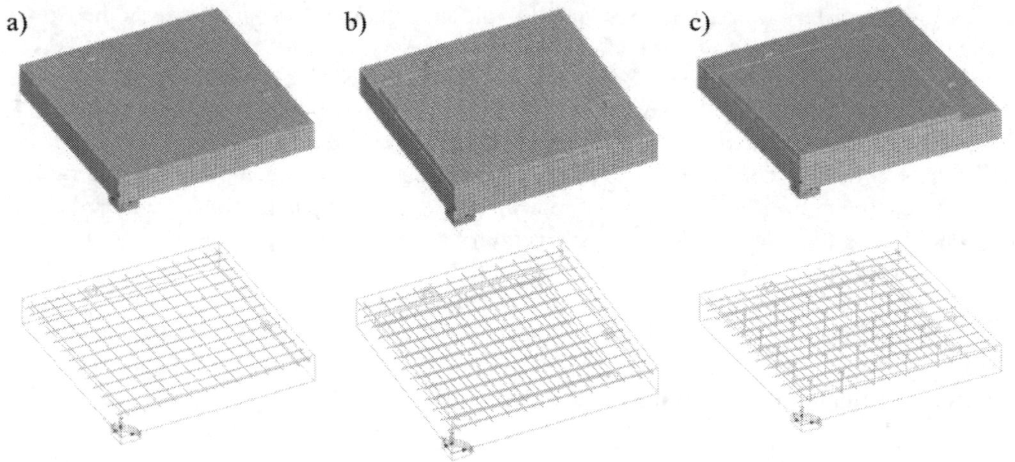


**Fig. 2.** Geometry of analysed fragment: a) fragment $M_0$ b) fragment $M_1$, $M_P$ c) fragment $M_{2b}$.

One modelled fragment marked as $M_P$ represents strengthened flat slab by concrete overlay, where interface concrete-concrete was considered as perfect connection. Both concrete and reinforcement characteristics were taken over previously mentioned models.

Load was applied by deformation gradually increasing in 0.1 mm steps on the circular column element and whole analysis was iterated by Newton-Raphson method in 80 analysis steps.

**Table 1.** Interface parameters.

| | | $M_{1a}$ | $M_{1b}$ | $M_{1c}$ | $M_{1d}$ | $M_{1e}$ | $M_{1f}$ | $M_{2b}$ |
|---|---|---|---|---|---|---|---|---|
| Friction coefficient | $\mu$ [-] | 0.9 | 1.2 | 1.5 | 1.5 | 0.9 | 0.3 | 1.2 |
| Cohesion coefficient | $c$ [MPa] | 2.17 | 3 | 3.75 | 2.5 | 3.5 | 2 | 3 |
| Tensile strength | $f_{t,i}$ [MPa] | 0.769 | 1.062 | 1.328 | 0.885 | 1.24 | 0 | 1.062 |
| Normal stiffness | $K_{NN}$ [x10$^4$ MN/m$^3$] | 1.195 | 1.650 | 2.063 | 2.848 | 3.991 | 20 | 1.650 |
| Tangential stiffness | $K_{TT}$ [x10$^4$ MN/m$^3$] | 1.195 | 1.650 | 2.063 | 1.375 | 1.926 | 20 | 1.650 |

## 3 Results from the NLFEA modelling

The resistance of the flat slab fragment was determined by failure of the flat slab. Failure mode of the flat slab was signalized by a sharp drop in force at the constant deformation as it can be observed from the following diagrams (Figure 3, Figure 4).

Original fragment of flat slab ($M_0$) achieved higher deflection in failure mode as all the strengthened fragments (Figure 3a). Behaviour of fragment $M_{1f}$ was similar to fragment $M_0$ and it was caused by partial delamination from the beginning of the loading process. When full delamination was achieved, the resistance of the flat slab was exceeded, which led to failure. From the beginning of the loading process, the concrete overlay behave as separate element and negatively influenced original flat slab by its weight. Different interface parameters in fragment $M_{1a\text{-}1e}$ did not change the results of loading force, neither deflection and results are quite similar, even when minimum friction according to maximum cohesion coefficient were used.

Figure 3b compares two models $M_{1b}$ and $M_{2b}$ with the same interface parameters, and the only difference in using shear connectors in fragment $M_{2b}$. Fragment $M_{2b}$ achieved higher resistance as fragment $M_{1b}$. It is an object of discussion which peak of the diagram represents failure. At first drop of the peak, shear connectors are able to stabilize concrete overlay from delamination. Fragment reports increasement in both loading force and deflection of the flat slab. Observation of diagram in Figure 3b, can lead to question, if it is possible achieve higher load force with higher deflection by using shear force ($M_{2b}$), when comparing to perfect connection ($M_P$).

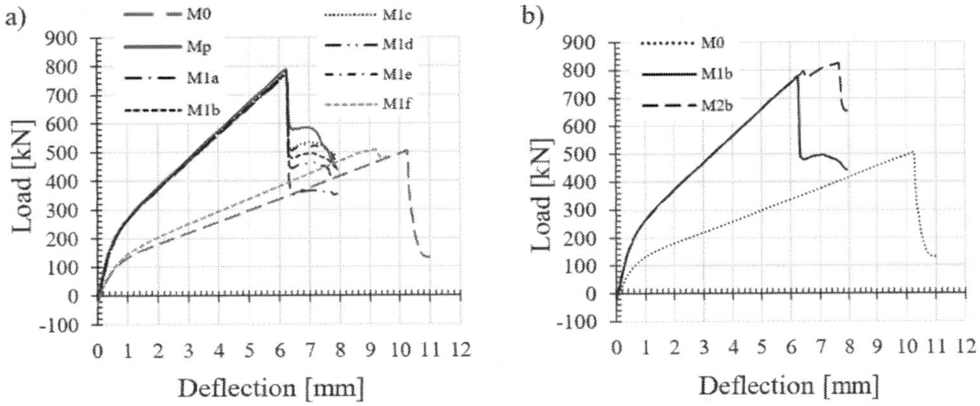


**Fig. 3.** Relationship between load force and deflection.

The major difference between each model is obtained by observing individual results from principal strains and deformation. Principal strain in fragment $M_{1b}$ just before failure indicates for bending failure (Figure 4a). After the delamination of concrete-concrete interface, brittle failure by punching shear appears (Figure 4b). Separate behaviour of the concrete overlay can be observed from Figure 5. From the beginning of loading process, cracks are concentrated along concrete-concrete interface. The behaviour of original flat slab does not depend on the influence of concrete overlay even when reaching failure mode (Figure 5a). Concrete overlay did not significantly contribute to punching shear capacity of the original flat slab.

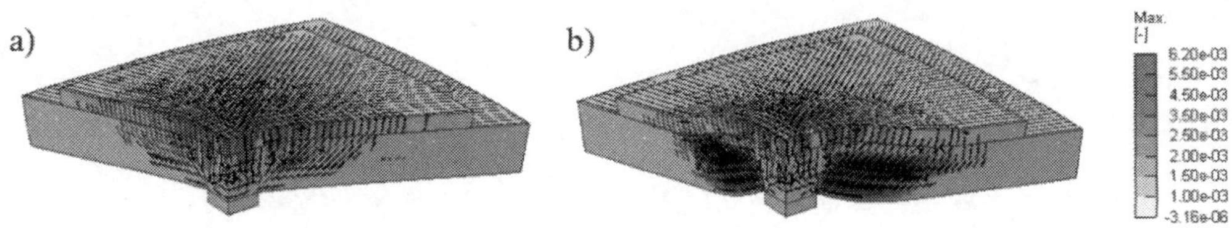


**Fig. 4.** Principal total strain of fragment $M_{1b}$.

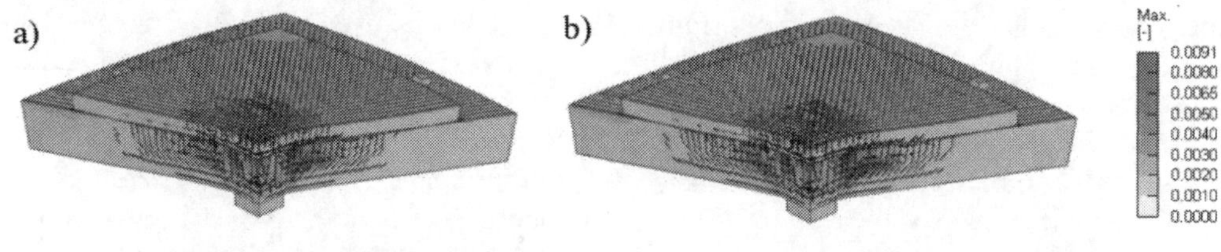


**Fig. 5.** Principal total strain of fragment $M_{1f}$.

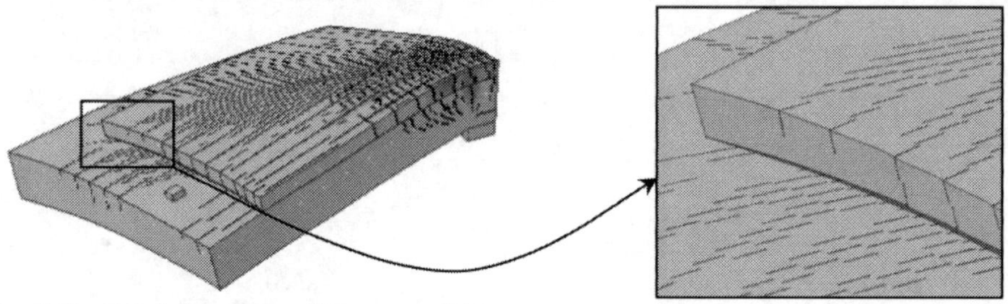

**Fig. 6.** Partial debonding of fragment $M_{2b}$.

## 4 Conclusions

Results of non-linear analytical model show the importance of the interface parameters. Some interface parameters can be obtained by experimental testing of the specimens, e.g. pull out tests. Fragment $M_{1f}$ represents a slip contact in concrete-concrete interface, because of low friction coefficient. Without roughening the surface of flat slab, increase in the punching shear capacity is negligible. Fragment $M_{2b}$ provides good results, when observing the behaviour of the flat slab model. After drop of the peak in diagram (Figure 3b), shear connectors are activated, which increases the stiffness of the flat slab and allows higher values of deflection with increased punching shear capacity of flat slab.

First set of fragments namely from $M_{1a}$ to $M_{1e}$ reach very similar or the exact values of punching shear capacity and also deflection. However, deformation in some load steps before failure could remind delamination, but software did not notice the loss of cohesion. Figure 6 shows this phenomenon at the moment when loading force reaches the punching shear capacity of original slab, which represents 65 percent of punching shear capacity of modelled fragment $M_{2b}$.

Results from this parametric study will serve as calibration models to experimental study of tested flat slabs.

This work was supported by the Scientific Grant Agency VEGA under the contract No. VEGA 1/0310/22.

## References

1. L. Kadlec, V. Cervenka, *Uncertainty of Numerical Models for Punching*. In proceedings of fib Symposium Copenhagen, May 18-20, Denmark
2. Ľ. Kormošová, J. Halvonik, T. Augustín, *Uncertainties in nonlinear modelling of punching,* in Engineering mechanics 2020, pp. 294-297, (2020)
3. CEB/FIP; CEB Bulletin No. 213/214: *CEB-FIP Model Code 90*, Fédération Internationale du Béton, pp. 460, (1993)

4. V. Červenka, L. Jendele, J. Červenka, *ATENA Program Documentation Part 1 Theory*, Atena, 2012: pp. 1-282, (2012)
5. M.E. Mohamad, et al., *Friction and cohesion coefficients of composite concrete-to-concrete bond,* in Cement and Concrete Composites 56, pp. 1-14, (2015)
6. COMITÉ EURO- INTERNATIONAL DU BÉTON *fib Model Code 2010, Final draft – Volume 2* (2012). 148–162 s. ISBN 9782883941069
7. H.D.P. FERNANDES, *Strengthening of flat slabs with reinforced concrete overlay – Analysis and development of the solution.* Universidade Nova de Lisboa (2019)

# Predicting pipeline life after an emergency repair

*Serhii* Palchyk[1*]

[1]Odesa State Academy of Civil Engineering and Architecture, 65029 Didrihson st., 4, Ukraine

**Abstract.** The operating conditions of main pipelines are analysed and it is established that fatigue and corrosion damage are the main causes of their destruction. The paper provides a substantiated answer to the question of the effectiveness of repairs using overlay elements. The number of load cycles that will take place before the destruction of the emergency section after repair with the use of overlay elements is determined by an example.

## 1 Introduction

The reliability of pipelines largely determines the continuity of industrial operations. According to statistical data, the number of pipeline failures in Ukraine tends to increase.

Failures occur mainly due to corrosion deterioration and ageing of pipelines, imperfect design solutions, factory defects in pipes, defects in construction and installation and repair works, by the fault of operating personnel and other reasons. A range of defects on the pipeline walls, group or continuous corrosion sores reduce the pipeline's load-bearing capacity.

Pipeline rupture accidents are relatively rare, but even a minor pipeline rupture can cause enormous damage due to environmental pollution, possible explosions and fires and loss of life. This is why maintaining the integrity of the pipeline's line section is one of the main challenges in pipeline transport.

Prompt and high quality elimination of the damage is of great importance, as pipeline failures are often accompanied by high economic losses and a long-term deterioration of the quality of environment.

The easiest and most common way to deal with emergencies on operating pipelines is to use various types of overlays. Modern overlay repair techniques have a drawback in reducing the safe operating life of pipelines and some of them can be used only as a temporary solution.

The existing methods for prediction of pipeline life including cyclic loading are based on the known Coffin-Manson low-cycle fatigue equation. In this work the method of low-cycle durability estimation of pipes with damage after emergency repair is applied for prediction of service life of gas pipelines.

[*] Corresponding author: spalchyk@ukr.net

## 2 Current state of the researched issue

Numerous internal technical improvements have been developed and applied to avoid most failures in the day-to-day operation of onshore and offshore facilities.

Ukraine's gas transmission system has been in operation for an average of 30 to 52 years, depending on the age of its components. During its operation, a significant part of the main gas pipelines and process equipment have either exhausted their service life or have been subject to repeated maintenance and overhauls. Some have even become obsolete. The specified service life of gas pipelines and gas transmission system facilities requires significant investments to improve reliability and maintain them in a technically sound condition, as well as to comply with technological regulations for gas pumping.

Various organizations exist around the world to record pipeline failures. The purpose of these organizations is to systematize and analyse the causes of failures in gas pipelines.

In 1982, six European gas transmission system operators took the initiative to collect data on unintentional gas releases from transmission pipeline systems. This cooperation was formalized through the establishment of EGIG (European Gas Pipeline Accident Data Group). The aim of this initiative was to provide a broad basis for calculating safety indicators for pipeline systems in Europe, thus providing a reliable picture of the number, frequency and nature of incidents. Today, EGIG brings together seventeen European gas transmission system operators and has an extensive database of pipeline incidents collected since 1970.

According to the EGIG report, the causes of accidents on European gas pipelines are as follows (Fig.1):

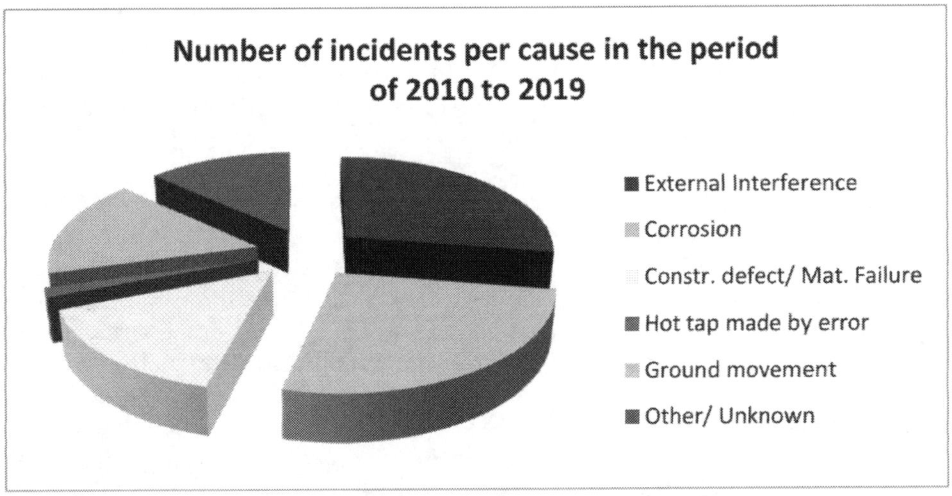


**Fig. 1.** Number of incidents per cause in the period of 2010 to 2019 (according to the report of EGIG [1]).

The diagram shows that one of the most common mechanical failures in a range of gas pipelines is caused by corrosion of equipment and systems.

A variety of causes and types of failures that are caused by corrosion have been investigated in the literature. Corrosion is a significant threat to pipeline systems, which can lead to leaks. Different types of approaches to analysis of corrosion causes have been studied in [2] for mechanical systems. Their findings have been used to predict accidents. Different types of accident assessments have been applied with regard to their specific features for different types of damage. Several scientific methods have been proposed to control the occurrence of cracks due to environmental influences with service life assessment criteria depending on the operating conditions [3].

Calculations for the strength of structural elements under long-term cyclic loads were given in [4], where defect-free materials and fatigue failure in the classical sense of this phenomenon were assumed. Experimental studies are also known, based on the results of which researchers constructed limit diagrams of the cyclic strength of structural elements with cracks, i.e. their residual cyclic strength [5, 6].

In studies [7], an automatic pipeline emergency assessment system has been developed, allowing the operator to interact with it. The capabilities of the emergency assessment system can be upgraded in conventional systems by obtaining data using spectral analysis and special instrumentation. In [8], authors have described various causes of pipe wall thinning and developed methods for repair and prediction of wall thinning rate based on measurement data. As a result of painstaking work, scientists from different countries were able to summarize and combine a variety of prediction methods for different types of pipes used in gas pipeline systems. A model for assessing the technical condition of submarine pipeline systems has been proposed in [9].

J.Z. Sikorska, M. Hodkiewicz, L. Ma considered the possibility of applying a three-level process business model for engineering problems. The proposed method allows selecting a suitable model to describe a three-level failure model [10]. In [11] an approach of plastic pipe fractures prediction using finite element method for 24-pipe with ring-shaped crack has been proposed. The construction of models based on statistical data for emergency assessment has been proposed in [12].

A simple model of crack growth in predicting accidents in gas transmission system using Bayesian interference and finite element method was proposed in [13]. On the basis of probabilistic methods a branch standard for prognosis of technical condition of mechanical systems was developed in Ukraine, and it is applicable to pipes as well [14]. Nowadays there are many new researches of tubing defects and pipeline emergency forecasts on the basis of statistical data with the help of different methods, taking into account various factors [15 - 21]. But the issue of operational reliability and durability of gas pipelines is very important, and therefore requires further detailed and in-depth study using new scientific achievements.

In Ukraine, important contributions to the problem was made by scientists such as A.B. Aybinder, P.P. Borodavkin, A.A. Korshak, V.V. Kurochkin, N.A. Malyushin, V.Y. Grudz, D.F. Timkiv, L.S. Shlapak, V.M. Ivashin, and E.I. Krizhanivsky, B.S. Bilobran. It should be noted that the research papers [22, 23] investigated the influence of operation parameters of a non-gas pipeline on its flow capacity, and the works [24, 25] describe the stressed - deformed state of line part of gas pipelines, problems of forecasting durability of pipelines due to low cycle stresses were considered. The Institute of Electric Welding named after E.O. Paton studied the problems of forecasting durability of pipelines taking into account low cycle stresses. The Paton Electric Welding Institute developed a method of welding multilayer inserts into a pipeline section made of single-layered pipes. When this section is destroyed by internal pressure, this method makes it possible to stop the avalanche crack propagation process in the multilayer inserts.

The process of stress corrosion cracking in existing main pipelines is spread over decades. Today, as the internal pressure and diameter of pipes increase, this process is becoming "younger", causing failure after 20 years or less of pipe service.

For the crack formation process to occur, the following factors must be present simultaneously

- sufficient cyclic stresses;
- a corrosive environment;
- a material that is incapable of resisting corrosion (pipe steels are such materials).

The process of crack initiation and development can be divided into three stages: initiation, propagation (stable slow and rapid catastrophic) and arrest. At each of these stages, once the fracture area is detected, it is subject to repair. One of the main repair methods is the use of overlay elements. Such repairs are considered temporary. Below we show that as a result of repair using overlay elements, the pipeline service life can be increased.

## 3 Residual life prediction

The Coffin-Manson low-cycle fatigue equation is the basis for existing methods for predicting the cyclic service life of pipelines. This equation relates the initial basic parameters – plastic strain amplitude $\varepsilon_a$ and relative contraction $\psi$. In terms of being able to determine in practice an estimate of local value for $\varepsilon_a$ for pipeline components is problematic due to the difficulty of finding the radius of curvature of most vertices of defects. This method is therefore appropriate for structural elements for which we can determine local values at the defect site $\psi$ and $\varepsilon_a$. These difficulties lead to the fact that the regulated methods of calculation of low-cycle durability of pipelines, in general, should be referred to the category of theoretical ones, at least - for the objects of magistral pipeline transport. The above-said predetermines finding of other approaches to estimation of pipeline durability.

Let's consider laws of multi-cycle and low-cycle serviceability of metal with reference to pipe steels, taking into account peculiarities of stressed state of metal of structural elements of pipelines.

Metal fatigue will be understood as a degree of damageability of metal structure under repeated action of stresses. Stresses can be normal, tangential or equivalent. It should be taken into account that the limit stresses repeatedly applied to the most common carbon and low-alloy steels are several times less than the destructive stresses under single loading.

Most of the proposed low cycle fatigue equations relate the number of cycles to failure $N$, the plastic strain amplitude $\varepsilon_{nl}$ and total strain amplitude $\varepsilon_{np}$. They are Orowan equation:

$$N\,\varepsilon_{nl} = const; \tag{1}$$

equation of Coffin-Manson:

$$N^{m_u}\varepsilon_{nl} = C_u, \tag{2}$$

where $m_u$ and $C_u$ are constants. Constant $C_u$ is related to limit ductility of the material:

$$C_u = \frac{1}{2}\varepsilon_{np} = \frac{1}{2}\log\left(\frac{1}{1-\psi}\right), \tag{3}$$

where $\psi$ – relative contraction at the breakage.

Substituting the summary strain amplitude $\varepsilon_a = \varepsilon_{nl} + \varepsilon_y$ in the equation of Coffin-Manson, where $\varepsilon_y$ is an elastic strain amplitude, the equation of durability can be re-written in the following form:

$$\varepsilon_a = \frac{1}{4}\sqrt[\frac{1}{m_u}]{N}\cdot\varepsilon_{np} + \frac{\sigma_{-l}}{E}. \tag{4}$$

Where $\sigma_{-l}$ is a fatigue limit. Sometimes, this equation can be represented in relative strain coefficients:

$$\sigma_a^* = \varepsilon_a \cdot E = \frac{1}{4}\sqrt[\frac{1}{m_u}]{N}\cdot\varepsilon_{np} + \sigma_{-l}. \tag{5}$$

Degree $m_u$ in these formulas depends on the mechanical characteristics of the metal. Steels with a high index $m_u$ are characterized by a lower ratio of tensile strength to yield strength. For example, when increasing $\sigma_в$ from 700 to 1400MPa coefficient $m_u$ changes from $0.50$ to $0.65$. This relation is described (approximately) by a formula: $m_u=0.5+0.0002(\sigma_в -700)$. Let us note that parameter $m_u$ correlates with a coefficient $m$. When $m < 0.15$ $m_u = 0.2$. When $m > 0.15$ parameter $m_u$ depends linearly on the coefficient of strain hardening: $m_u = 0.2 + 2.4 (m - 0.125)$ [26].

Manson, based on strain and force criteria, proposed a more general low cycle fatigue equation in amplitudes of total strain:

$$\varepsilon_a = SN^{-s} + TN^t,$$

where $S$, $s$, $T$ *and* $t$ are material constants, while for many materials they are as follows:

$$s = 0.12; S = 3.5\frac{\sigma_в}{E}; t = 0.6; T = log\frac{1}{1-\psi}.$$

Longevity of an element is defined by integrating the equation

$$N_{mp} = \int_{h_0}^{h_{kp}} \frac{dh}{C_\sigma \cdot (\Delta K_{i\varepsilon})^{n_\sigma}}, \tag{6}$$

where $h_0$ and $h_{kp}$ - initial and critical crack depths. The equation (6) can be represented in the following shape:

$$N_{mp} = N_0 \cdot K_n. \tag{7}$$

Where $K_n$ in the first approximation equals to $h_0/h_{kp}$, $N_0$ is defined by the equation (7) when the initial value of coefficient of strain intensity $N_{mp}K_{i\varepsilon}^{(0)}$:

$$N_0 = \frac{h_{kp}-h_0}{C_\sigma \left(K_{i\varepsilon}^{(0)}\right)^{n_\sigma}}. \tag{8}$$

Taking into account $K_n$ and $N=t/v$ this formula can be re-written as follows:

$$t_p = \frac{\delta_0 \cdot \eta_0 (n_h - 1)}{v \cdot n_h \cdot C_\sigma \left(K_{i\varepsilon}^{(0)}\right)^{n_\sigma}}, \tag{9}$$

where $t_p$ is a time till an element is destroyed; $\delta_o$ is a thickness of element; $n_h = h_{kp}/h_o$; $v$ – frequency of load cycles.

Let's turn our attention to a practical example of calculations. Let a constructive element to function under periodic strain cycle with a maximum load $\sigma_{max}$, which equals to working load $\sigma_p$. Then, the working load equals to $\sigma_p=0.67\sigma_i$, namely $\sigma_p=201MПa$. Then a coefficient of strain hardening $m = 0.23$, and relative contraction $\psi = 52$ %.

With the specified initial data, the cyclic fracture durability parameters will be $n_\sigma= 1.23$ and $C_\sigma= 1.116 \cdot 10^{-4}$. According to diagnostic data, the initial depth of the fracture-like defect $h_0=3.8mm$. Then we find the critical defect depth corresponding to the operating load $\sigma_p$. The steel 17ГС is a plastic material with a parameter of fracture resistance $\alpha_{тр}$ [26] equals to 1 ($\alpha_{тр}=1.0$). This suggests that an element (pipe) with an extended crack-like defect will fracture at pressures that produce average stresses in the weakened model section close in magnitude to the time resistance of the metal $\sigma_в$. In other words, the strength of such elements will depend proportionally on the degree of weakening of the wall:

$$\sigma_{HB} = \sigma_{B}^{0M}\left(1 - \frac{h}{\delta_0}\right). \tag{10}$$

Let us substitute $\sigma_{HB}$ with a value of work load $\sigma_p$, we get $h_{kp} = 9.8mm$. The relative depth of the defect is $\eta_{kp} = 0.71$, and the initial value $\eta_0 = 0.275$.

Then, let us find a coefficient of strain intensity $K_{i\varepsilon}$. Under $\eta_0 = 0.275$ according to GOST 25.506-85 (State standard), let us find a correction function $Y_5 = 2.85$. Then, a coefficient of strain intensity will be equal to

$$K_i = 201\sqrt{0.0038} \cdot 2.85 \approx 35MPa\sqrt{M}. \tag{11}$$

Namely:

$$K_{i\varepsilon} = \left(\frac{35}{300}\right)^{\frac{2}{1+0.23}} \approx 0.0304\,. \tag{12}$$

Taking into account the value found for $K_{i\varepsilon}$, let's use the formula (12) in order to find $N_{\text{тр}}$:

$$N_{mp} = \frac{3.8}{9.8}\frac{(9.8-3.8)\cdot 10^{-3}}{1.116\cdot 10^{-4}\cdot(0.0304)^{1.23}} \approx 1531\text{cycles}.$$

This number of cycles is valid, if the starting condition is a zero number of cycles: $P_{min}=0$; $P_{max}=P_p$, where $P_p$ is a work load ($P_p=3.35$MPa). A reduction in amplitude of the strain rate factor by a factor of two results in an almost 2.5 - fold increase in durability. At frequency of loading cycles $\eta$ = 365 cycles /year, we obtain a calendar time estimate till destruction $t_p$= 4.2year. Thus, the parameters of the kinetic equation of low-cycle crack resistance have been estimated and formulas for calculating the lifetime of elements with crack-like defects have been obtained.

## 4 Conclusions

The article analyses the operating conditions of magistral pipelines and establishes that fatigue and corrosion-fatigue damage are the main causes of their destruction. The existing methods for predicting the durability of pipelines require improvement, as they cannot fully take into account all operational factors.

The use of a system of comprehensive monitoring of the technical condition of magistral welded pipelines and timely repair of its parts will reduce the level of premature corrosion wear of components, increase the overhaul period of technological equipment and increase the level of its safe operation.

The article sought to answer the question of how successful the use of overhead elements is in pipeline repair. A specific example is used to demonstrate an algorithm for determining the residual service life of a pipeline after repair.

Thus, the article solved an urgent scientific and technical problem by using the Coffin-Manson equation to prove the effectiveness of repairing gas pipeline defects with the help of overlay elements and established that the number of cycles that will pass before the defective section of the pipeline is destroyed is several years.

## References

1. Gas pipeline incidents, 11th Report of the European Gas Pipeline Incident Data Group (period 1970 – 2019) (egig.eu/report)
2. P.V. Varde, M.G. Pecht, *Life Prediction*. In: Risk-Based Eng. An Integr. Approach to Complex Syst. Ref. to Nucl. Plants. Singapore: Springer Singapore, 115–39 (2011)
3. P.L. Andresen, F.P. Ford, *Life prediction by mechanistic modeling and system monitoring of environmental cracking of iron and nickel alloys in aqueous systems*. Mater Sci Eng A **103**, 167–84 (1988)
4. A. Carpinteri, *Handbook of fatigue crack propagation in metallic structures*, Amsterdam: Elsevier (1994)
5. P. Forrest, *Fatigue of metals*, 9781483160733, (1962)
6. R.B. Heywood, *Designing Against Fatigue*, 0412068206, (1961)
7. W.G. Papadimitriou, *Autonomous Remaining Useful Life Estimation*, 20080004839, (2008)
8. Nishiguchi, F. Inada, M. Takahashi, B. Ogawa, T. Inagaki, T. Ohira, et al. *A review: Japanese pipe wall thinning management based on JSME rules and recent R&D studies performed to enhance the rules*, E-J Adv Maint **2**, 14–24 (2010)
9. P. Vaidya, M. Rausand, *Remaining useful life, technical health, and life extension*, Proc Inst Mech Eng Part O J Risk Reliab **225**, 219–31 (2011)
10. J.Z. Sikorska, M. Hodkiewicz, L. Ma, *Prognostic modelling options for remaining useful life estimation by industry*, Mech Syst Signal Process **25**, 1803–36 (2011)
11. Y. Li, K. Hasegawa, P. Hoang, B. Bezensek, *Prediction Method for Plastic Collapse of Pipes Subjected to Combined Bending and Torsion Moments*. J Press Vessel Technol **134** (2010)
12. X-S. Si, W. Wang, C-H. Hu, D-H. Zhou, *Remaining useful life estimation – A review on the statistical data driven approaches*, Eur J Oper Res **213**, 1–14 (2011)
13. A. Coppe, M.J. Pais, R.T. Haftka, N.H. Kim, *Using a Simple Crack Growth Model in Predicting Remaining Useful Life*, J Aircr 2012 **49**,1965–73 (2012)
14. National standard of Ukraine. *Evaluation and prediction reserve resource (service term) technical systems*. DSTU. UkrNDNC, Kyiv (2015).
15. H. Rezaei, B. Ryan, I. Stoianov, *Pipe Failure Analysis and Impact of Dynamic Hydraulic Conditions in Water Supply Networks,* Procedia Eng 2015 **119**, 253–62 (2015)
16. M.S. El-Abbasy, A. Senouci, T. Zayed, F. Mirahadi, L. Parvizsedghy, *Artificial neural network models for predicting condition of offshore oil and gas pipelines,* Autom Constr **45**, 50–65 (2014)
17. M. Mahmoodian, V. Aryai, *Structural failure assessment of buried steel water pipes subject to corrosive environment*, Urban Water J **14**, 1023–30 (2017)
18. R. Farmani, K. Kakoudakis, K. Behzadian, D. Butler, *Pipe Failure Prediction in Water Distribution Systems Considering Static and Dynamic Factors*. Procedia Eng 2017 **186**, 117–26 (2017)
19. L.S. Lahmadi, N. Terrissa, A. Zerhouni, *Data-driven method for estimating the remaining useful life of a Composite Drill Pipe*, IEEE, (2018)

20. R. Tavakoli et al. *Remaining useful life prediction of water pipes using artificial neural network and adaptive neuro fuzzy inference system models*. Texas: UTA Libraries, (2018)
21. B. Cai B, X. Shao, Y. Liu, X. Kong, H. Wang, H. Xu et al. *Remaining useful life estimation of structure systems under the influence of multiple causes: Subsea pipelines as a case study*. IEEE Trans Ind Electron **67**, 5737-47 (2020)
22. M.D. Serediuk, *Study of the peculiarities of the hydraulic characteristics of a non-isothermal oil pipeline*, Exploration and development of oil and gas deposits **1**, 6, 96-100 (2003)
23. O.M. Mandryk, *Analysis of the causes of emergencies and destruction of main gas pipelines*, Scientific Bulletin of NLTU of Ukraine **25.1**, 155-162 (2015)
24. V. Grudz, *Typical calculations of reliability indicators of gas and oil supply systems. Complex industry methodology*, Ivano-Frankivsk, Fakel, 76 p., (2009)
25. O. Stepova, I. Rassoha, L. Blazhko, O. Hanoshenko, *Calculation of Lifetime of Steel Oil Pipelines with the Account of Corrosive Environment Affect*. Proceedings of the 2nd International Conference on Building Innovations. ICBI 2019. Lecture Notes in Civil Engineering, Springer **73** (2020)
26. M.D. Serediuk, J.V. Yakymiv, V.P.Lisafin, *Pipeline transport of oil and oil products*, Ivano-Frankivsk, IFNTUOG, 517 p. (2002)

# PV panels as a renewable energy source for residential building – an economical comparison before and during the energy crisis

*Michal* Brandtner[1*], *Vaclav* Sotolar[1], *Adam* Bohacek[1] and *Vaclav* Venkrbec[1,2]

[1] Brno University of Technology, Faculty of Civil Engineering, Veveri 331/95, 60200 Brno, Czech Republic
[2] UiT The Arctic University of Norway, Faculty of Engineering Science and technology, Lodve Langesgate 2, 8514 Narvik, Norway

**Abstract.** The application of renewable energy sources is a current trend today due to the limited sources of fossil. A substantial part of these energy sources is solar radiation. The development of electricity prices was significantly affected by two global crises, the COVID-19 pandemic and the energy crisis associated with the Russia-Ukraine conflict. The essence of the article is to compare the economic sense and appropriateness of investing in a photovoltaic power plant system and its return over time, considering the prediction of the development of electricity prices between 2020 and 2023 and subsequently also with a 15-year price prediction. The article deals with investment, energy consumption and financial evaluation of a PV system with PV panels installed on the ventilated façade of a residential building in the Czech Republic. The conclusions of the article show that the increase in investment in the PV panels themselves between 2020 and 2023 is a crucial factor for the return of the entire system. Predictive models of the development of electricity prices and associated liquidity of the investment also play a significant role in this issue.

## 1 Introduction

European countries have agreed on a new 2030 Framework on climate and energy, which sets a target for the share of renewables to reach at least 27 % of the gross final energy consumption of the European Union by 2030 [1]. The development of renewable energy sources is a promising concept to tackle climate change and reduce greenhouse gas emissions [2]. It also aims at reducing energy dependency between countries [3], which has been a current characteristic goal for most European countries in recent years, among other reasons due to the energy crisis resulting from the ongoing Russia-Ukraine conflict [4].
The development of electricity prices was significantly influenced by two global crises, the COVID-19 pandemic and the energy crisis associated with the Russian-Ukrainian conflict [4]. On the basis of global crises, the price of electricity is rising. However, many developed European countries want to be independent from Russia and other great powers, so they are

* Corresponding author: 159111@vutbr.cz

looking for a solution to the energy crisis in renewable sources [5]. The production of these systems very often ends up depending on another commodity that the country is unable to produce on its own.

Photovoltaic panels are used for the production of electrical energy, which is produced thanks to light radiation, but not thermal radiation. The most famous photovoltaic panels are made of silicon cells (crystalline cells), in which direct current is generated. The produced current is led by special cabling to the voltage converter. Direct current is converted into alternating current. From there, the electric current is distributed to the battery storage of electrical energy, or to the home electrical grid. In the event of a surplus of electricity, energy can be supplied to the distribution electricity network.

Photovoltaic energy production is one of the fastest growing energy sources, which is influenced by the amount of sunlight and the temperature of the photovoltaic panel. The efficiency of photovoltaic panels decreases with increasing temperature. Article [6] deals with the experimental investigation of photovoltaic panel performance using phase change materials. These materials can absorb a significant amount of heat when used on the back of the panel, increasing their output power.

PV cell is a device that converts solar energy directly into electricity. PV cells are made from materials we call semiconductors. The basic material of the semiconductor is silicon. However, it has the disadvantage that in its raw state it is a very poor conductor of electricity. For this reason, silicon is combined with other elements like phosphorus and boron. Semiconductors created by mixing silicon and phosphorus have an excess of electrons. These semiconductors are referred to as N-type semiconductors. Semiconductors that are created by mixing silicon and boron, on the other hand, are deficient in electrons. Places where electrons are missing are called holes. These holes have a positive electric charge, so we call these semiconductors P-type semiconductors. If we connect an N-type and a P-type semiconductor together, the electrons begin to spontaneously move toward the P-type semiconductor and the holes begin to move toward the N-type semiconductor. In places where where these semiconductors are connected is a barrier that prevents electrons and holes from moving across the entire semiconductor. When the PV cell is irradiated with sunlight, this barrier is overcome, and the electrons penetrate into the P-type semiconductor and the holes into the N-type semiconductor. This creates an electric voltage on the PV cell [7]. The following kinds of PV panels are currently known: (a) Silicon crystalline solar panel; (b) Thin film solar panels; (c) Perovskite solar cells; (d) Organic solar panel; e) Photovoltaic double-shell panels (PV-DSF); (f) Building Integrated Photovoltaics (BIPV); (g) Photovoltaic façades [7, 8, 9].

The history of photovoltaic panels dates back to the 50s of the last century, when scientists first discovered the photovoltaic effect. The first solar-powered building was built in 1979, and the use of PV panels on building façades has grown steadily since then. Nowadays, the use of photovoltaic panels on building façades is becoming increasingly popular and the market is growing rapidly. This trend is driven by the growing interest in renewable energy sources as well as efforts to reduce energy costs and reduce the carbon footprint of buildings. In recent years, building-integrated photovoltaics (BIPV), where PV panels are integrated into the design of the building rather than being added as an afterthought, has become more popular. This approach allows for more aesthetically pleasing designs and has helped to dispel the notion that PV panels are unattractive and detract from the appearance of a building.

In addition, the development of smart grid technology will allow building owners to better manage energy consumption and integrate their PV panels into the larger energy grid, further reducing costs and increasing the efficiency of the technology.

From the above, it can be logically deduced that façade systems are currently unique and have not been subjected to the necessary studies. The aim of this article is to explore one such solution, in this case on a specific case study of an apartment building in the Czech Republic.

This paper fills a gap in the literature and contributes to the overall expansion of knowledge in the field of photovoltaic systems.

The following chapters of the paper are structured as follows: chapter 2 presents the methods and background facts about the case study, chapter 3 summarizes the findings and discusses them, and is followed by chapter 4, which summarizes the main highlights and provides conclusions.

## 2 Methods

### 2.1 Parameters for installing PV panels on the façade

PV panels can be placed on the façade for architectural reasons or because they cannot be placed on the roof structure for other reasons. The best efficiency of photovoltaic panels is usually when installed at an inclination of 30° to 40°. Another important aspect in the placement of photovoltaic panels is the orientation to the cardinal directions as shown in Figure 1 [10].

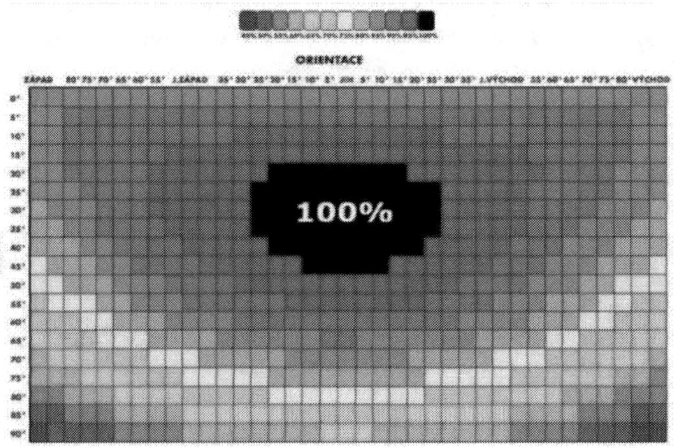


**Fig. 1.** Effect of orientation to cardinal directions and inclination of PV panels on their performance [10].

The ideal placement is on the south façade, while the north façade is not suitable. On the façade, PV panels can be placed at 90° and serve as the top layer of a ventilated. Panels mounted completely vertically show an efficiency loss of about 25 % compared to panels at an optimum inclination of 35°. This placement is most common in cases where the aim is not to disturb the architecturally determined mass of the building, as visible in Figure 2. Another option is to install the panels at a lower angle, which usually increases the efficiency of energy production, but at the detriment of poorer aesthetics and possible shading of the building interior. Such an installation is suitable, i.e. for office spaces. The PV plant has a total area of 53.6 m². The most important value in the calculation is also the average daily sunshine. In recent years, the average daily sunshine was 8 hours, however, only 5 hours of sunshine has been considered, due to the lower electricity production and the earlier eventual financial return of the PV plant. In the case study presented below, option b) was chosen, i.e. the placement of the PV panels at 90°, and only on the south façade of the building.

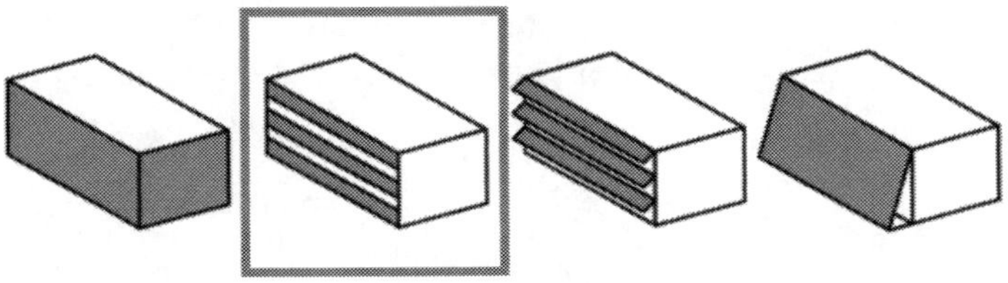


**Fig. 2.** Options for placing photovoltaic panels on the façade of a building with a selected option for a case study [11].

## 2.2 Reference building for case study and calculation of electricity consumption

The case study considers the possibility of using solar energy for the needs of an apartment building consisting of three floors above ground and one underground floor. 3D model of building is visible in Figure 3. There are three apartments on each floor and Table 2 shows the reference set of domestic appliances and their average running time per day. The basic parameters of the PV system are shown in Table 1. The PV panels are fixed to a support grid, which is further anchored within the south façade of the building to the supporting structure. The electricity generated will be stored in LiFePO4 batteries and in case of surplus electricity will be transferred to the distribution grid.

**Fig. 3.** 3D model of case study reference building [authors own work].

**Table 1.** Basic parameters of PV panels.

| | | | |
|---|---|---|---|
| **Dimensions** | Atypical | **Nominal voltage** | 43.4 V |
| **Cell type** | Mono-crystalline | **Max. load current** | 9.22 A |
| **Nominal panel output** | 400 Wp | **Voltage at idle** | 52.7 V |
| **Weight per 1 m$^2$** | 12 kg | **Max. system voltage** | 1000 V |
| **Warranty period** | 25 years | **Total area of PV panels** | 53.6 m$^2$ |
| **Max. panel efficiency** | 19.4 % | **Average daily hours of sunshine** | 5 hours |

Table 2 shows the electricity consumption per 1 year. Each housing unit consumes approximately 23.6 kWh per day and 212.0 kWh of electricity is consumed per day in total.

The total electricity consumption of the apartment building for 1 year is 77,388 kWh. Then, the electricity production of the PV power plant per 1 year is shown in Table 3.

**Table 2.** Electricity consumption of an apartment building.

| Appliance type | Average power input [kW] | Appliance use time [hours] | Daily consumption [kWh] |
|---|---|---|---|
| Household lighting | 0.80 | 2.5 | 2.0 |
| Electric boiler, volume 200 l | 2.20 | 1.0 | 2.2 |
| Electric hair dryer | 2.30 | 0.15 | 0.3 |
| Combined fridge | 0.30 | 1.0 | 0.3 |
| Electric kettle | 2.20 | 0.25 | 0.6 |
| Microwave oven | 0.90 | 0.25 | 0.2 |
| Built-in oven | 3.30 | 0.5 | 1.7 |
| Electric induction cooktop | 6.50 | 1.0 | 6.5 |
| Electric cooker hood | 0.08 | 1.0 | 0.1 |
| Dishwasher | 0.80 | 1.5 | 1.2 |
| Washing machine | 1.00 | 1.0 | 1.0 |
| Clothes dryer | 4.50 | 1.0 | 4.5 |
| Wi-Fi router | 0.01 | 24.0 | 0.2 |
| LCD TV | 0.20 | 2.0 | 0.4 |
| **Total electricity consumption of 1 apartment unit per day [kWh]** | | | 23.6 |
| **Number of apartment units in the apartment building [pcs]** | | | 9 |
| **Total electricity consumption of the apartment building per day [kWh]** | | | 212.0 |
| **Total electricity consumption of the apartment building per 1 year [kWh]** | | | 77.388 |

**Table 3.** Electricity production of the PV power plant per 1 year.

| | |
|---|---|
| **Generated power of one panel per 1 m2 [Wp] per hour** | 400 |
| **PV panel efficiency at 90° inclination [%]** | 70 |
| **Actual power of one panel per m² [Wr] per hour** | 280 |
| **Total PV panel area [m2]** | 53.6 |
| **Electricity produced per hour [kW]** | 15.01 |
| **Average daily sunshine time [hours]** | 5 |
| **Total electricity produced per day [kWh]** | 75.04 |
| **Total electricity produced per year [kWh]** | 27 389.6 |

## 3 Results and discussion

The objective of the case study was to find out how long it takes for a PV plant to be profitable. That is, how long it takes to return to the investor the initial cost that was invested. Table 4 shows the calculation of the cost of a PV power plant for the façade of an apartment building. The calculation takes into account the administrative work, the actual installation and delivery of the PV plant to the final site and its installation. The total cost of the PV power plant is € 27,627 in 2020 and € 43,073 in 2023 as shown in Table 4. No state or other subsidies were considered.

The return calculation for the PV system implemented in 2020 is shown in Table 5. There is a visible decrease in efficiency over time. Also, the change in the unit market price of electricity has a substantial role in the return on investment. This parameter has been taken for the study from the Energy Brainpool 2022 forecasting model [12], which assumes a rapid decrease in the market price of electricity after 2030. This results in a slower return on investment in the single years after 2030. This trend is also evident in the sniff chart in Figure 6. Energy consumption is assumed to be constant in future years. This parameter is also questionable given the tendency of end users to behave responsibly and sustainably and to save electricity.

**Table 4.** Total cost of the PV power plant for the apartment building in 2020 and in 2023.

| Item | Pcs | Unit | Unit cost in 2020 [EUR] | Unit cost in 2023 [EUR] | Cost of item in 2020 [EUR] | Cost of item in 2023 [EUR] | Difference (2023-2020) [EUR] |
|---|---|---|---|---|---|---|---|
| PV power plant project documentation | 1 | set | 133 | 626 | 133 | 626 | 493 |
| PV panel | 45 | pcs | 246 | 441 | 11 080 | 19 861 | 8 782 |
| Solar cable 6 mm$^2$ | 5 | 100 m | 95 | 144 | 473 | 720 | 246 |
| Connectors for PV panels | 90 | pcs | 1 | 1 | 68 | 128 | 60 |
| PV panel anchoring system | 132 | pcs | 5 | 20 | 650 | 2 683 | 2.033 |
| Combined voltage inverter | 1 | pcs | 337 | 1 487 | 337 | 1 487 | 1.150 |
| Voltage regulator | 2 | pcs | 57 | 109 | 114 | 218 | 104 |
| Battery | 1 | pcs | 11 932 | 13 281 | 11 932 | 13 281 | 1 349 |
| Installation costs of the PV power plant | 1 | set | 2 841 | 4 069 | 2 841 | 4 069 | 1 228 |
| Total cost of the PV power plant for the apartment building | | | | | 27 672 | 43 073 | 15 446 |

**Table 5.** Returns calculation of the PV power plant for the apartment building, constructed in 2020.

| Years of service | Efficiency [%] | Generated electricity [kWh] | Unit market price per 1kWh [EUR] | Market value of electricity produced in a given year [EUR] | Electricity consumption in a given year [kWh] | Market price of electricity consumed in a given year [EUR] | Returns [EUR] |
|---|---|---|---|---|---|---|---|
| **2020** | 70 | 27 389.6 | 0.1894 | 5 187.42 | 77 388.0 | 14 656.82 | 5 187.42 |
| **2021** | 69 | 26 998.3 | 0.2201 | 6 817.08 | 77 388.0 | 14 656.82 | 12 004.50 |
| **2022** | 68 | 26 607.0 | 0.2525 | 6 718.28 | 77 388.0 | 14 656.82 | 18 722.78 |
| **2023** | 67 | 26 215.8 | 0.2475 | 6 488.29 | 77 388.0 | 14 656.82 | 25 211.07 |
| **2024** | 66 | 25 824.5 | 0.1722 | 4 445.92 | 77 388.0 | 14 656.82 | 29 656.99 |
| **2025** | 65 | 25 433.2 | 0.1263 | 3 210,94 | 77 388.0 | 14 656.82 | 32 867.93 |
| **2026** | 64 | 25 041.9 | 0.0976 | 2 443.01 | 77 388.0 | 14 656.82 | 35 310.94 |
| **2027** | 63 | 24 650.6 | 0.0803 | 1 980.45 | 77 388.0 | 14 656.82 | 37 291.39 |
| **2028** | 62 | 24 259.4 | 0.0780 | 1 893.33 | 77 388.0 | 14 656.82 | 39 184.73 |
| **2029** | 61 | 23 868.1 | 0.0735 | 1 753.22 | 77 388.0 | 14 656.82 | 40 937.95 |
| **2030** | 60 | 23 476.8 | 0.0689 | 1 616.70 | 77 388.0 | 14 656.82 | 42 554.64 |
| **2031** | 59 | 23 085.5 | 0.0666 | 1 536.76 | 77 388.0 | 14 656.82 | 44 091.40 |
| **2032** | 58 | 22 694.2 | 0.0643 | 1 458.62 | 77 388.0 | 14 656.82 | 45 550.03 |
| **2033** | 57 | 22 303.0 | 0.0620 | 1 382.28 | 77 388.0 | 14 656.82 | 46 932.30 |
| **2034** | 56 | 21 911.7 | 0.0608 | 1 332.88 | 77 388.0 | 14 656.82 | 48 265.18 |
| **2035** | 55 | 21 520.4 | 0.0597 | 1 284.38 | 77 388.0 | 14 656.82 | 49 549.56 |
| **2036** | 54 | 21 129.1 | 0.0585 | 1 236.77 | 77 388.0 | 14 656.82 | 50 786.33 |
| **2037** | 53 | 20 737.8 | 0.0574 | 1 190.07 | 77 388.0 | 14 656.82 | 51 976.40 |

From Figure 4 it can be seen that the investment made in 2020 is discarded after 4 years of operation, when revenues of 25 thousand € have been achieved in this year, on the contrary, according to Figure 6, the investment relayed in 2023 is not discarded even after 15 years of operation. The unit price of electricity plays a key role in this difference, and the increased cost of procurement also plays a relatively significant role. The total cost of acquiring an identical PV plant has increased from €27 672 to €43 073 over three years, an increase of €15 401, an increase in input costs of 55.6 %.

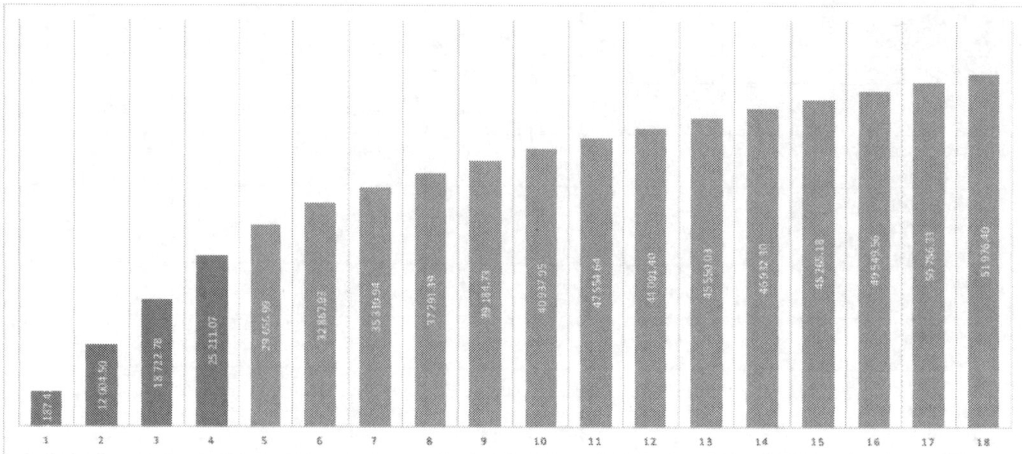


**Fig. 4.** Model of financial return on PV power plant in € – constructed in 2020.

Analogically, as the return for 2020 was calculated, the return for the PV plant implemented in 2023 was calculated in the same way, as can be seen in Table 6 and Figure 5.

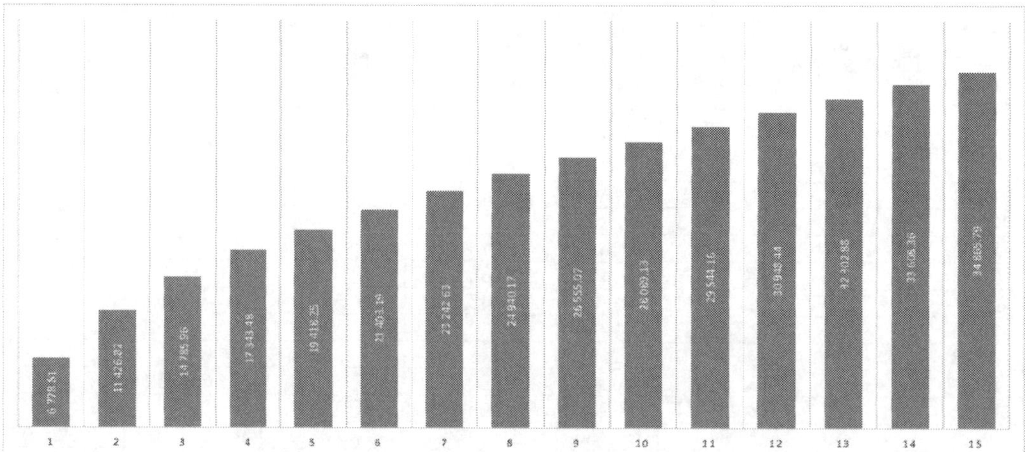


**Fig. 5.** Model of financial return on PV power plant - constructed in 2023.

**Table 6.** Returns calculation of the PV power plant for the apartment building, constructed in 2023.

| Years of service | Efficiency [%] | Generated electricity [kWh] | Unit market price per 1kWh [EUR] | Market value of electricity produced in a given year [EUR] | Electricity consumption in a given year [kWh] | Market price of electricity consumed in a given year [EUR] | Returns [EUR] |
|---|---|---|---|---|---|---|---|
| **2023** | 70 | 27 389.6 | 0.2475 | 6 778.81 | 77 388.0 | 19 153.21 | 6 778.81 |
| **2024** | 69 | 26 998.3 | 0.1722 | 4 648.01 | 77 388.0 | 19 153.21 | 11 426.82 |
| **2025** | 68 | 26 607.0 | 0.1263 | 3 359.14 | 77 388.0 | 19 153.21 | 14 785.96 |
| **2026** | 67 | 26 215.8 | 0.0976 | 2 557.53 | 77 388.0 | 19 153.21 | 17 343.48 |
| **2027** | 66 | 25 824.5 | 0.0803 | 2 074.76 | 77 388.0 | 19 153.21 | 19 418.25 |
| **2028** | 65 | 25 433.2 | 0.0780 | 1 984.95 | 77 388.0 | 19 153.21 | 21 403.19 |
| **2029** | 64 | 25 041.9 | 0.0735 | 1 839.44 | 77 388.0 | 19 153.21 | 23 242.63 |
| **2030** | 63 | 24 650.6 | 0.0689 | 1 697.53 | 77 388.0 | 19 153.21 | 24 940.17 |
| **2031** | 62 | 24 259.4 | 0.0666 | 1 614.90 | 77 388.0 | 19 153.21 | 26 555.07 |
| **2032** | 61 | 23 868.1 | 0.0643 | 1 534.07 | 77 388.0 | 19 153.21 | 28 089.13 |
| **2033** | 60 | 23 476.8 | 0.0620 | 1 455.03 | 77 388.0 | 19 153.21 | 29 544.16 |
| **2034** | 59 | 23 085.5 | 0.0608 | 1 404.28 | 77 388.0 | 19 153.21 | 30 948.44 |
| **2035** | 58 | 22 694.2 | 0.0597 | 1 354.43 | 77 388.0 | 19 153.21 | 32 302.88 |
| **2036** | 57 | 22 303.0 | 0.0585 | 1 305.48 | 77 388.0 | 19 153.21 | 33 608.36 |
| **2037** | 56 | 21 911.7 | 0.0574 | 1 257.43 | 77 388.0 | 19 153.21 | 34 865.79 |

A comparison of the two realisations, i.e. in 2020 and 2023, can be seen in the graph in Figure 6. The trend curves for the incremental return on investment in each year can be seen here. Although the logarithmic progression of the curves is not too different, the upfront investment costs play a significant role, where these costs shift the curve upwards on the y-axis and thus delay the amortisation of the investment in subsequent years.

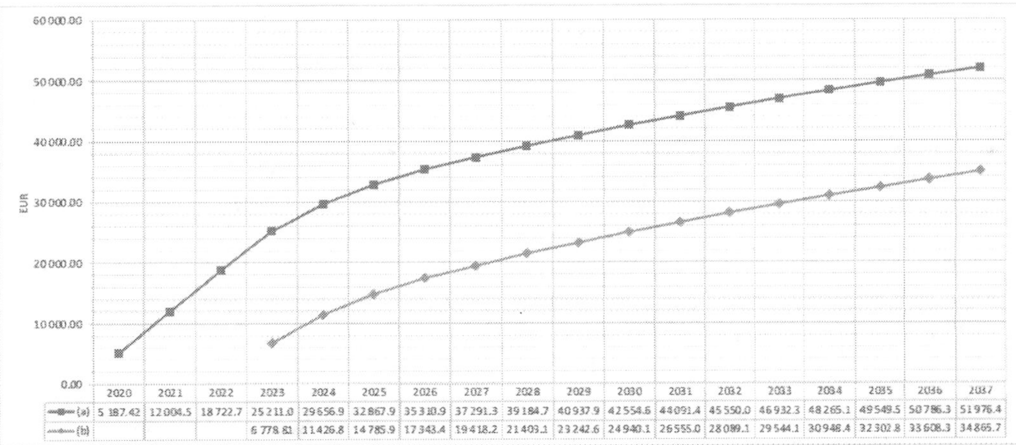


**Fig. 6.** Comparison of the return on investment of the PV power plant: (a) constructed in 2020; (b) constructed in 2023.

The results also showed how much electricity can be generated by a PV plant on the façade of an apartment building. Furthermore, the study shows the differences in initial costs and returns of a PV plant when implemented in 2020 and in 2023. An important aspect in the calculation is the power of one PV panel. The proposed PV panel of 400 Wp is mounted at an inclination of 90°. This inclination is undesirable in terms of PV performance, dropping to 70 % of full power. PV plants have an average lifetime of 20-25 years. In the calculation of electricity production, a 15-year lifetime was considered due to the reasonable financial return of PV plants being between 10-15 years on building façades. A decrease in energy production with each year of 1 % is considered as a gradual degradation of PV panels. Results showed that with each year used, the electricity production of PV façade on the apartment building decreases by 39.31 kWh. From 5 year to 10 year, the electricity production decreases by 9782.0 kWh. The comparison with the electricity produced by PV façade in 1 year and the electricity consumption of apartment building in 1 year is around one third.

## 4 Conclusion

The aim of the paper was to examine the role of the energy crisis in the context of the Russia-Ukraine conflict. The conflict occurred shortly after the global pandemic COVID-19. It was found that the unit price of electricity plays a relatively significant role in the amortisation of the initial investment in a PV system for a smaller 9-unit apartment building. The increased purchase cost is also a key factor. It increases by 55.6 % between 2020 and 2023. The total cost of purchasing an identical PV system has increased from €27 672 to €43 073 over three years.

Based on the results of the case study, the following highlights can be summarised:

• A key starting point for early returns is the initial cost of a PV system.

• The unit cost of electricity is a key parameter affecting the returns from a PV system. This can be a side effect of global crises.

• Global crises have only a secondary impact on PV system returns, such as pandemics or wars. Their duration is usually short and the calculation of returns is over a longer time horizon.

• The amortisation of the upfront investment in a PV system is significantly affected by local and supranational legislation regulating electricity prices. This is therefore one of the secondary effects of the global crises and wars.

At a time when electricity prices are on the rise in many EU countries, the investment in a PV system and other renewable energy sources is a matter of concern for many property owners in order to avoid future problems with energy costs and is a key issue in the operation of the building. This article fills a gap in the literature of the last few years, which have been affected by crises of various origins. In addition, it provides a guide for the owners and investors of new buildings on how to approach the financial equilibrium in the implementation of a PV power plant on the façade of an apartment building. The article also describes and reflects on current trends and parameters that influence the market.

This research acknowledges the support received from Brno University of Technology, Faculty of Civil Engineering; and UiT The Arctic University of Norway, Faculty of Engineering Sciences and Technology, for providing basic infrastructure and laboratory facilities.

## References

1. C. A. Horowitz, Int. Leg. Mater. **55**, 4 (2016)
2. D. Ş. Armeanu, G. Vintilă, Ş. C. Gherghina, Energies **10**, 3 (2017)
3. D. N. Madsen, J. P. Hansen, Renew. Sust. Energ. Rev. **114** (2019)
4. R. E. Mbah, D. Wasum, Adv. Soc. Sci. Res. J. **9**, 3 (2022)
5. Independent, but still very close to Russia. Met. Powder Rep. **64**, 4 (2009)
6. S. Adibpour, A. Raisi, B. Ghasemi, A. R. Sajadi, G. Rosengarten, Renew. Energy. **165** (2021)
7. V. Poulek, M. Libra, Elektro, Magazine for Electrical Engineering **3** (2010)

8. S. Preet, M. K. Sharma, J. Mathur, A. Chowdhury, S. Mathur, J. Build. Eng. **32** (2020)
9. J. Han, L. Lu, H. Yang, Y. Cheng, Energy Procedia **158** (2019)
10. Energie Snadno [in Czech] (2023)
11. V. Matajs, Solar Experts [in Czech] (2015)
12. A. Schmitt, Energy BrainBlog (2022)

# Cohesion and adhesion of a sealant joint subjected to compressive stress

*Adam* Bohacek[1*], *Michal* Brandtner[1] and *Jiri* Slanhof[1]

[1] Brno University of Technology, Faculty of Civil Engineering, Veveri 331/95, 60200 Brno, Czech Republic

**Abstract.** This article is devoted to the evaluation of the suitability of the use of sealants when applied to a cement-based material and its subsequent exposure to compressive stress. Aquapanel cement board is chosen as the underlying material. In general, cementitious materials are characterized by a high concentration of dust particles on the surface, which prevent the correct adhesion of the sealant to the substrate, and therefore their sealing is often problematic. The sealants presented in this research are selected based on the manufacturer's recommendations for suitability for sealing cement-based materials. Base plates are made from cement board, which are subsequently sealed according to the procedures established by the ČSN EN ISO 11432 standard. The samples produced in this way are then cycled and their resistance to pressure stress is subsequently tested. The results of the test performed in this way are evaluated by measuring and visually evaluating the adhesion and cohesion of the cemented joint. Based on the results, it is possible to recommend all of the tested sealants for sealing cementitious base materials that are exposed to compressive stress.

## 1 Introduction

The issue of cementing individual materials is still a current topic in the field of construction today. The main functions of the cemented joint are to allow a certain range of movement of the substrates, preventing the entry of water, air and other substances into the structure. Furthermore, it provides thermal and sound insulation, possibly also resistance to fire [1] [2]. Sealed joints are exposed to a large number of external influences. Therefore, it is necessary to choose a sealant based on the effects against which it should be resistant. This information is provided by the sealant manufacturer as a recommendation for sealant use. However, this recommendation does not always correspond to reality, and the cemented joint is able to retain its properties and not degrade even after exposure to these influences [3].

In this article, the author specifically deals with the sealing of a problematic cement-based material, the adhesion and cohesion of a cemented joint after its exposure to compressive stress.

## 2 Material

For the purposes of the research published in this contribution, the materials for its implementation are selected. First, the base material is selected. Aquapanel cement board is chosen as the underlying material. In addition, sealants are selected, which are intended by the manufacturer for sealing the selected base material.

---

* Corresponding author: bohacek.a@fce.vutbr.cz

The Aquapanel cement board is chosen because it is problematic when sealing. This sealing problem is caused by the presence of fine dust particles on the surface of the board, which disrupt the cemented joint formed by the base material and the applied sealant.

Sealants are selected based on several criteria. One of them is the manufacturer's recommendation on the suitability of use on the selected base material. Another criterion is availability for ordinary consumers and, last but not least, the price of the selected sealant. According to the manufacturer's recommendations, there is no need to use a primer for selected sealants.

## 3 Methods

The methods used in this contribution are based on the valid European technical standard with the designation ČSN EN ISO 11432 Building constructions – Sealants – Sealants – Determination of resistance to compression. This European technical standard defines the test sample and the exact test procedure.

The test sample defined by the standard consists of two base plates with plan dimensions of 30x50 mm and a thickness of 12.5 mm. These base plates are made of aquapanel cement board. It is also made up of wooden spacers with plan dimensions of 9x12 mm and a height of 50 mm. Two base plates and two wooden spacers delimit the space for the sealant with floor plan dimensions of 12x12 mm and a height of 50 mm. Five test samples are made for each sealant. During production, the regulations and recommendations of the manufacturer of the selected sealant are taken into account. Furthermore, the following measures defined by the technical standard are observed: exclude the formation of air bubbles, press the sealing sealant to the contact surfaces of the underlying bodies and smooth the surface of the sealant to the plane of the auxiliary bodies and spacers.

After the production of the test samples, the test samples are stored for 28 days at a temperature of (23±2) °C and a relative humidity of (50±5) % in order to allow the sealant to mature properly. Before starting the testing of the test samples, it is necessary to submit the test samples according to the procedure prescribed by the technical standard. This prescribed cycle of preloading test specimens is repeated three times. The preset is as follows:

1. 3 days in the oven at room temperature (70±2) °C
2. 1 day in distilled water at room temperature (23±2) °C
3. 2 days in the oven at room temperature (70±2) °C
4. 1 day in distilled water at room temperature (23±2) °C

For the research published in this contribution, a pressure test is used, the exact procedure of which is described in the already mentioned European technical standard. The test takes place at a temperature of (23±2) °C and a relative humidity of (50±5) %. First, the wooden spacers are removed and the test specimens are placed in the testing machine and compressed at a rate of (5.5±0.7) mm/minute to 75 % or 80 % of the initial length, or any other % value agreed upon by the parties involved. For this research, a value of 75 % of the initial length is chosen.

The result of this test is the recorded maximum force required to compress the test sample to a value of 75 % of the original length. Furthermore, it is the calculated secant modulus of elasticity, which is calculated from the relation:

$$\sigma = \frac{F}{s}$$

σ secant modulus [N/mm2]

F force at selected compression [N]
S initial cross-sectional area of the test specimen [mm2]

**Table 1.** Relation of compression value to initial length.

| Compression length relative to initial length [%] | Compression [%] | Length after compression [mm] |
|---|---|---|
| 75 | 25 | 9.0 |
| 80 | 20 | 9.6 |

## 4 Results

Five test samples are produced and subsequently tested for both selected sealants. The test results are shown below in table 2. For each test sample tested, the measured value of the maximum force required to compress to 75 % of the original length in Newtons is shown in the results table. The table also shows the average value of the maximum forces required for compression to 75 % of the original length for both selected sealants. The table of results also shows the calculated secant modulus of elasticity for the individual tested test samples and the calculated average value of the secant modulus of elasticity for the individual tested sealants.

**Table 2.** Test according to ČSN EN ISO 11432.

| Sealant | Test sample nr. | Max. force to compress to 75 % of original length [N] | Diameter of max. force for compression to 75 % of original length [N] | Secant modulus of elasticity [N/mm$^2$] | The diameter of the secant modulus of elasticity [N/mm$^2$] |
|---|---|---|---|---|---|
| Sealant I | 1 | 220.04 | 306.87 | 0.37 | 0.51 |
| | 2 | 343.67 | | 0.57 | |
| | 3 | 267.21 | | 0.45 | |
| | 4 | 306.47 | | 0.51 | |
| | 5 | 396.98 | | 0.66 | |
| Sealant II | 1 | 134.98 | 148.74 | 0.22 | 0.25 |
| | 2 | 155.10 | | 0.26 | |
| | 3 | 147.28 | | 0.25 | |
| | 4 | 159.46 | | 0.27 | |
| | 5 | 146.88 | | 0.24 | |

# 5 Analysis of results

It is clear from the results of the tests that for sealant I, a force greater than 300 N is needed to compress it to 75 % of the original length, which indicates a high resistance of the sealant against compression. For sealant II, it is necessary to develop a force of almost 150 N, i.e. half less than for sealant I. Therefore, we can say that sealant I is more resistant to compression than sealant II.

As for the failure of the test samples, there was no disruption of the cemented joint during testing for both tested sealants, which indicates their solid structure resistant to the pressure stress to which the test samples were subjected during testing.

The results of the maximum force for compression to 75 % of the original length of the individual tested samples are shown in Figure 1.

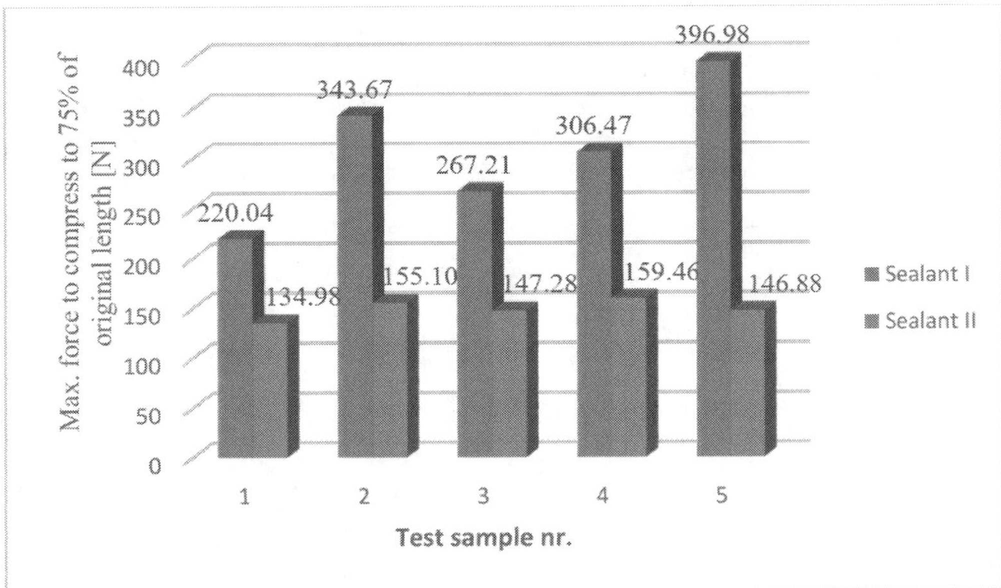


**Fig. 1.** Comparison of the results of the maximum force for compression to 75 % of the original length of both tested sealants.

# 6 Conclusion

Based on the results of the pressure test, it can be declared that both of the selected sealants met the conditions of the test, during which there was no violation of cohesion and adhesion of the tested test samples. Based on the pressure test, both selected sealants can therefore be recommended for sealing the selected cement base material.

The contribution was made with the support of junior specific research of the Faculty of Civil Engineering of the Brno University of Technology conducted under the designation FAST-J-22-7872.

## References

1. E. M. Petrie, *Handbook of adhesives and sealants*, New York (2009)
2. J. Klosowski, A. T. Wolf, *Sealants in construction*, United Kingdom (2016)
3. K. Lal Mittal, A. Pizzi, *Handbook of sealant technology*, United Kingdom (2009)

4. A.T. Wolf, *Movement capability of sealants*, Construction and Building Materials, USA (1991)
5. M. Z. L. Chew, X. Zhou, S. T. Tan, *On-site weathering of sealants under tropical conditions*, Construction and Building Materials (2004)
6. H. Stögbauer, A. T. Wolf, *The influence of heat ageing on one-part construction silicone sealants*, Construction and Buildings Materials (1991)
7. ČSN EN ISO 11432 *Building construction - Jointing products - Sealants - determination of compressive strength* Prague: Czech Standards Institute (2006)
8. J. Šlanhof. M. Šimácková, P. Liška, B. Nečasová, *Sealing possibilities of glass fibre reinforced concrete elements and fibre - Cement facade cladding panels*, Proceedings of an International Scientific Conference in Bratislava (2014)
9. P. Liška, B. Nečasová, J. Šlanhof, M. Šimáčková, *Determination of Tensile Properties of Selected Building Sealants in Combination with High-pressure Compact Laminate (HPL),* In: Procedia Engineering (2015)
10. J. Šlanhof, P. Liška, B. Nečasová, *Evaluation of test methods for testing of sealants, Advances and Trends in Engineering Sciences and Technologies II*, Proceedings of the 2nd International Conference on Engineering Sciences and Technologies (2016)

# The influence of the mass moisture content of the masonry on the choice of insulation system for historic buildings

*Kristian* Hutkai[1*], *Dusan* Katunsky[1]

[1]Technical university of Kosice, Faculty of Civil Engineering, Institute of Architectural Engineering, Vysokoskolska 4, 042 00 Kosice, Slovakia

**Abstract.** When discussing the energy efficiency of historic structures, increased humidity of the perimeter wall is a crucial problem. It greatly affects the choice of insulation system, because if the drying of the structure becomes impossible, permanently increased humidity of the structure can lead to serious damage to the perimeter structure. The contribution focuses on the interior insulation design of historical structures with various types of elevated humidity with the use of dynamic simulation techniques to evaluate the appropriateness of several thermal insulation materials. The degree of drying and the thermal-humidity response of the buildings after the historic wall has been insulated from the inside will be assessed based on the findings of HAM simulations.

## 1 Introduction

Depending on atmospheric conditions, each solid material contains a certain amount of water. The level of wetting depends on the temperature and relative humidity of the surrounding environment, porosity and pore structure of the material. The term material moisture can be characterized as the amount of water found in the pore structure of a dry material. It is expressed by the mass or volume ratio of water to the solid part of dry matter.

The source of increased humidity can be rainwater, subsurface, surface, underground, condensed, operational, built-in and soil moisture. It is soil moisture that poses the greatest risk of increased moisture in perimeter structures in historic buildings. Those buildings were built several decades or even centuries ago, when barriers against the penetration of soil moisture into the structures were not yet available. For this reason, the statement that the perimeter structures of listed buildings show a certain level of increased humidity may be true. At the same time, legal equirements on energy efficiency also force the construction industry to take a responsible approach to the renovation of buildings and to make their energy management as efficient as possible. Thermal protection of historic buildings could save up to 40% of total energy consumption and 36% of $CO_2$ emissions in Europe [1]. The high architectural and cultural value of these buildings, especially their exposed facade elements, prevent the use of thermal insulation in the most effective way - from the exterior

* Corresponding author: kristian.hutkai@tuke.sk

side. Therefore, the only solution to reduce the energy consumption of historical buildings is the use of insulating materials from the interior. This form of use of thermal insulation in combination with increased moisture in the structure can cause irreversible degradation processes due to the accumulation of moisture and cause moisture damage, frost damage, wood rot, e.g. in ceiling constructions, formation of mold and thermal bridges in the connections of the partition with the perimeter structure [2-7].

According to [8], masonry moisture can be classified as:

**Table 1.** Humidity classification according to STN P 73 06 10.

| **Humidity** | **Mass moisture content M%** |
|---|---|
| Very low | < 3 |
| Low | 3.0-5.0 |
| Increased | 5.0 – 7.5 |
| High | 7.5 - 10 |
| Very high | > 10 |

The paper investigates the effect of capillary active calcium silicate board and mineral wool with a vapour barrier layer on the inner surface of the structure and examines the heat-moisture response of a historical wall with levels of initial humidity according to STN P 73 06 10 - very low (3 M% - mass moisture content) and low (5 M%)

## 2 Methodology with input and boundary conditions

The experiment was performed by the WuFi simulation program [9], which enables the analysis of one-dimensional complex heat and water transfer in building materials under dynamic boundary conditions [10,11]. Numerical analysis took place over a period of 5 years in climate conditions of Košice according to [12].

The balance equations of heat and moisture transfer in the simulation tool are described as:

$$\frac{dw}{d\varphi}\frac{\partial\varphi}{\partial t} = \nabla.((D_\varphi\nabla\varphi + \delta_p\nabla(\varphi.p_{sat})) \tag{1}$$

$$\frac{dH}{dT}\frac{\partial T}{\partial t} = \nabla.(\lambda\nabla T) + h_v\nabla.(\delta_p\nabla(\varphi.p_{sat})) \tag{2}$$

Whereas the liquid water transport coefficient is defined as:

$$D_\varphi(w) = D_w(w).\frac{dw}{d\varphi} \tag{3}$$

Where: $dH/dT$ (J/(m$^3$ K)) is the heat storage capacity of the moist building material, $dw/d\varphi$ (kg/m$^3$) represents the moisture storage capacity of the building material, $\partial T/\partial t$ (K/s) Is the temporal change of the temperature, $\partial\varphi/\partial t$ (1/s) is the temporal change of the relative humidity, $\delta$ (kg/(msPa)) the water vapour permeability, $p_{sat}$ (Pa) the partial pressure of saturated watet vapour in the air, $h_v$ (J/kg) the latent heat of evaporation of water, $\lambda$ (W/(mK) the thermal conductivity, $T$ (°C) is the thermodynamic temperature, $\varphi$ (–) the relative humidity, $D_\varphi$ (kgm/s) liquid transport coefficient of the building material, $D_w$ (m$^2$/s) liquid diffusivity [9-11].

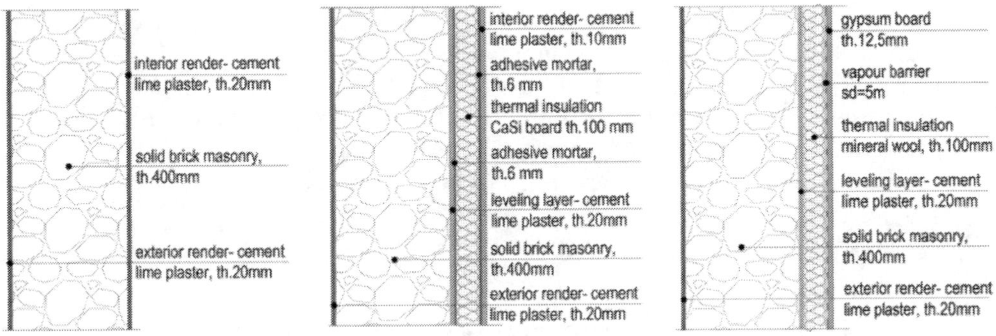


**Fig. 1.** Assessed compositions of solid brick masonry with internal insulations.

The initial condition in the 1st calculation scenario was considered 3 M% mass moisture content of the perimeter brick wall and in the 2nd calculation scenario 5 M% mass moisture content. Figure 1 shows the composition of evaluated structures - uninsulated brick wall, brick wall insulated with capillary active calcium silicate board and wall insulated with mineral wool in combination with vapor barrier layer with diffusion equivalent air layer thickness sd=5m.

## 3 Results

As part of the results, the course of mass moisture content of nodal points in the structure was analysed over a period of 10 years. These points were chosen on the basis of the most risky parts, namely on the surface of the masonry (or under the thermal insulation), in the 100 mm depth of the masonry from the interior and in the middle of the masonry.

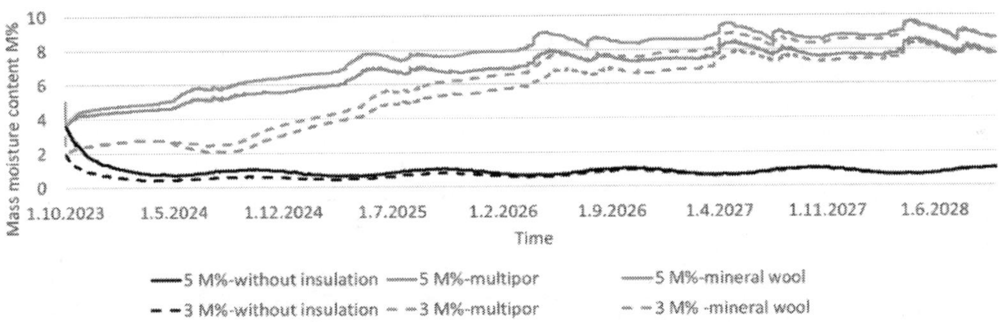


**Fig. 2.** Course of mass moisture on the surface of masonry at initial moisture of 3 M% and 5 M%.

Figure 2 shows the course of the mass moisture of the masonry under the layer of thermal insulation or plastering in the case of an existing composition. In the case of an uninsulated wall, a drying trend can be seen from an initial mass humidity of 3 M%, or 5 M% to approximately 1 M% after the first year. In the following period, it is possible to observe cyclically recurring wetting in the autumn period and drying in the spring months. Mass humidity reaches its maximum value in the month of October and minimum in April. After a five-year monitoring period, the waveforms merge and show identical mass moisture. After using the calcium silicate board, the mass moisture shows an increasing trend during the simulated period. In the summer months, slight drying can be observed, but it is negligible compared to the rising trend. After 4 years, the mass moisture stabilizes and the value oscillates between 7 and 8% mass moisture. After the application of mineral wool in combination with a vapor barrier layer with an equivalent diffusion thickness of sd=5m, the drying of the structure towards the inside is limited and a steeper trend of wetting of the structure is evident. After 2.5 years of the simulated period, in the case of mineral wool with an initial moisture content of 3 M%, the mass moisture will exceed the value of a calcium-silicate board with an initial moisture content of 5 M%. It is clear from this that the structure, which includes a calcium-silicate board, shows a more favourable course of mass moisture.

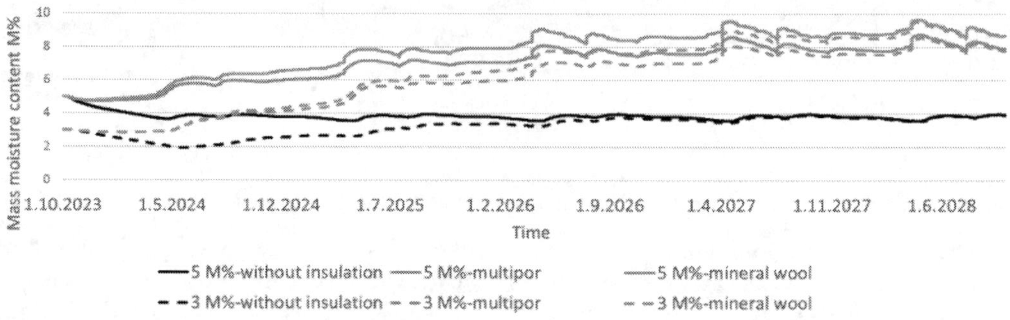


**Fig. 3.** Course of mass humidity at a depth of 100 mm from the interior at initial humidity of 3 M% and 5 M%.

Figure 3 shows the curves of mass moisture at a depth of 100 mm in the structure. The development of mass moisture is similar to that on the surface of masonry. The mass humidity of the wall at this depth is influenced not only by the parameters of the internal environment, but also by the conditions of the external environment. This tendency is demonstrated by the case of an uninsulated wall, where the average mass humidity is approximately three times higher than on the surface of the masonry. At this depth, the internal warm air has less influence on the mass humidity level and is more dependent on the external air parameters. The development of mass moisture when using thermal insulation is similar to that on the surface of masonry. Calcium-silicate board even in the case of an initial moisture content of 5% after a period of 5 years shows a lower final moisture content than mineral wool with an initial moisture content of 3%.

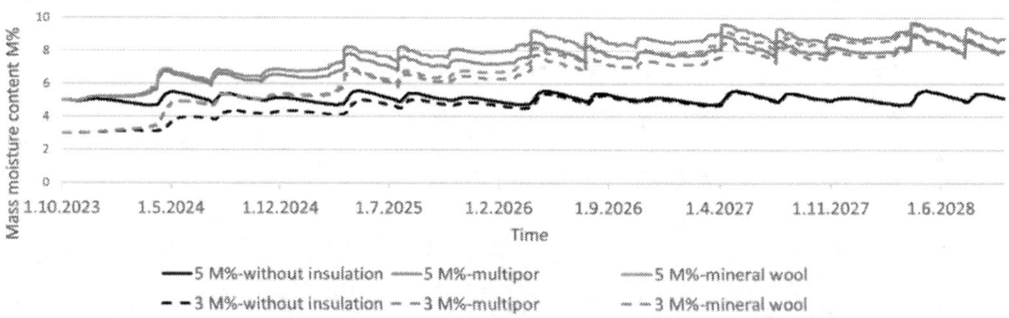


**Fig. 4.** Course of mass moisture in the middle of the masonry at initial moisture of 3 M% and 5 M%.

Figure 4 shows the curves in the middle of a 400 mm thick wall. In this part of the masonry, the parameters of the internal environment influence the course of mass moisture to a negligible extent. The development of mass humidity depends on the external parameters of the environment - on the amount and intensity of precipitation and solar radiation. The mass humidity for a period of 5 years with an uninsulated wall is around 5 M%, which represents, according to [8], the upper limit of a low level of humidity. In the case of calcium-silicate board, this value is at the level of 8 M% and in the case of mineral wool 9 M%. According to [8], these values are already at a high level.

## Conclusion

From the results, it can be concluded that thermal insulation boards with open porosity (capillary activity) are a safer solution for reducing their energy consumption than vapor-tight insulation systems for the perimeter walls of listed buildings, where any level of moisture content is assumed. These boards with their pore structure are able to receive water and gradually release it towards the interior, thus preventing excessive accumulation of water. In the case of a vapor-tight insulation system of mineral wool, drying towards the interior is limited and a larger amount of water in the constructions is evident. The vapor barrier layer should have a sufficiently high diffusion resistance to prevent internal condensation during the heating season. At the same time, during the cooling season (in the summer months) it must have a sufficiently low diffusion resistance factor to prevent summer condensation. It is possible to state that calcium silicate board in terms of water accumulation has a more favourable effect than mineral wool even with a higher initial mass humidity. However, it should be noted that any intervention supporting the increase of the thermal insulation capacity of the structure automatically disrupts its thermal and moisture behaviour. It is necessary to find a compromise between the degree of insulation and the accumulation of water in the structure in order to eliminate the risks of any malfunctions as much as possible. At the same time, if the wall is too wet, it is necessary to use an effective drying method before insulating it.

The overall thermal-moisture behaviour of insulation systems on the internal surface is closely related to the parameters of the internal environment (temperature and relative humidity), therefore, in the next phase of research in this issue, the most favourable conditions of the internal environment will be quantified and defined to maximize the potential of insulation systems for the purposes of internal insulation.

This research is supported by VEGA 1/0626/22 Design and evaluation of building structures and the indoor environment of buildings for demanding conditions.

## References

1. A. Blumberga, K. Kašs, E. Kamendere, G. Žogla, A. Kamenders, D. Blumberga, ... & E.J. de Place Hansen. Robust Internal Thermal Insulation of Historic Buildings-State of the art on historic building insulation materials and retrofit strategies. (2019)
2. S.P. Bjarløv, G.R. Finken, and T. Odgaard, Retrofit with interior insulation on solid masonry walls in cool temperate climates–An evaluation of the influence of interior insulation materials on moisture condition in the building envelope, Energy Procedia, **78**, 1461-1466, (2015)
3. E. Brandt, T. Bunch-Nielsen, G. Christensen, C. Gudum, M.H. Hansen, E.B. Møller. SBianvisning **224**: Fugt i bygninger. Statens Byggeforskningsinstitut, Hørsholm. (2009).
4. J. Toman, A. Vimmrova & R. Černý. Long-term on-site assessment of hygrothermal performance of interior thermal insulation system without water vapour barrier. Energy and Buildings, **41**(1), 51-55. (2009).
5. J. Zhao, J. Grunewald, U. Ruisinger & S. Feng. Evaluation of capillary-active mineral insulation systems for interior retrofit solution. Building and Environment, **115**, 215-227. (2017)
6. R. Walker & S. Pavía. Thermal performance of a selection of insulation materials suitable for historic buildings. Building and environment, **94**, 155-165. (2015)
7. E. Vereecken, L. Van Gelder, H. Janssen & S. Roels. Interior insulation for wall retrofitting–A probabilistic analysis of energy savings and hygrothermal risks. Energy and Buildings, **89**, 231-244. (2015)

8. STN P 73 06 10- hydroizolácie stavieb
9. https://wufi.de/en/software/wufi-pro/
10. M. Krus: Moisture Transport and Storage Coefficients of Porous Mineral Building Materials. Theoretical Principles and New Test Methods Fraunhofer IRB Verlag, 1996
11. H. M. Künzel H. M. (1995). Simultaneous heat and moisture transport in building components. One-and two-dimensional calculation using simple parameters. IRB-Verlag Stuttgart, **65**. (1995)
12. http://climate.onebuilding.org/WMO_Region_6_Europe/SVK_Slovakia/index.html

# A review of building envelope constructions designed for demanding operating and climatic conditions

*Zuzana* Dicka[1*], *Dusan* Katunsky[1], and *Erika* Dolnikova[1]

[1]Institute of Architectural Engineering, Faculty of Civil Engineering, Technical University of Kosice, 042 00 Kosice, Vysokoskolska 4, Slovakia

**Abstract.** Under the term demanding climatic conditions in connection with the topic, we can primarily imagine the excessive overheating of buildings in summer and on the other hand the exposure of the building to low temperatures in winter. Considering the extreme climatic conditions, the design of the building requires an individual approach. First of all, it is necessary to pay increased attention to the envelope structures such as roof and perimeter walls. The building can also be affected by other extreme conditions such as solar radiation and associated excessive lighting, i.e. glare in the work environment. This topic is increasingly coming to the fore due to climate change. Summer seasons are getting warmer and this problem is getting even worse. Even in locations where this issue has never been dealt with before. The article summarizes the conducted research related to the issue of overheating and excessive lighting of buildings. It also informs about how researchers in Slovakia and different locations around the world deal with this problem.

## Introduction

As we all know, we have been facing climate change in recent decades. This topic is becoming more and more important nowadays. Anthropological activity worsens living conditions all over the world [1]. However, the effects of these changes can mitigate by measures in various sectors, and construction is one of them. It is the built-up urban environment that contributes to climate change. As a result of the long-term change in the outdoor climate, buildings may overheat, and the associated disruption of indoor comfort may occur. Therefore, it is necessary to reconsider the concept of the building from the initial design [2]. It is essential to design buildings that take on the dynamics of the climate throughout their life cycle. One of the main aspects of building design is the choice of materials and the structure of the building envelope. We should pay a special attention to the physical properties of these materials. It is also necessary to consider the characteristic climate-forming factors besides it's locality. However, we must not forget the critical part of the building, namely its glazed components. Precisely the unwanted solar gains through transparent surfaces cause unpleasant overheating in summer [3].

* Corresponding author: zuzana.dicka@tuke.sk

## 1 Factors influencing the summer overheating of buildings

The most significant factor influencing the overheating of buildings is precisely climate change. It causes increasingly frequent continuous series of warm days, which significantly affect the thermal comfort in the interior of buildings. There may be so-called thermal stress, which occurs in an environment with an excess, or vice versa, with a lack of heat [4]. It can harm the health of building occupants and pose health risks, especially for the most vulnerable groups of people. Climate change does not develop linearly in all locations. Cities overheat faster than remote rural areas [5]. Therefore, another essential factor is the location and density of buildings.

The intensity of air exchange and increasingly higher requirements for the airtightness of buildings have an impact on summer overheating. An airtight building eliminates the heat loss produced by the heating system, but on the other hand, it does not remove the accumulated heat. For this reason, heat gains increase [6]. The outcomes of the study conducted by Laouadi et al. showed that highly insulated and airtight buildings are more prone to overheating than older ones that are less insulated and airtight [7].

Other factors are the thermotechnical properties of envelope structures and materials, their thermal accumulation, and thermal inertia. The temperature change on the inner surface depends on the heat-accumulating properties of the materials. Thermal attenuation indicates how many times the temperature amplitude of the inner surface is smaller than that of the outer air [8]. Němeček, in his doctoral thesis, dealt with the influence of the internal thermal accumulation of structures and their summer thermal stability. Emphasis on comparing wooden and masonry buildings. From the simulation study by Němeček and Kalousek, we could conclude that the accumulation mass has a lower impact than ventilation. The accumulation capacity of structures in the interior has a somewhat smaller influence, as much as the small thickness of the internal structure effectively participates in the accumulation of heat within the daily cycle [9].

Last but not least is the influence of the shape of the building envelope. We can partially solve the problem of overheating and energy efficiency already at the initial design phase when we decide on the overall form of the building. The most significant influence is the ratio between the building's surface area and volume (i.e. building's shape factor). In a study by Košir et al., the authors present the connection between the shape of the building, the orientation, and the area of the windows concerning the building's energy needs. They concluded that for the considered Central European climate, an elongated building shape is more suitable than a compact one [10]. A study from Great Britain by Lavafpour et al. investigates the geometric implications of facade design to mitigate overheating in energy-efficient houses. The authors designed a self-shading facade with the shape of an optimal sloping wall, which protects the building from excessive sunlight in the summer. It could potentially serve as an effective shielding device [11].

### 1.1 Factors influencing thermal stability in summer

Internal thermal stability is significant for the assessment of thermal comfort in summer and winter. It describes the temperature characteristics of the interior space of the building concerning the external conditions. Summer thermal stability of the room is an increasingly topical problem. Calculation assessment applies for the so-called Critical room – space with the expected highest heat load [12]. For the object or room not to overheat, it is necessary to ensure the correct thermal stability design. We can consider a room to be thermally stable, if it's temperature remains within the permitted values for a specified time interval [13]. Factors affecting thermal stability in the summer are structural, architectural, and dispositional solutions of the building. Also, the orientation of the building to the cardinal

points, especially the orientation of the main facade, significantly affects the heat losses and gains of the building. It influences the overall temperature regime of the internal environment [14]. Other factors that affect thermal stability are [15]:
- modification of external surfaces of the building envelope
- segmentation of the outer surface
- the ratio of opaque and glazed parts of the building envelope
- shading of glazed parts
- heat gains through transparent parts of the building envelope
- heat gains through opaque parts of the building envelope
- ventilation of the building
- the measure of the object's accumulative capacity
- internal heat gains from people and appliances

## 2 Solar gains through transparent surfaces

The potential source of overheating is unwanted heat gains through transparent structures. From the beginning of construction, openings were omitted in the perimeter constructions so daylight could penetrate the interior [16]. The windows are the most vulnerable part of the building. Compared to non-transparent perimeter constructions, they are up to five times weaker from a thermotechnical point of view. Excessive solar gain from windows is one of the most common causes of building overheating [17]. Němeček found in his work that solar gains through the translucent fillings of the openings have the greatest influence on the highest temperature achieved in the interior [9]. The optimal orientation of the glass facade is to the south, southeast, or southwest. Other directions require smaller glazing areas or none at all. It's about the ideology of Socrates' solar house, which perfectly illustrates the principles of passive use of solar radiation. Solar gains through the all-glazed southern facade are desirable in winter but, on the other hand, cause excessive overheating in summer. It causes increased energy costs for cooling systems, which we currently want to avoid (Figure 1.) [18].

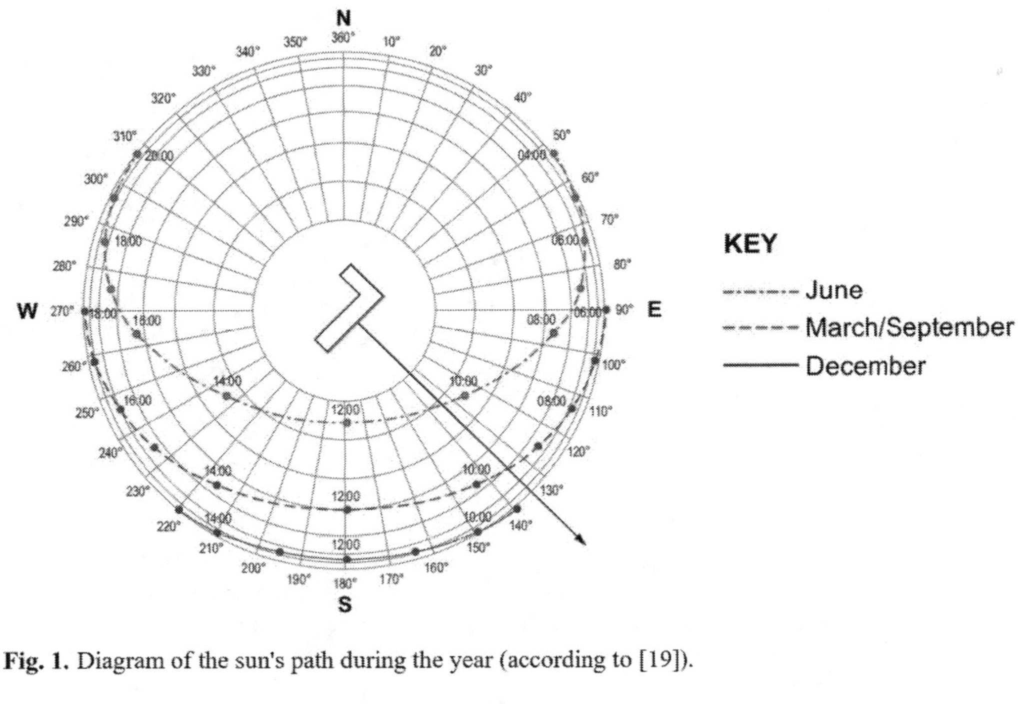


**Fig. 1.** Diagram of the sun's path during the year (according to [19]).

In recent years, the application of all-glass systems as building envelopes has expanded. The curtain wall has become an icon of developing cities. These structural elements are often the subject of analysis, and many researchers see the enormous energy-saving potential [20]. Apart from the disadvantage of summer overheating, these fully glazed systems also have a positive side. They ensure sufficient natural lighting in the working or living space. Natural daylight and workplace lighting are essential to optimal working conditions and employee health [17].

## 3 Elimination of solar gains through transparent surfaces

Transparent parts of the perimeter structure of the building are highly exposed, and their thermal insulation and other properties affect the quality of the building's interior space. The need for new technologies and approaches to improve their quality parameters is constantly increasing [21]. Solar gains through transparent surfaces can be eliminated by a generally known solution, namely shading devices. Another significant elimination factor of thermal gains is the correct selection of the type of window and window glazing. The choice of adequate elements is one of the most effective strategies for minimizing solar gains [17].

### 3.1 Shading elements

We already encounter the shading element in the ideology of Socrates' sunny house, which also points to the problem of summer overheating of buildings (Figure 2.). To prevent summer sunlight from entering the interior, Socrates proposed extending the roof on the south facade of the house. Roof slopes to the north. The slope angle depends on the latitude of the location and the knowledge of the geometry of the sun's rays [18].

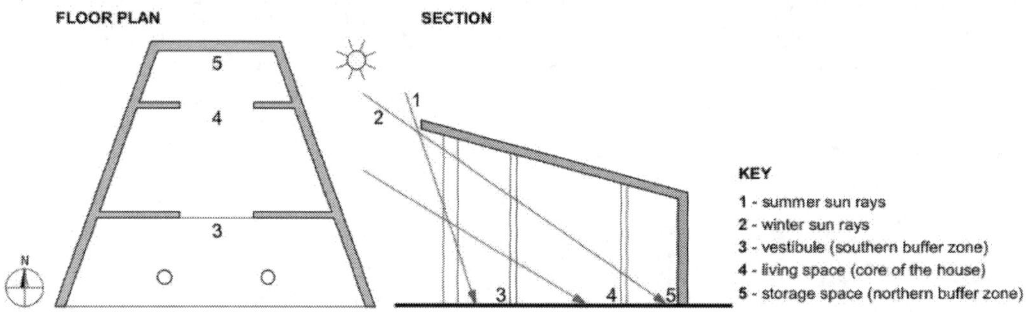


**Fig. 2.** The principle of Socrates' solar house (according to [18]).

Shading systems are currently an integral part of architecture on buildings with significant glass surfaces. It also regulates direct light and prevents glare and unwanted reflection of sunlight in the interior. High brightness in the visual field can disturb the visual comfort of building users [22]. Nowadays, we know a large number of shading elements. We can divide them into different points of view. According to their location, we can subdivide them into indoor (blinds, curtains, window films, etc.) and outdoor (shutters, outdoor blinds, sun shades, awnings, pergolas, etc.). Shading elements can be stationary, movable, and combined, depending on the control type, manual and automatic. Other divisions are, for example, material used, colour, surface treatment or light transmission. We generally know that external shading achieves higher efficiency than internal ones [23]. In the study by Kraus et al., the authors evaluated the influence of shading elements on the thermal comfort of commercial space in the summer season in the Central European climate

(Czech Republic). Their work aimed to study the effect of solar gains and their impact on thermal comfort. They assessed a total of 9 variants of internal and external shading. With the help of calculations and simulations, they achieved the most favourable results when applying an external semi-transparent colour film to the entire glazing area. The indoor air temperature decreased by 5.17 °C. In second place, the automatically controlled external blinds were closed, reducing the internal air temperature by 3.56°C. The outside screen blackout blinds with automatic control, in dark colour also stood up. Internal blinds with manual control fared the worst, with a temperature drop of only 1.48°C [24].

Shading elements do not always have to be part of the building. Also, the surrounding objects and landscape elements of the building locality environment affect shading. A study by Hwang et al. found that vegetation in an urban landscape can dramatically reduce a building's energy costs. Trees cast shade on buildings, reducing the internal temperature during warm seasons [25]. In another study by Chagolla et al., the authors dealt with measurements and simulations of the effect of trees shade on the internal temperature and thermal load of a test house in the climate zone of Mexico. Measurements and simulations present one warm and one cold week. The results show that the annual energy saving was increased due to the shade of trees by up to 76.6%. Therefore, shading by trees can significantly reduce indoor air temperature in warm climates and achieve thermal comfort [26].

### 3.2 Glazing of transparent constructions

The researcher Usta and Zengin focused on energy simulations of four different types of glazing in the climatic zone of Istanbul. Glazing depends on the thickness of the glass and the air gap. The type of glazing mainly affects the building in terms of internal thermal comfort. The study proves that building's energy demand can decrease by 25% by using a suitable glazing material [17]. In Slovakia, the critical period of overheating of buildings is the interval from April to September. During this period, solar radiation increases and causes significant heat gains in the interior of buildings. In his study, researcher Masaryk dealt with designing a system that would allow absorption or reflection of the heat-bearing part of the spectrum of solar radiation while allowing sufficient visible light into the interior of the building. The principle consists of creating a liquid layer on a transparent surface with a suitable additive. The best results reached black acrylic paint, which achieved the performance of white colour but with 1/3 of its volume [27]. Researchers Buratti et al. compared double glazing with aerogel insulation with single and double glazing in three different climates. The simulations apply to school buildings, where the glazing area was 30% of the building surface. The results show that using aerogel glazing reduces solar gain by 73% compared to single glazing and by 56% compared to double glazing [28]. Researcher Zhou also deals with aerogel glazing, and in his study, he devotes himself to a holistic review of the application of aerogel in building glazing (Figure 3.). Passive solutions, as a type of new glazing with more advanced materials, represent a more flexible way for practical applications. The study provides an in-depth analysis of the aerogel material, its production, and construction applications. The results show the potential of this material as a promising candidate for the energy efficiency of buildings [29].

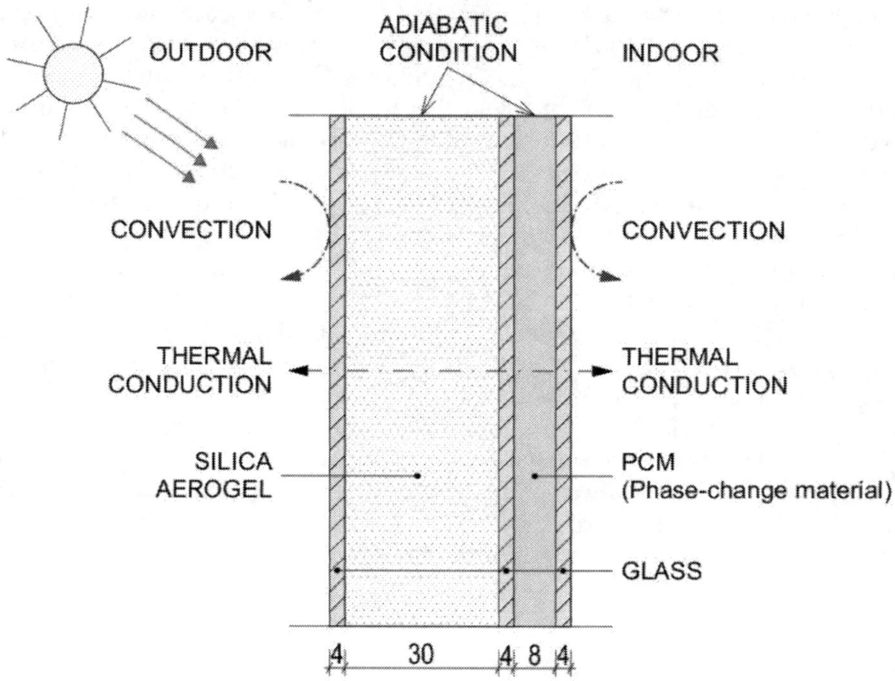


**Fig. 3.** Structural configuration of a PCM and aerogel-integrated window glazing system (according to [29]).

## Conclusion

Due to the climate changes in recent years, we can assume an increasingly frequent occurrence of exceedingly warm days everywhere in the world. For this reason, we should pay more attention to the theme of the summer overheating of buildings [30]. However, we should not only focus on how to mitigate climate change but also on how to face its consequences. It is necessary to focus on finding a long-term efficient solution for protecting buildings from overheating. Nevertheless, we can eliminate some of the problems and help improve thermal stability by improving insulation, better quality windows and last but not least, shading elements and a shading system. Correctly designed and right-implemented sun protection creates maximum visual and thermal comfort in the interior. It leads to improved productivity, performance and above all, the health of building users [24].

This paper was elaborated with the financial support of the research project VEGA 1/0626/22 of the Scientific Grant Agency of the Ministry of Education, Science, Research and Sports of the Slovak Republic and the Slovak Academy of Sciences.

## References

1. B. Metz, *Controlling Climate change*, Cambridge University Press, 376, (2010), ISBN 978-0-521-76403-2
2. C. Baglivo, *Dynamic Evaluation of the Effects of Climate Change on the Energy Renovation of a School in a Mediterranean Climate*, Sustainability, **13 (11)**, (2021)
3. O. Zhukova, N. Nehoda, *Forecast of changes in climate factors of the city of Kyiv and their impact on the life cycle of buildings*, Ecological safety, and envir. protection, **3 (43)**, (2022)

4. R. Králiková, *Thermal Comfort and Working Environment*, in Proceedings of the 17th International Conference, Bratislava, March 14-15, ESE-**33** (2017)
5. E. A. Mathez, J. E. Smerdon, *Climate change 2*, New York, Columbia University Press, 2018, 520, ISBN 9780231172837
6. A. Sdei, M. Mcevoy, *The effects of insulation and air-tightness in reducing the overheating risk of retrofitted social housing*, in Proceedings of the PLEA2016 Los Angeles – 32nd International Conference on Passive and Low Energy Architecture, 11-13 July, (2016)
7. A. Laouadi, M. Bartko, M. A. Lacasse, *A new methodology of evaluation of overheating in buildings*, Energy and Buildings, **226**, (2020)
8. I. Chmúrny, *Thermal protection of buildings,* Bratislava: JAGA Group, 230, (2003), ISBN 80-88905-27-3
9. M. Němeček, M. Kalousek, *The effect of internal heat accumulation of passive house structures on their summer thermal stability,* Doctoral thesis - Brno University of Technology, (2018)
10. M. Košir, T. Gostiša, Ž. Kristl, *Influence of architectural building envelope characteristics on energy performance in Central European climatic conditions*, Journal of Building Engineering, **15**, 278-288 (2018)
11. Y. Lavafpour, S. Sharples, B. Gething, *The impact of building form on overheating control: a case study of Larch House*, Architectural Science Review, **63**, 467-480, (2020)
12. F. Kulhánek, *Construction physics II. Building thermal technology* , ČVUT, 144, (2006), ISBN 80-01-03408-9
13. J. Hirs, J. Mohelníková, R. Yao, *Winter and summer thermal stability and room daylighting*, in Proceedings of the 7th International Conference in Building and Environment SuDBE (2015)
14. M. Bielek, *Building and energy*, VIDAS, Banská Bystrica, 214, (1995), ISBN 80-85306-07-7
15. D. Katunský, *Construction physics*, TUKE, (2013), ISBN 978-80-553-0972-9
16. I. Augusta, *Building thermal technology for everyone, Institute of Building Information*, Praha, 63, (1991), ISBN 80-85380-01-3
17. P. Usta, B. Zengin, *An evaluation of the glazing type impact on building energy performance through a building simulation*, Journal of Energy Systems, **6(1)**, 1-17, (2022)
18. dennisrhollowayarchitect.com [online] „Socrates Sun-Tempered House“, available: http://www.dennisrhollowayarchitect.com/SocratesHouse.html, last accessed 2022/5/27
19. coltinfo.sk [online] „Colt- shading and facade systems “, available: https://www.coltinfo.sk/na-stiahnutie.html, last accessed 2023/2/24
20. M. A. Memari, *Curtain Wall System*, American Society of Civil Engineers, 217, (2013), ISBN 978-0-7844-7768-7
21. A. Puškár et al., *Windows, glazed walls, doors, gates*, Jaga group, s.r.o., 266, (2008), ISBN 978-80-8076-062-5
22. D. Katunský et al., *Acoustics and daylighting*, TUKE, Košice, 105, (2015), ITMS 26110230093
23. J. Hraška, *Shading technique of buildings*, Spektrum STU, 102, (2020), ISBN 978802750066
24. M. Kraus, I. J. Senitková, L. Kučerová, *The impact of solar shading elements on thermal comfort*, in AIP Conference Proceedings, **2488**, (2022)
25. W. Hwang, P. E. Wiseman, V. Thomas, *Simulation of Shade Tree Effects on Residential Energy Consumption in Four U.S. Cities*, CATE, **9**, (2016)
26. M. Chagolla et al., Effect *of Tree Shading on the Thermal Load of a House in a Warm Climate Zone in Mexico*, Proceedings of the ASME IMECE 2012 November 9-15, Texas, USA (2012)
27. M. Masaryk, *Adaptive Shading System as a Building Heat Load Reduction*, in Proceedings of the MATEC Web of Conferences **369**, (2022)
28. C. Buratti et al., *Analysis of Nano Silica Aerogel Based Glazing Effect on the Solar Heat Gain and Cooling Load in a School under Different Climatic Conditions*, Environ. Sci. Proc. (2021)
29. Y. Zhou, *Artificial neural network-based smart aerogel glazing in low-energy buildings: A state-of-the-art review*, iScience, **24**, (2021)
30. K. Stejskalová, N. Vavřínová, *Assessment of the Summer Thermal Stability of the Attic Room Using Two Different Software*, Civil and Environmental Engineering, **16**, 360-369, (2020)

# The use of a general method for calculating the resistance of strengthened concrete columns

*Jan* Pecka[1]*and *Katarina* Gajdosova[1]

[1] Faculty of Civil Engineering, Slovak University of Technology in Bratislava, Radlinského 11, 810 05 Bratislava, Slovak Republic

**Abstract.** Columns are one of the basic types of vertical load-bearing elements. In buildings, they are used where there is a requirement to create free spaces. This results in an effort to minimize the cross-section dimensions of the column and to maximize the use of its resistance. The need of the increase in resistance is the reason for columns strengthening. The choice of a suitable strengthening method must take into account the way the column is loaded. Some methods are universal, the others are more effective in the case of predominant compressive, or predominant bending loading. In presented research, the attention is focused on strengthening slender reinforced concrete columns by progressive methods - FRP (fibre reinforced polymer) and FRC (fibre reinforced concrete) applications. For slender columns, second-order effects must be taken into account. The current Eurocode lists three methods of analysing these effects. The simplified methods proposed for the analysis of a reinforced concrete column cannot be applied directly to the calculation of the resistance of a column strengthened by any of the progressive methods. The article is devoted to the calculation of column resistance using a general method that allows the contribution to column resistance from strengthening to be considered.

## 1 Introduction

The requirement to increase the resistance of an existing column can result from various reasons, the required increase in resistance can be achieved by one of several methods of column strengthening. When choosing a strengthening method, it is important to know the way the column is loaded so that the optimal method can be chosen.

In general, strengthening methods can be divided into two groups. The first achieves an increase in resistance by adding a material involved in load transfer. The second group includes methods using the confinement effect, which increases the concrete strength of the existing column. Strengthening by confinement is suitable for short columns stressed mainly by an axial compressive force. We can include e.g. jacketing with FRP sheets, wrapping with mortars reinforced with steel mesh or strengthening with steel caging.

For columns loaded mainly by bending moments, as well as for slender columns, strengthening by confinement is less effective. Methods based on the addition of material are

* Corresponding author: jan.pecka@stuba.sk

more suitable. With the classic method of strengthening by adding concrete reinforcement and layer of concrete, both flexural and compressive resistance increases, but there is a significant increase in the cross-section dimensions. Progressive methods include the application of CFRP (Carbon Fibre Reinforced Polymers) strips both by the NSM (Near Surface Mounted) and EBR (Externally Bonded Reinforcement) method, which enables an increase in bending resistance with a minimal increase in cross-section. Another option is to jacket the column with a layer of FRC (Fibre Reinforced Concrete) or UHPFRC (Ultra-High Performance FRC). The jacketing can actively participate in load transfer and at the same time it acts as a confinement on the column core. Unlike jacketing with traditional concrete with the addition of reinforcement, strengthening with thinner layers of 3-4 cm is possible with FRC, or UHPFRC.

For slender columns, it is necessary to consider second-order effects. In the currently valid Eurocode (EC2), three methods for the analysis of second-order effects are given [1]. Two simplified methods: 1. Method based on nominal stiffness, which is based on the estimation of bending stiffness by a model taking into account the effect of cracks, physical nonlinearities and creep. 2. The method based on the nominal curvature, which is based on the calculation of the deflection based on the buckling length and the estimation of the maximum curvature. None of the simplified methods directly allow to consider the strengthening by progressive materials. The third method is the general method, which is the only one that allows the strengthening with the progressive materials to be considered when calculating the resistance of the column.

The presented study is focused on increasing the resistance of slender concrete columns using progressive methods, application of NSM CFRP and jacketing with FRC/UHPFRC. The aim of the paper is to verify the calculation of the resistance of strengthened columns by the general method on already realized experiments. In the future, the calculation by the general method will be used in the design of the own experiment.

## 2 General Method

The general method makes it possible to consider geometrical nonlinearities in addition to material nonlinearities. It allows to analyse any cross-section shape, boundary conditions, any stress-strain diagram, uniaxial, or biaxial bending moment. The general method is based on following assumptions [2]:

- Linear strain distribution in a cross-section;
- Equal strains in materials at the same level of the cross-section;
- Given stress-strain relationships for materials.

### 2.1 Calculation algorithm

For the analysis of slender columns by the general method, the calculation method A was chosen, described in dissertation of Čuhák (2013) [3]. The result of the calculation is the N-M load curve of the column for the specified eccentricity of the first order.

Based on the assumptions of the calculation that the strain in the cross-section is linear, it is possible to define the strain at any level of cross-section based on the values of strains at its edges ($\varepsilon_o$, $\varepsilon_h$). The column cross-section is divided into parts, for which the dimension and the position in the cross section are determined (Fig. 1). The stresses in the individual parts are determined based on the defined stress-strain diagrams of materials and the position in the cross-section.

In the cross-section loaded by the external forces $N_E$, $M_{Ey}$ the equilibrium of forces must apply: $\sum N=0$, $\sum M=0$. The strains $\varepsilon_o$, $\varepsilon_h$ are unknown. The non-linear character of stress-strain diagrams requires the use of one of the optimization methods for the solution. Newton's method is used for calculation of strains corresponding to the applied external load.

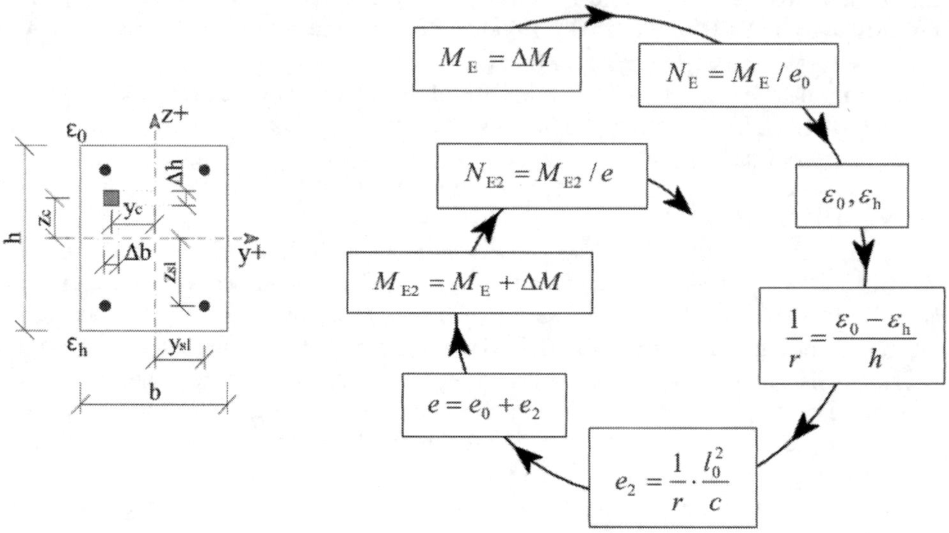


**Fig. 1** Cross-section division scheme and principle of cyclic calculation.

Calculation method A itself is a cyclic calculation, where in each round of the cycle the acting moment is increased by the value *ΔM*. The result is the N-M loading curve of the column. The calculation starts by $\varepsilon_0=0$, $\varepsilon_h=0$, $N_E=0$, $M_E=0$. The first order eccentricity of force $e_0$ is defined. An increment of the moment $\Delta M$ is added to the acting moment $M_E$. Subsequently, through the eccentricity of the first order, the corresponding axial force $N_E$ is calculated. The deformations $\varepsilon_0$, $\varepsilon_h$ are calculated from the equilibrium condition of forces in the cross-section using Newton's method. Based on the deformation of the cross-section, the curvature is calculated using the equation (1). Based on the curvature, it is possible to calculate the eccentricity of the second order according to (2) [1]. The total eccentricity is determined by the sum of the eccentricities of the first and second order. Subsequently, the next round of calculation is entered. An increment is added to the bending moment. Using the resulting eccentricity from the previous cycle, the axial force is calculated, then strains are determined, etc. The calculation ends at the moment when the ultimate strain is reached in one of the materials and its failure occurs.

$$\frac{1}{r} = \frac{\varepsilon_0 - \varepsilon_h}{h} \tag{1}$$

$$e_2 = \frac{1}{r} \cdot \frac{l_0^2}{c} \tag{2}$$

In addition to the N-M curve calculated using the general method, an interaction diagram of the critical cross-section (ID) is also constructed for the purpose of presenting the resistance of the column cross-section. The conditions for failure of the cross-section are the same as the conditions for ending the cyclic calculation of the N-M curve. On the compressed side, it is the achievement of the ultimate strain of the material in the most stressed fibre of the cross-section. On the tensile side, it is the achievement of the ultimate strain of the reinforcement or the CFRP strips, depending on condition which occurs first.

## 2.2 Stress-strains diagrams

Material nonlinearities are taken into account by defined stress-strain diagrams. For concrete, a parabolic, alternatively block parabolic diagram defined according to EC2 [1] was used. The values of the material characteristics were taken from the literature. A bilinear stress-strain diagram with an ascending branch is considered for the reinforcement. For CFRP strips, the stress-strain diagram with linear elastic behaviour until failure is chosen. The compressive strength of the CFRP is not neglected. For FRC, a parabolic stress-strain diagram is used in the pressure area. In the area of tensile stresses, the diagram used is based on RILEM TC 162-TDF. The stress-strain diagrams are shown in Fig. 2.

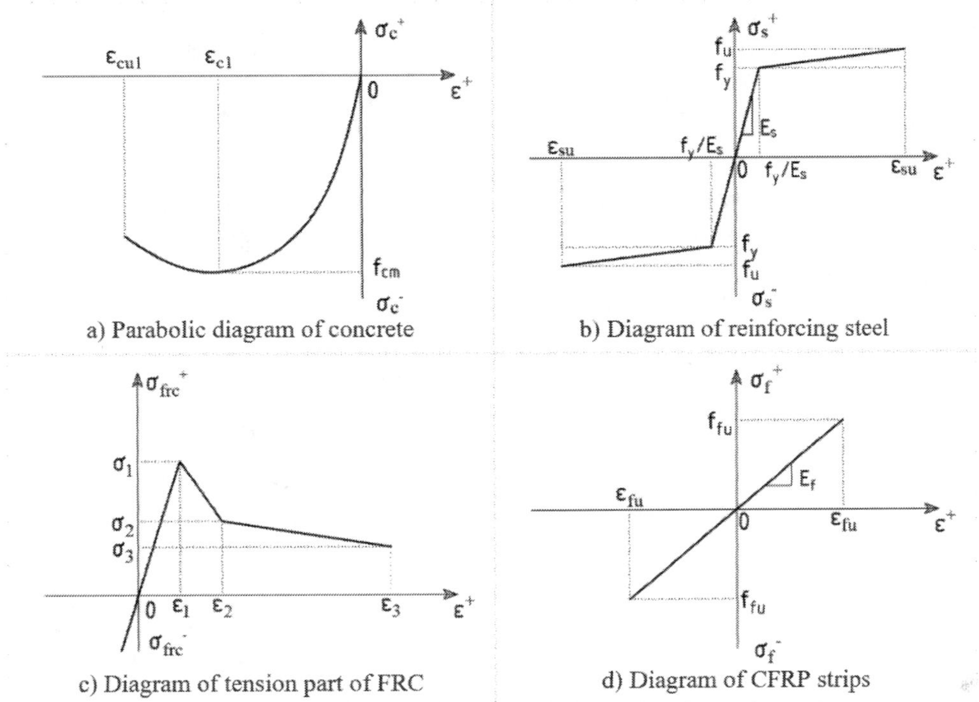


**Fig. 2** Stress-strains diagrams of materials.

# 3 Verification of the calculation method

The calculation of the resistance of strengthened columns by the analytical general method was chosen for the higher availability of software in which the calculation can be written.

For the purposes of verification of the calculation method, experimentally tested specimens of columns with a rectangular cross-section with hinged supports at both ends were chosen from literature review. The specimens are loaded with a combination of axial force and uniaxial bending moment. The following chapter is devoted to the comparison of the calculated results with the results obtained from the experimental measurements. The calculated results are graphically presented using N-M column curves along with the ID of the critical section. For comparison, the points corresponding to the experimentally determined resistance are plotted graphically - marked _*E* (Experimental) and the calculated resistance marked _*A* (Analytical). The mode of failure of the specimens is also compared. The designation of the specimens is taken from the literature. Other designations used: *N* (Non-strengthened), *S* (Strengthened), *CS* (Cross-Section).

### 3.1 Experimental work by Gajdošová

The first experimental work, chosen for comparison, was performed by Gajdošová (2010) [4]. Full-scale slender rectangular RC columns were strengthened either with longitudinal NSM CFRP strips, transverse CFRP wraps, or a combination of both strengthening methods. The cross-section of the columns was 150 x 210 mm with a length of 4.00 m. The specimens were longitudinally reinforced with 8 steel bars of 10 mm in diameter, with transverse reinforcement with diameter of 6 mm with 20 mm cover, all reinforcement was of B500B class. During test, the specimens were hinged supported on both ends with an eccentricity of 40 mm. For comparison, only reference columns and columns strengthened with NSM CFRP strips were selected. When strengthening by confining, taking into account the great slenderness of the column and significant eccentricity, the increase in resistance was insignificant [4]. The material characteristics were considered according to the literature. The results are compared in Table 1. and graphically shown in Fig. 3.

**Table 1.** Comparison of experimental results from Gajdošová and analytical solution.

| **Experimental** | | | | **Analytical** | | | |
|---|---|---|---|---|---|---|---|
| **Specimen** | $N_{R,E}$ [kN] | $M_{R,E}$ [kNm] | Failure mode | $N_{R,A}$ [kN] | $M_{R,A}$ [kNm] | Failure mode | Comparison of $N_R$ [%] |
| **C1, C2** | -286.0 | -23.5 | Loss of stability | -261.4 | -21.3 | Loss of stability | -9.4 |
| **C3, C4** | -323.0 | -26.8 | Loss of stability | -290.4 | -24.5 | Loss of stability | -11.2 |

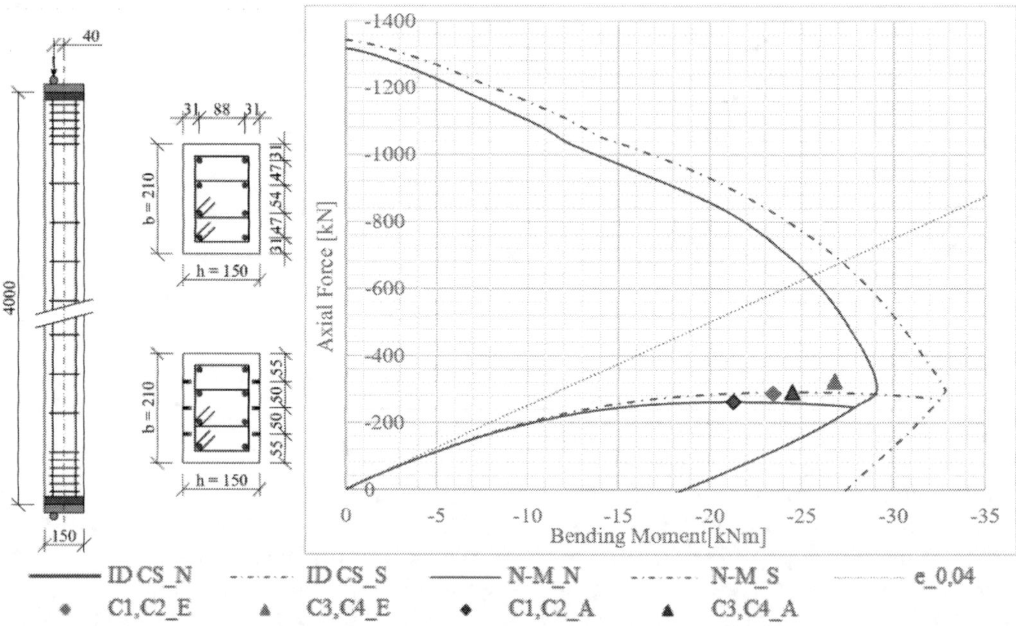


**Fig. 3** The graphic comparison of experimental results from Gajdošová and analytical solution.

The results of the analytical calculation show the same failure mode of the specimens as in the experimental verification, namely by loss of stability. The calculated maximum compressive force at the failure of the specimens is lower by 9.4% for the unstrengthened columns and by 11.2% for the strengthened columns. The calculated bending moments are approximately 10% lower than experimentally determined. We consider the given results to be a good match.

### 3.2 Experimental work by Kendický

The second selected experimental work was performed by Kendický (2017) [5]. In his work, he focused on the general method of calculating the resistance of slender reinforced concrete columns made of high-strength concrete. The experimental program consisted of 6 specimens S2-1 to S2-6. Cross-section dimensions were 150 x 240 mm with a length of 3.84 m. During experimental investigation, the specimens were hinged supported on both ends with an eccentricity of 40 mm. The longitudinal reinforcement consists of 4 steel bars of diameter of 14 mm, one in each corner of the column. The transverse reinforcement was of diameter of 6 mm with a 20 mm cover. The average cylindrical strength of concrete was 87 MPa and the modulus of elasticity 37 GPa. For comparison, the specimens S2-2 and S2-4 are chosen, which achieved the smallest and the largest value of the axial force at failure and the average value of all tested specimens. In the experimental investigation, all specimens failed due to loss of stability. The results are compared in Table 2. and graphically shown in Fig. 4.

**Table 2.** Comparison of experimental results from Kendický and analytical solution.

| Experimental | | | | Analytical | | | |
|---|---|---|---|---|---|---|---|
| **Specimen** | $N_{R.E}$ [kN] | $M_{R.E}$ [kNm] | $e_{2.E}$ [mm] | $N_{R.A}$ [kN] | $M_{R.A}$ [kNm] | $e_{2.A}$ [mm] | Comparison of $N_R$ [%] |
| **S2-2** | -378.8 | -28.6 | 35.5 | | | | -1.3 |
| **S2-4** | -346.0 | -26.6 | 36.8 | -374.1 | -29.6 | 39.1 | 8.1 |
| **Average** | -361.2 | -27.6 | 35.7 | | | | 3.6 |

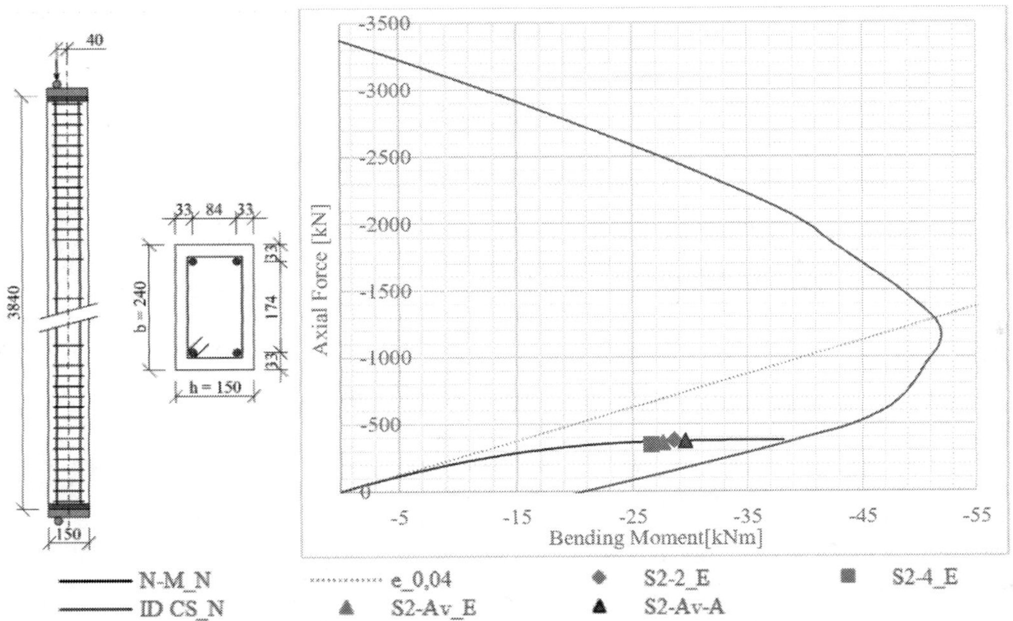


**Fig. 4** The graphic comparison of experimental results from Kendický and analytical solution.

The analytically calculated N-M curve of the column reaches its maximum inside the cross-section ID. Column failure occurs due to the loss of stability, as it was the case with experimentally tested specimens. The individual experimental results are close to the analytical N-M curve, but they achieved a little bit lower bending resistance as assumed analytically. The difference in the average maximum compressive force for all specimens is of 3.6% lower than the analytically calculated value. Which can be considered as a very good agreement between experimental verification and analytical calculation.

### 3.3 Experimental work by Khorramian and Sadeghian

The third selected experimental work was carried out by Khorramian and Sadeghian (2017) [6]. The specimens with square cross-section of 150×150 mm and a length of 500 mm were tested under concentric and eccentric compressive loading up to failure. Eccentricities of 0, 10, 20, and 30% of the cross-section height were used, i.e. 0, 15, 30, and 45 mm. Nine of the specimens were strengthened with 4 NSM CFRP strips of dimensions of 10 x 1.2 mm (N-ex-y). These were analytically compared, the block-parabolic diagram for concrete was used. The modes of failure were: concrete spalling in compression (CS), concrete crushing in compression by the strain 0.0035mm/mm (CC), compressive FRP crushing (CFC). The results are compared in Table 3 and graphically shown in Fig. 5.

**Table 3.** Comparison of experimental results from Khorramian and Sadeghian and analytical solution.

| **Experimental** | | | | **Analytical** | | | |
|---|---|---|---|---|---|---|---|
| **Specimen** | $N_{R.E}$ [kN] | $M_{R.E}$ [kNm] | Failure mode | $N_{R.A}$ [kN] | $M_{R.A}$ [kNm] | Failure mode | Comparison of $N_R$ [%] |
| N-e0-1 | -686.9 | - | CS | -850.4 | 0 | CC | 23.8 |
| **N-e0-2** | -862.1 | - | CS | | | | -1.4 |
| **N-e10-1** | -660.7 | - | CC→CFC | -669.2 | -10.4 | CC | 1.3 |
| N-e10-2 | -556.0 | - | CC→CFC | | | | 20.4 |
| **N-e10-3** | -661.6 | - | CC→CFC | | | | 1.1 |
| **N-e20-1** | -537.0 | - | CFC→CC | -510.9 | -15.7 | CC | -5.1 |
| **N-e20-2** | -552.9 | - | CFC→CC | | | | -8.2 |
| **N-e30-1** | -408.0 | - | CC→CFC | -369.3 | -17.0 | CC | -10.5 |
| **N-e30-2** | -389.8 | - | CFC→CC | | | | -5.6 |

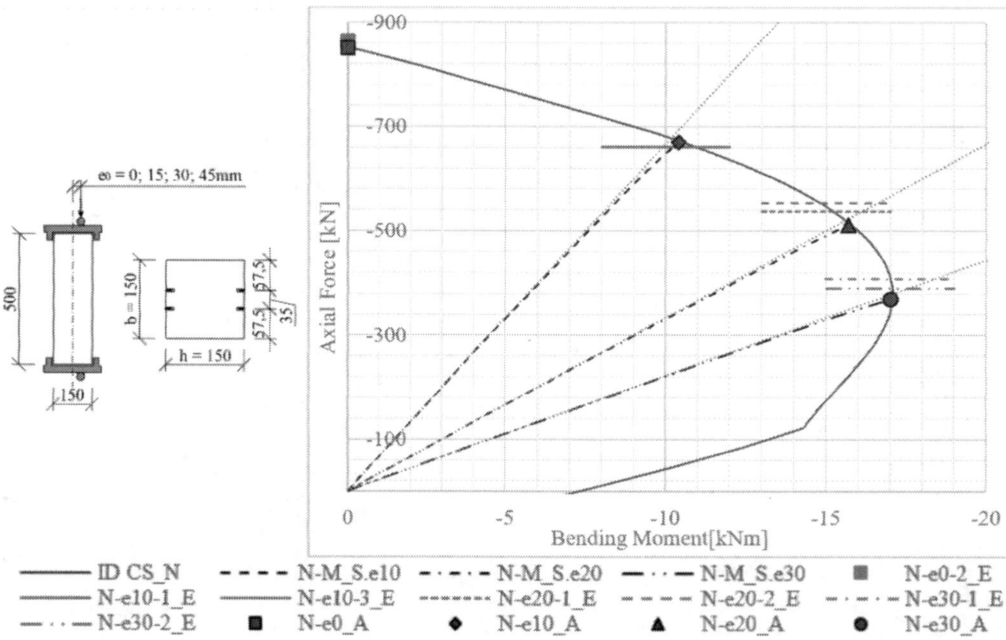


**Fig. 5** The graphic comparison of experiment from Khorramian and Sadeghian and analytic. solution.

The comparison can be considered as verification of ID of cross-section. Experimental results differ from analytical results up to -10% as in the comparison with Gajdošová. Due to the large differences in results of columns with same parameters, it is not appropriate to take the results of N-e10-2 and N-e0-1 into account. All the others results are in a good match.

### 3.4 Experimental work by Vavruš

The last, fourth experimental work chosen for the verification was performed by Vavruš (2020) [7]. In his work, he dealt with the strengthening of slender columns by jacketing with concrete. A total of 6 columns with a cross-section of 160 x 160 mm and a length of 2.50 m were tested. The longitudinal reinforcement of the column was 4 Ø10 mm, the transverse reinforcement was of diameter of 8 mm. All specimens were tested in a horizontal position with an eccentricity of 100 mm. Two specimens were reference (SN_01 and SN_02). Two were strengthened by jacketing with concrete layer of 35 mm in width with the addition of concrete reinforcement 4 Ø10 mm (SB_01 and SB_02). Two specimens were strengthened by jacketing with FRC layer of 35 mm in width (SV_01 and SV_02). The results are compared in Table 4 and graphically shown in Fig. 6.

**Table 4.** Comparison of experimental results from Vavruš and analytical solution.

| **Experimental** | | | | **Analytical** | | | |
|---|---|---|---|---|---|---|---|
| **Specimen** | $N_{R.E}$ [kN] | $M_{R.E}$ [kNm] | $e_{2.E}$ [mm] | $N_{R.A}$ [kN] | $M_{R.A}$ [kNm] | $e_{2.A}$ [mm] | Comparison of $N_R$ [%] |
| **SN_01** | -149.5 | -18.0 | 20.6 | -126.1 | 15.8 | 25.3 | -18.6 |
| **SN_02** | -147.5 | -18.5 | 25.4 | | | | -17.0 |
| **SV_01** | -410.6 | -49.8 | 21.2 | -400.6 | 46.5 | 16.1 | -2.5 |
| **SV_02** | -359.3 | -42.9 | 19.5 | | | | 11.5 |
| **SB_01** | -597.4 | -68.9 | 15.4 | -503.8 | 59.0 | 17.1 | -18.6 |
| **SB_02** | -517.1 | -64.1 | 23.9 | | | | -2.6 |

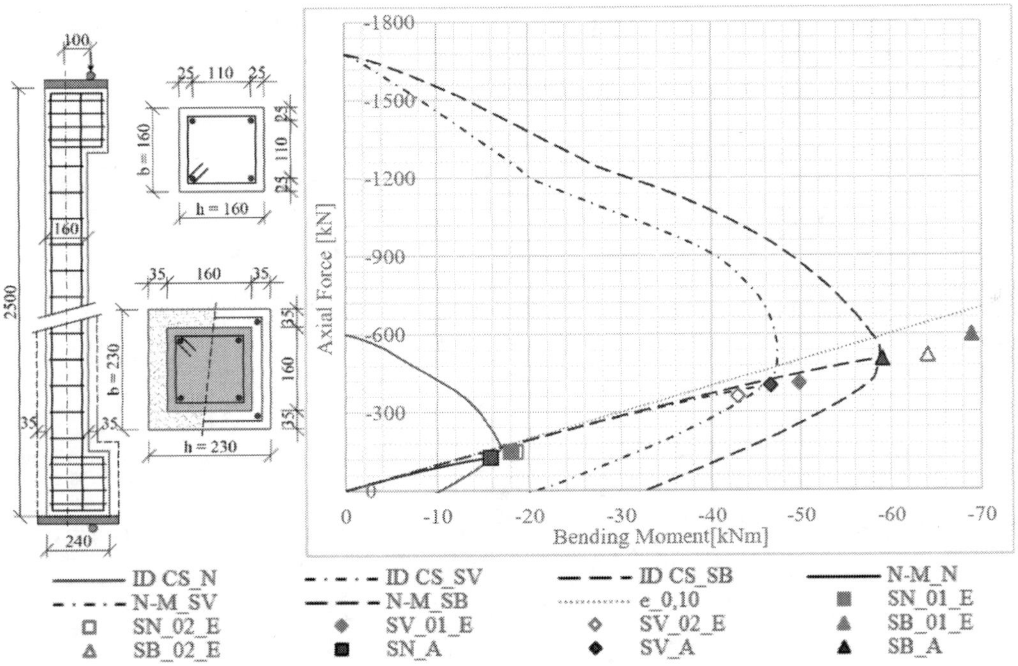


**Fig. 6** The graphic comparison of experimental results from Vavruš and analytical solution.

The failure of the specimens in the analytical calculation occurred due to crushing of the concrete on the compressed side, as by experimental specimens. The biggest differences in calculated and experimental resistance are 17.0 and 18.6% for reference specimens. The smallest difference is 2.5% and 11.5% by strengthening with FRC jacketing. With regard to the range of experimental values, we consider the given agreement to be satisfying.

## 4 Conclusion

The issue of strengthening existing structures is becoming more and more important, whether for reasons of increasing demands on structures, their neglected maintenance or an ecological issue. When strengthening slender columns, it is necessary to consider second-order effects. The simplified methods for the analysis of second-order effects given in the EC2 do not allow directly taking into account the strengthening by progressive methods. For this reason, a general method was chosen to consider the effects of column strengthening.

As a part of the verification of the general method - method A, the resistances of the experimentally tested column specimens were calculated. The results of the comparison of the analytical solution with the experimental results are as follows:

Using a general method, it is possible to calculate the resistance of slender columns made of normal as well as high-strength concrete.

The general method makes it possible to take into account the contribution of strengthening by methods based on the application of CFRP strips, jacketing by the classic concrete layer, or jacketing by the layer of FRC to resistance.

Between the experimentally measured maximum compressive forces and the values determined by the analytical method, the difference was a maximum of 18,6 % (comparison in Table 4) with the fact that most of the values were within 10% and close to this value, which can be evaluated as a very good agreement.

The calculation takes into account whether the column will fail due to loss of stability or due to failure of the critical section. In the future, it is possible to add additional boundary conditions for failure of the critical cross-section.

The verified general method will be used in the design of the experiment aimed at strengthening slender columns. by applying CFRP strips using the NMS method and jacketing by layer of FRC alt. UHPFRC.

This work was supported by the Scientific Grant Agency VEGA under the contract No. VEGA 1/0358/23.

## References

1. STN EN 1992–1–1 + A1 2015, *Eurocode 2: Design of concrete structures - Part 1-1: General rules and rules for buildings,* pp. 190, (2015)
2. B. Westenberg, *Second order effects in slender concrete structures,* pp. 98, KTH, Stocholm, (2004)
3. M. Čuhák, *Nonlinear analysis of concrete structures. Biaxial bending with axial force*, dissertation, pp. 110, SvF STU, Bratislava, (2013)
4. K. Gajdošová, *Strengthening of slender reinforced concrete columns with the use of fibre reinforced polymers*, dissertation, pp. 144, SvF STU, Bratislava, (2010)
5. P. Kendický, *Numerical and experimental analysis of slender concrete columns*, dissertation, pp. 130, SvF STU, Bratislava, (2017)
6. K. Khorramian, P. Sadeghian, *Strengthening concrete columns using NSM CFRP laminates*, in Proceedings of the 6th Asia-Pacific Conference on FRP in Structures, APFIS 2017, 19-21 July 2017, Singapore (2017)
7. M. Vavruš, *Use of high-quality concrete to strengthen existing load-bearing elements*, dissertation, pp. 152, SvF UNIZA, Žilina, (2020)

# Extracting the geometry of selected structural elements from BIM models

*Aset* Madiev[1,*], *Jan* Erdelyi[1], and *Richard* Honti[1]

[1] Department of Surveying, Faculty of Civil Engineering, Slovak University of Technology in Bratislava, Vazovova 5, 812 43, Bratislava, Slovakia

**Abstract.** The use of information technology is currently a prerequisite for the successful implementation of any project in various spheres of human activity. Building information modelling (BIM) allows you to use a virtual model of a building, covering the entire life cycle of a building, starting from design, and ending with its demolition. However, as in any field, BIM technology has advantages and disadvantages. Since the models are often detailed and complex 3D models, the extraction of the coordinates of characteristic points for geodetic purposes is more difficult as in the case of 2D drawings. In this article, we provide practical information on extracting geometric parameters of objects from a BIM model in the Industry Foundation Classes (IFC) format. Three objects (wall, structural opening, and column) were selected as examples. The results obtained showed that the applied approach provides reliable data on the geometry of the objects used.

## 1 Introduction

The use of information technology is currently a prerequisite for the successful implementation of any project in various spheres of human activity. Building Information Modeling (BIM) is one of the most promising developments in the architecture, engineering, and construction (AEC) industries. With BIM technology, an accurate virtual model of a building is constructed digitally. When completed, the computer-generated model contains precise geometry and relevant data to support the construction, fabrication, and procurement activities required to realize the building. When implemented appropriately, BIM facilitates a more integrated design and construction process that results in better quality buildings at lower cost and reduced project duration [1].

Nowadays the most widely used file exchange format for BIM is the Industry Foundation Classes (IFC). This format has gained its popularity due to an open specification for data exchange in the field of construction and facility management [2]. The developer of this format is the buildingSMART alliance, formerly known as the International Alliance of Interaction (IAI). The central principle of this alliance is the openBIM concept – the principle of openness and unification of data for all project participants, regardless of the software used

*Corresponding author: aset.madiev@stuba.sk

to solve a particular task. The concept implies transparent, open data, avoiding loss and duplication of information. In addition, the IFC format has the following features:

1. The ability to manually configure the attributes of objects. Due to the openness for editing the IFC file, it is possible to fine-tune any entity to suit your needs without using a BIM platform [3].

2. Reading (viewing) projects. The IFC format is most often used as a visualization of the project through viewing programs for the purpose of checking the original project. At the same time, after importing, the project file is initially closed for writing and cannot be accidentally changed.

3. Support and development prospects. The format is being developed by the joint efforts of an alliance of the largest CAD companies with the support of universities around the world.

4. IFC is a data exchange format and an interface for converting entities of one software into entities of another. As a result of the double conversion, despite the OpenBIM concept, there are errors and data loss due to imperfections in the algorithms for importing and exporting BIM platforms [4].

Currently, the newest version of IFC is IFC 4.3.x dev, however the IFC 4.4.0 is already being developed [5]. A BIM model exported into the IFC format was used in this work as an example (Fig. 1). The aim of the article is to extract geometric information from an IFC format file to use the obtained data in the execution of selected geodetic works.

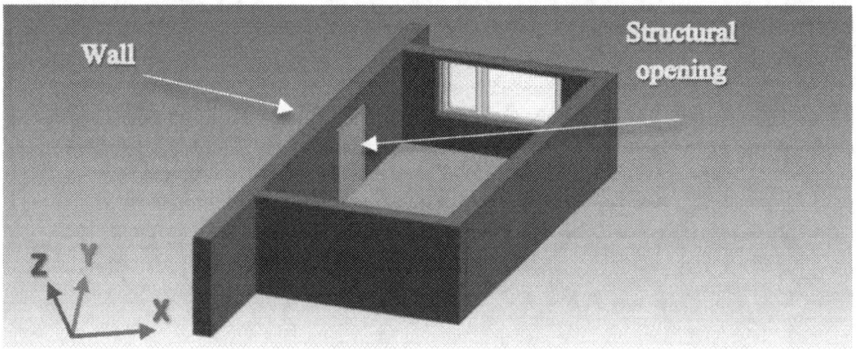


**Fig. 1.** A BIM model imported into the IFC format.

## 2 Wall geometry extraction from IFC

According to the definition by the International Organization for Standardization (ISO), a wall is a vertical structure, usually, stone or concrete, that restricts or separates the structure and performs load bearing or retaining functions [6].

In contrast to the definition given by [6], the definition of a wall in buildingSMART is supplemented by the following: "... Walls are usually vertical or almost vertical flat elements, often designed for bearing structural loads. At the same time, the wall should not be a carrier ..." [7]. There is a division into IfcWallStandartCase and IfcWall in the buildingSMART catalogue. So, IfcWallStandartCase is used for all occurrences of walls that have a constant thickness along the wall path and where the thickness parameter can be fully described by a set of material layers. These walls are always represented geometrically using the SweptSolid geometry if a three-dimensional geometric representation is assigned. IfcWall, is used for all other wall occurrences for walls with varying thickness along the wall path (for example, polygonal walls), or walls with non-rectangular cross-sections. The information provided in IfcWall contained the following information stored as text: #518=IFCWALL('3EiGoFx85EV8Kh5UhCp_73',#42,'BasicWall:SN.257_WALL_RCO_250:443268',$,'BasicWall:SN.257_WALL_RCO_250',#478,#514,'443268',.NOTDEFINED).

The information stored in the line is the following:

– #518 – the sequence number of the line in the IFC file.

– IFCWALL – internal data type.

– '3EiGoFx85EV8Kh5UhCp_73' – IFC–GUID is a globally unique identifier that provides a way to uniquely identify an object.

– #42 – Information about the file. This includes information about the organization and the user who created this object.

– 'BasicWall:SN.257_WALL_RCO_250:443268' – the name of the object specified by the user.

– $ - a symbol indicating the absence of an optional element.

– 'BasicWall:SN.257_WALL_RCO_250' – the type of object specified by the user.

– #478 – information about the location of the object.

– #514 – information about the shape of the object.

– 443268 – object tag.

– NOTDEFINED – preferred object type.

To trace the entire connection of storing all geometry information, it is necessary to follow the links from the initialization of the object to the final line with data about the coordinate system used. The initialization string is located at position #478, has the name IfcLocalPlacement and contains two links: #136 and #477 (both are IFCLocalPlacement). From the [7], we understand that IfcLocalPlacement indicates the location of the object. In this case, the location consists of two components: information about the relative object of which the current object will be constructed (PlacementRelTo), as well as information about the object itself (RelativePlacement). If there is no PlacementRelTo, it means that the object is located absolutely (without connecting to another object). Based on this definition, it was assumed that the information contained in line #136 and all subsequent references within it refers to information about a relative object. And the information contained in line #477 refers to information about the location of the object itself (Fig. 2).

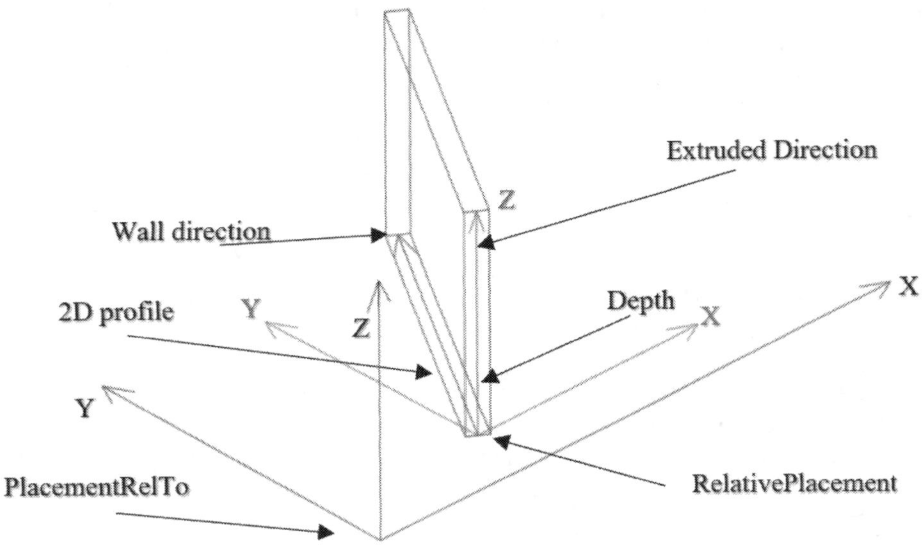


**Fig. 2.** Definition of IfcLocalPlacement in a three–dimensional coordinate system (XYZ (red)– the coordinate system of the object itself (RelativePlacement). XYZ (black) – the coordinate system of a relative object (PlacementRelTo) Wall direction – the direction of the wall or a unit vector.).

Following the link to the number #136, we could find that it also has the name IfcLocalPlacement and contains an identical set of parameters (#136 =

IFCLocalPlacement(#33,#135). At the end of this iteration, we come to line #146 (#146= IFCAxis2Placement3D(#6,$,$), which represents the location and orientation for placing elements in three-dimensional space. The object is characterized by a point having coordinates in a two- or three–dimensional rectangular Cartesian coordinate system (IfcCartesianPoint), two axes (IfcDirection), with one of the axes representing the Z-axis direction of a local coordinate system, and the second - the X-axis direction of local coordinate system. It is also worth noting that the IfcDirection components are not normalized, so if a unit vector is needed, it should be normalized first and used only after. If there are no IfcDirection elements, as in our case, this means that the axes are set by default and equal to X-axis [1., 0., 0.,], Y-axis [0., 1., 0.,], Z-axis [0., 0., 1.,]. The location of the object itself contains an identical set of parameters (a point and two vectors). After determining the information about the location of the object, it is necessary to determine the remaining elements to be able to recreate the desired object independently. Therefore, the next step for analysis is information about the shape and dimension of the object.

All information about the shape of the object is presented in the IfcProductDefinitionShape. In our case, this description is presented in line #514 (#514= IfcProductDefinitionShape($,$,(#483,#512))). Shape-related information includes shape representation, related representation information, and assignment to representation layers or elements of a topological representation for connectivity systems. Based on the example presented, the information about the form was divided into two links (#483 and #512, with the same name IfcShapeRepresentation). Despite the similarity of the names, in the first case (#483) was presented information about the object axis, and in the second (#512) - information about the shape of the object. Since information about the shape of the object itself is considered the most important, the second case of IfcShapeRepresentation (#512) will be considered here, but their data structure is the same, so all explanations can be applied to the first case (#483).

#512 IfcShapeRepresentation - represents the concept of a specific geometric representation of a product or product component in the context of a specific geometric representation (definition from buildingSMART). And, indeed, if we look at the contents of line #512 (#512= IfcShapeRepresentation(#105,'Body','SweptSolid',(#502))), we can see the following parameters:

1. #105 is the first element of the string containing information about the subcontext. Since each object has several instances of IfcRepresentation subtypes, each of which is assigned to different geometric representations, then using the information about the subcontext, you can choose the most appropriate representation for displaying the geometric shape of the object, depending on the target view and scale. This element contains attributes that represent information about the form of representation:

    1.1. Parent context. The most important attribute containing information about the coordinate system, accuracy, dimension of spatial coordinates and true north.

    1.2. Information about the scale.

    1.3. Updated information about the target form.

    1.4. User-defined form (this parameter may be missing).

2. Information about the object concerning which all information is provided (Axis, Body, etc.).

3. Representation types for shape representation.

4. And the most important element is the location. In our case, this value corresponds to the string #502 IfcExtrudedAreaSolid.

From the definition provided by [7], IfcExtrudedAreaSolid is the body of an elongated area defined by covering a bounded flat surface. Having considered the structure of the element (#502= IFCExtrudedAreaSolid(#500,#501,#20,2590.31254649137)), we can observe the following parameters:

1. IfcArbitraryClosedProfileDef – defines an arbitrary two-dimensional profile for the use within the swept surface geometry, the swept area solid or a sectioned spine. It is given by an outer boundary from which the surface or solid can be constructed.
2. The direction in which the surface should be swept.
3. The distance the surface is to be swept (wall height).

As a result of the above procedures, the following data presented in Table 1.

**Table 1.** Extracted geometry information of the Wall from IFC.

| Name | Value |
|---|---|
| PlacementRelTo | For the entire BIM model, this coordinate system was unified and had the following characteristics [0,0,0] - coordinates of the reference point represented as [X, Y, Z]; [1,0,0], [0,1,0], [0,0,1] – the direction of the vectors X, Y, Z, respectively. |
| Relative Placement coordinate system | The coordinates of the reference point are [4715.1401476308, -3629.42855196745, -120]. The direction of vector Z [0,0,1], the direction of the unit vector [0.532833356621246, 0.846220192427323 ,0]; |
| 2D profile coordinates | #977= (1570.3558653296, -125.); #979= (1820.35730921413, -125.); #981= (6852.24049489, -125.); #983= (7102.2420493596, -125.); #985= (7242.24291986258, -125.); #987= (7753.68813497982, -125.); #989= (7754.94535999682,125.); #991= (6892.90502698309,125.); #993= (6752.90502698309,125.); #995= (5675.68018319826,125.); #997= (5535.67635462137,125.); #999= (3702.4115197198,125.); #1001= (3562.40814192163,125.); #1003= (1.18712323781781,125.). For each point, the format was (X, Y). |
| Swept direction | The direction in which the surface should be swept = [0,0,1]. |
| Wall height | 2590.31254649137 |

## 3 Geometry extraction of structural opening from IFC

According to the definition from [7], IfcOpeningElement defines an opening, recess or slot reflecting voids. It represents a void within any element that has physical manifestation. There are two different types of opening elements:

– An opening, where the thickness of the opening is greater or equal to the thickness of the element.

– A recess or niche, where the thickness of the recess is smaller than the thickness of the element.

To capture the differences, the inherited ObjectType attribute is used, which specifies "Opening" for the first type and "Recess" for the second type of detached elements. If the ObjectType value is set to NOTDEFINED, "Opening" is assumed. The information provided in IfcOpeningElement was as follows: #1343= IFCOPENINGELEMENT ('069cAVrUT5WBODNYwCoEJm',#42,'Basic Wall:SN.257_WALL_RCO_250:444020',$, 'Opening',#1342,#1336,'444020',.OPENING.). The only difference from the wall's characteristics is the 3rd item ('Basic Wall:SN.257_WALL_RCO_250:444020'), indicating the element in which this opening is located.

Since the description of IfcOpeningElement is identical to the description of IfcWall, the structure of extracting information about the location of the opening is identical to the structure of extracting information about the location of the wall, except for one thing. Since the openings are "inside" the wall, they have two PlacementRelTo objects. The first PlacementRelTo object for openings will be the wall PlacementRelTo object, and the second will be the wall RelativePlacement. At the same time, openings also have their own Relative Placement (Fig. 3).

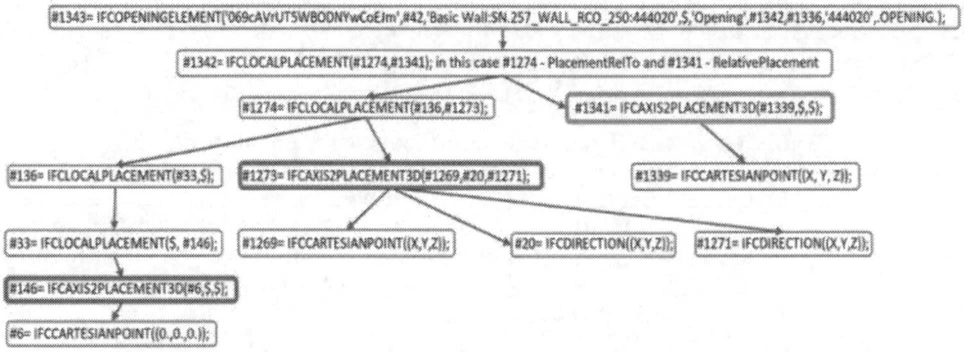


**Fig. 3.** The structure of the location information in structural opening. PlacementRelTo is highlighted in a red rectangle. RelativePlacement – into a green rectangle.

Unlike the almost identical structure of location information with IfcWall, the structure of information about the shape of the opening has more differences. All information about the shape of the object is also presented in the IfcProductDefinitionShape. In our case, this description is presented in line #1336 (#1336= IfcProductDefinitionShape($,$,(#1332))). Due to the entry of the opener into the wall, there is no need for information about the axes here to represent the shape and space of the object. Further, the description of the object's shape has an identical IfcWall structure, up to the moment of the shape's location.

IfcRectangleProfileDef is an entity used for describing the location of the object's centre, which in our case is in line #1314. According to [7], this attribute defines a rectangle as a profile definition used by the geometry of the sweeping surface or the solid body of the sweeping area. The attribute itself includes the following parameters:

1. Name.
2. location (described by IfcCartesianPoint and IfcDirection).
3. The length of the rectangle in the direction of the X and Y axes.

A feature of this attribute is the change in the direction of the axes. When an opening is located, the directions of the main axes of the object are changed relative to the relative coordinate system used for the entire BIM model (since the direction of the axes of the wall coordinate system coincides with the direction of the axes of the relative coordinate system). Thus, the X-axis of the structural opening corresponds to the Z-axis of the wall, the Y-axis of the structural opening corresponds to the X-axis of the wall, and the Z-axis of the structural opening corresponds to the Y-axis of the wall (Fig. 4).

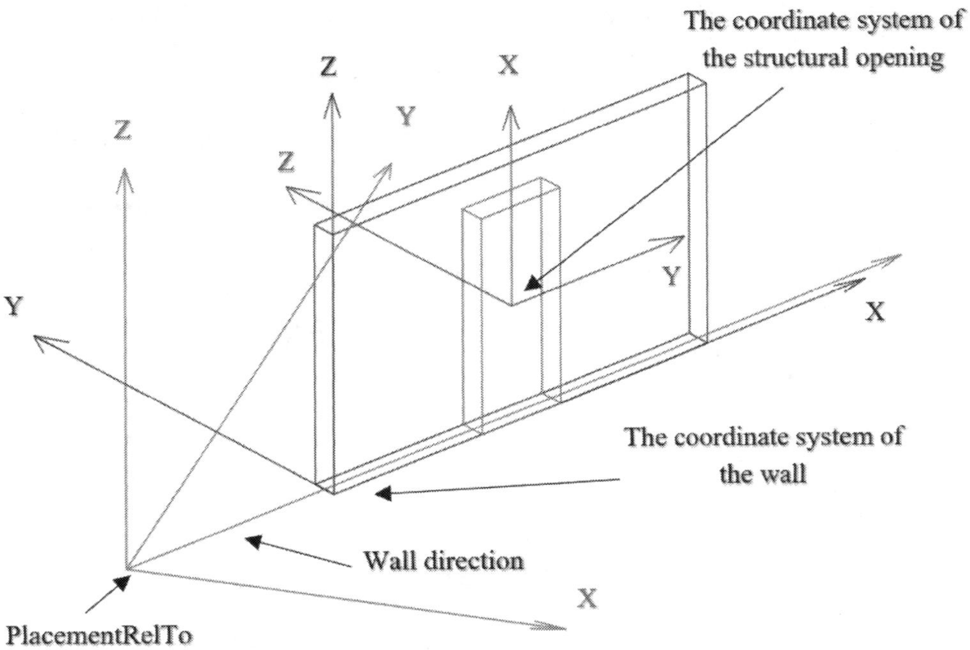


**Fig. 4.** Scheme of the IfcOpeningElement axes relative to the wall coordinate system.

The result of the performed work is presented in Table 2.

**Table 2.** Extracted information from structural opener.

| Name | Value |
|---|---|
| RelativePlacement coordinate system, described by IfcCartesianPoint [X, Y, Z] | [1548.79055287, 70, 0] |
| The centre of the opener, described by IfcCartesianPoint [X, Y, Z] | [1075, 350] |
| The extent of the rectangle in the direction of the x-axis and y-axis | 2150 and 900 |

## 4 Column geometry extraction from IFC

From the definition [7], IfcColumn is a vertical structural element that is often aligned at the intersection of the structural grid. The information provided in IfcColumn was as follows: #409=IFCCOLUMN('2l2PK1jcD8jhsjs8kucjTh',#42,'Family1:Family1:206432',$,'Family1: Family1',#408,#402,'202859',.COLUMN.).

As can be seen from the example, the structure of the columns is identical to the structure of the walls, so the data extraction process is identical to the data extraction process from IfcWall. The only difference is the representation of the object's shape. If the column has a cylindrical shape, the IfcCircleProfileDef attribute is used, describing the shape of the column using a point (IfcCartesianPoint), direction (IfcDirection) and radius. If the column has a rectangular shape, the IfcRectangleProfileDef attribute is used, describing the shape of the column using a point (IfcCartesianPoint), direction (IfcDirection) and the length of the rectangle in the direction of the X and Y axes. At the same time, the height, regardless of the column's shape, is represented in a separate attribute (IfcExtrudedAreaSolid).

Thus, the extracted data, stored in the line for a column has the following form:

– PlacementRelTo coordinate system. For the entire BIM model, this coordinate system was unified and had the following characteristics [0,0,0] - coordinates of the reference point represented as [X, Y, Z]; [1,0,0], [0,1,0], [0,0,1] – the direction of the vectors X, Y, Z, respectively. In our case the data was stored in #408 → #407 → #405.

– RelativePlacement. The coordinates of the reference point are [-173.102273899521, -320.272744411158, 0]. In our case the data was stored in #408 → #146 → #33 → #32.

– 2D profile coordinates (0, - 4.92417986287059E-13.). In our case the data was stored in #402 → #400 → #398 → #390 → #387 → #386 → #384 → #383 → #381.

– Radius. 304.800 In our case the data was stored in #402 → #400 → #398 → #390 → #387 → #386 → #384.

– The direction in which the surface should be swept (column height) = [0,0,1]. In our case the data was stored in #402 → #400 → #398 → #390 → #387 → #386 → #20.

– Column height – 4000 mm. In our case the data was stored in #402 → #400 → #398 → #390 → #387 → #386.

The column restored in the AutoCAD environment based on the extracted data had the form (Fig. 5).

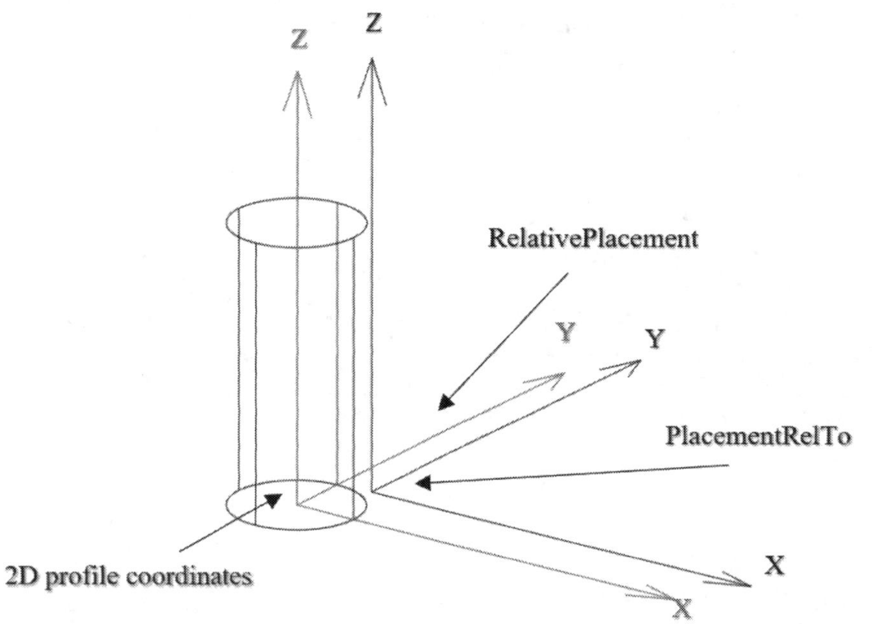


**Fig. 5.** Column restored based on extracted data.

## 5 Conclusion

The aim of this study was to export geometric information from a BIM model in to IFC format. In this model, a wall, a construction opening, and a column were chosen as examples. The search and extraction of the data were done manually. The results showed that the applied approach provides reliable data on the geometry of the objects used. The obtained data can be used to check the geometric parameters of construction objects. However, to completely use such data, it is necessary to automate the extraction procedure and as a further study we are going to use Matlab software to solve this task.

This publication was created with the support of the Scientific Grant Agency of the Ministry of Education, science, research and sport of the Slovak Republic and the Slovak Academy of Sciences for the project VEGA-1/0272/22.

## References

1. C. Eastman, P. Teicholz, R. Sacks, K. Liston, BIM Handbook: A Guide to Building Information Modeling for Owners, Managers, Designers, Engineers and Contractors, Wiley Publishing, 2008
2. R. Honti, J. Erdélyi, *Possibilities of BIM data exchange*, in Proceedings of $18^{th}$ International Multidisciplinary Scientific GeoConference, SGEM, 2-8 July 2018, Albena, Bulgaria (2018)
3. S. Azhar, Leadersh. Manag. Eng. **11**, 12 (2011)
4. T. H. Kolbe, A. Donaubauer, *Semantic 3D City Modeling and BIM.* In Urban Informatics, Cham, Switzerland, 2021
5. ISO 16739-1:2018 Industry Foundation Classes (IFC) for data sharing in the construction and facility management industries – Part 1: Data scheme, 2018
6. ISO 6707-1:1989 Building and civil engineering – Vocabulary – Part 1: General terms, 1989
7. "buildingsSMART" [Online]. Available: http://www.buildingsmart-tech.org/%20specifications [Accessed 12.03.2023]

MATEC Web of Conferences **385**, 01017 (2023) https://doi.org/10.1051/matecconf/202338501017
*Young Scientist 2023*

# Analysis of the thermal and technical parameters of the vegetated roof and its effect on the energy demand – Case study A

*Ján* Domanický[1*], *František* Vranay[2]

[1-2]Technical university of Kosice, Faculty of Civil Engineering, Institute of Architectural Engineering, Vysokoškolská 4, 042 00 Kosice, Slovakia

**Abstract.** ENERGY!!! In this energy era, when the development of energies and the development of their prices is a daily topic that affects people's lives, it is necessary to direct our thinking towards the reduction of global energy demands. In my industry, it's reducing buildings' energy needs, specifically for heating and cooling. It is known from many researches and studies how it is possible to reduce the energy demand of the building in the area of constructions by insulation, or suitable design of the envelope or in the area of Building services, by using renewable energy sources or by using a suitable source and type of heating and cooling. Therefore, I took a different direction in my research. PASSIVE COOLING. Passive so that we don't fight against energy with energy, but that we use the amount of it that falls on the object in the largest possible amount and eliminate the rest with suitable design and suitable elements. I focus on VEGETATION and its potential in this direction. My work has several chapters, and in this article I am dealing with one of the first, namely the analysis of the thermal engineering parameters of the constructions in case study A.

## 1 Introduction

### 1.1 Energy efficiency of building

The topic of global warming is more than actual these days. The media often reports on this problem and people actively perceive it. The weather itself is significantly affected by this phenomenon, and this has a great impact on the population. One of the main, if not the primary, impact of climate change is urbanization [1] – the movement and centralization of the population into cities. It is possible to monitor its increase in the long term, and forecasts do not indicate the opposite. Every year, this phenomenon has an upward trend in most states, although it fluctuates in some cases. Slovakia is an example of this. According to data from the study [2], the development of the urban way of life has the following values.

* Corresponding author: jan.domanicky@tuke.sk

**Table 1.** The value of urbanizm [2].

| Place | Year | | |
|---|---|---|---|
| | **1990** | **2014** | **2050** |
| World | 43% | 54% | 66% |
| Europe | 70% | 73% | 82% |
| East Europe | 68% | 69% | 78% |
| Slovakia | 56% | 54% | 63% |

In Fig. 1 according to [3] it can be seen that already in 2007 the world exceeded the level of urbanism by 50%.

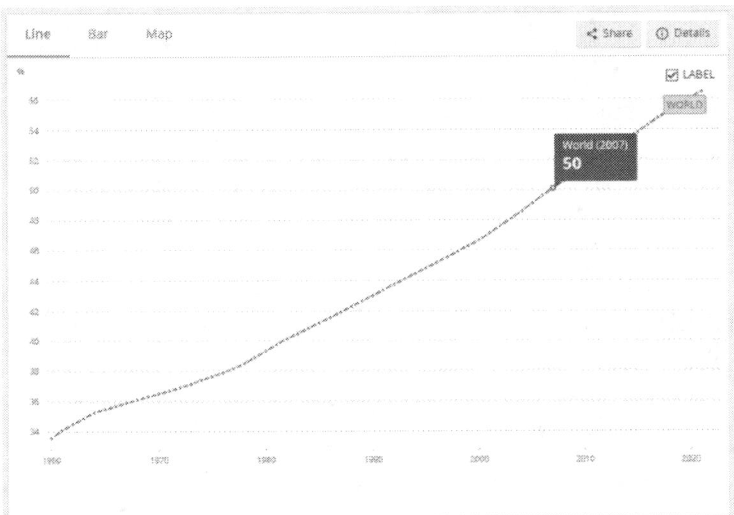


**Fig. 1.** Rising trend of urbanizm [3].

This trend has an impact on several aspects of the climate crisis. Therefore, we got into one big climatic chain. Urbanism causes an increase in the temperature of cities, thereby strengthening one of the significant problems of cities - UHI (Urban Heat Islands). It also increases the local need for energy for heating and cooling, and this means an increase in the production of greenhouse gas emissions. Today's goal is the exact opposite and the reduction of energy consumption.

This article is part of the research in my dissertation, where one of the main goals is to find ways in our climate to prevent the energy fight in buildings with additional energy as much as possible. Of course, it is not possible to completely eliminate it, but it is necessary to find ways to reduce it as much as possible. One of the options (in addition to proper architectural design, shielding, appropriate building envelope, etc.) is the use of passive heating and cooling elements. In my work, I focus on the application of vegetation elements and its effect on heating and cooling, specifically on the interior part.

### 1.2 Vegetation elements

Today, there is no doubt that vegetation elements, especially vegetation roofs, have many advantages, such as reducing rainwater runoff, enriching urban biodiversity, mitigating UHI, improving air quality, reducing thermal oscillation on the roof surface - extending the life of the roof membrane, reducing noise levels and reducing energy consumption. This article deals with the latter aspect [1].

During the summer season, due to the rising temperature in the exterior and subsequently in the interior, the demand for cooling increases. The calculated indoor temperature in Slovakia during the winter period is 20°C and 26°C in the summer. However, it is known that people often think the opposite. They require 20°C in summer and 24°C and more in winter. Therefore, the demand for heating and cooling is increasing. Since cooling is often provided by air conditioning units, this results in an increase in the exterior temperature, whether at the local or urban level, and subsequently supports the UHI issue. It is known that through evapotranspiration, a large amount of solar radiation can be converted into latent heat, which does not cause an increase in temperature [4]. This is another reason to use passive cooling elements.

Why the roof? In our climate zone, it is problematic to maintain vegetated exterior facades. Also, these structures - horizontal building structures are exposed to high thermal stress during the summer season, and require significant measures to prevent heat loss in the winter season [5].

As the benefits are proven, some states have decided to include their applications in cities and support them. It was legally incorporated by Toronto in 2009 [6]. Other cities, such as e.g., supported this issue. Chicago.

## 2 The influence of vegetation elements on heat flow and heat conductivity coefficient

Several factors affect how well the vegetation roof can meet the building's energy needs. Among them are the thickness and thermal conductivity of the substrate, volumetric humidity, height of vegetation, leaf efficiency, emissivity and their coverage, stomatal resistance - the product of linear discontinuous functions of leaf water potential and solar flux, as well as the type and thickness of the roof insulation and from the climatic climate [1].

As for the effect of the vegetated roof on the heat flow, or for the thermal and technical parameters of the roof, which I would use in energy balances, I drew from texts [5,7]. This case study deals with the analysis during the heating period.

In the study [7], 3 types of roofs (differing in insulation thickness) were considered. For our climatic zone and our country, it is necessary to consider a well-insulated roof.

Table 2. Shows the values of the heat transfer coefficients of the given structures either with or without a vegetated roof for three levels of insulation [7].

1. uninsulated roof
    a. 25 cm concrete
    b. 10 cm concrete
2. slightly insulated roof
    a. 15 cm of concrete + 5 cm of insulation + 15 cm of concrete,
    b. 5 cm concrete + 5 cm insulation + 5 cm concrete
3. well-insulated roof
    a. 15 cm of concrete + 15 cm of insulation + 15 cm of concrete,
    b. 10 cm of concrete + 10 cm of insulation + 10 cm of concrete

**Table 2.** Values of heat conductivity coefficients [7].

| Roof | | Heat conductivity coefficient [W/m².K] | | Saving | |
|---|---|---|---|---|---|
| | | Without vegetation | With vegetation | W/m².K | % |
| 1 | a | 7.76 | 1.73 | 6.03 | 77.7 % |
| | b | 18.18 | 1.99 | 16.19 | 89.1 % |
| 2 | a | 0.74 | 0.55 | 0.19 | 25.7 % |
| | b | 0.8 | 0.59 | 0.21 | 26.3 % |
| 3 | a | 0.26 | 0.24 | 0.02 | 7.7 % |
| | b | 0.4 | 0.34 | 0.06 | 15.0 % |

## 3 Direction of research

The article captures case study A, where I chose to start with an unrealistic object of a warehouse-type hall with dimensions of 100x200x10m (width x length x height) - part A1. In this case, the roof area makes up 43.48% of the building's heat exchange envelope. For the principle assessment, I considered roof 3a, which is the closest to our conditions.

Subsequently, when choosing an object, or other buildings, it is necessary to consider the factor of the building's shape and thus the ratio of the individual areas of the building's envelope to the vegetation area. This article features a building with a large roof-to-wall ratio. For comparison, I added an apartment building with dimensions of 14.76x31.66x23.1m (width x length x height) - part A2, where the roof area is only 15.13% of the building area. In fig. 2 and 3 are the dimensions of the two compared objects.

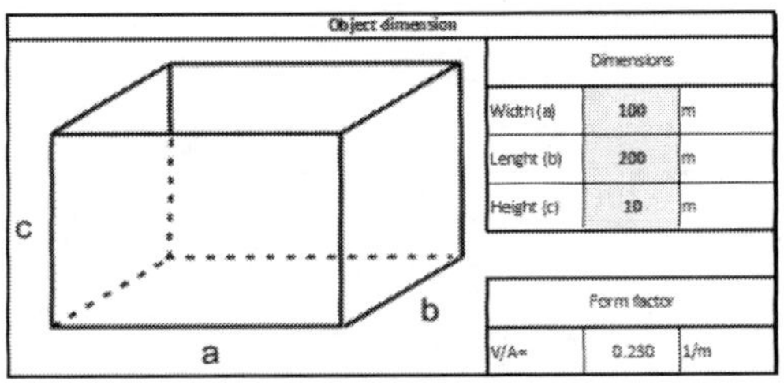


**Fig. 2.** Hall dimensions – part A1.

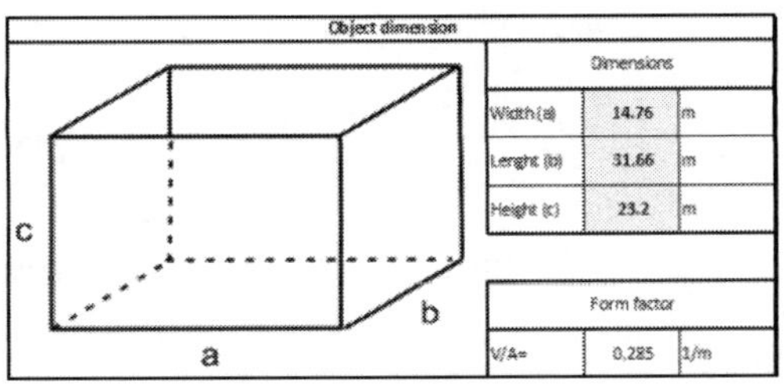


**Fig. 3.** Residential dimensions – part A2.

After selecting the object, the surfaces of the perimeter structures (floor, walls and roof) were compared. These relationships can be seen in table 3 and table 4.

**Table 3.** Area ratio of packaging structures – part A1.

| The ratio of the area of the walls to the roof | | |
|---|---|---|
| Area | m2 | % |
| Walls | 6000 | 13.04% |
| Roof | 20000 | **43.48%** |
| Object | 46000 | 100.00% |

**Table 4.** Area ratio of packaging structures – part A2.

| The ratio of the area of the walls to the roof | | |
|---|---|---|
| Area | m2 | % |
| Walls | 2153.888 | 69.74% |
| Roof | 467.3016 | **15.13%** |
| Object | 3088.4912 | 100.00% |

The thermal and technical parameters of the structures are the same for both parts, so that only the roof structures can be accurately compared. Part A1 is shown in table 5 and 6. Part A2 in table 8 and 9.

**Table 5.** Calculation of heat losses by constructions without a vegetated roof – part A1.

| Heat loss of an object without a vegetation element | | | | | | | |
|---|---|---|---|---|---|---|---|
| | A | U | θi | θe | Δθ | Qi | α |
| | m2 | W/m2.K | °C | °C | K | W | % |
| SO1 | 6000 | 0.144 | 20 | -13 | 33 | 28465.80 | 12% |
| PDL | 20000 | 0.191 | 20 | 3 | 17 | 65022.56 | 27% |
| S - without vegetation | 20000 | 0.222 | 20 | -13 | 33 | 146779.52 | 61% |
| | | | | | | **240267.88** | |

**Table 6.** Calculation of heat losses by constructions without a vegetated roof - part A1.

| Heat loss of an object with a vegetation element | | | | | | | |
|---|---|---|---|---|---|---|---|
| | A | U | θi | θe | Δθ | Qi | α |
| | m2 | W/m2.K | °C | °C | K | W | % |
| SO1 | 6000 | 0.144 | 20 | -13 | 33 | 28465.80 | 12% |
| PDL | 20000 | 0.191 | 20 | 3 | 17 | 65022.56 | 28% |
| S - with vegetation | 20000 | 0.2053 | 20 | -13 | 33 | 135477.50 | 59% |
| | | | | | | **228965.86** | |

**Table 7.** Total object heating balance - part A1.

| Part | S - without veg. | S - with veg | Saving | |
|---|---|---|---|---|
| | W | W | W | % |
| Construction | 146779.52 | 135477.50 | 11302.02 | 7.70% |
| Object | 240267.88 | 228965.86 | 11302.02 | 4.70% |

In part A1, a significant reduction in heat loss through the roof structure can be seen. This influence is also displayed in the overall balance of the object - Table 7. The balance of part A is shown in table 10.

**Table 8.** Calculation of heat losses by constructions without a vegetated roof – part A2.

| Heat loss of an object without a vegetation element | | | | | | | |
|---|---|---|---|---|---|---|---|
| | A | U | θi | θe | Δθ | Qi | α |
| | m2 | W/m2.K | °C | °C | K | W | % |
| SO1 | 2153.888 | 0.144 | 20 | -13 | 33 | 10218.69 | 67% |
| PDL | 467.3016 | 0.191 | 20 | 3 | 17 | 1519.26 | 10% |
| S - without vegetation | 467.3016 | 0.222 | 20 | -13 | 33 | 3429.52 | 23% |
| | | | | | | **15167.46** | |

**Table 9.** Calculation of heat losses by constructions without a vegetated roof - part A2.

| Heat loss of an object with a vegetation element | | | | | | | |
|---|---|---|---|---|---|---|---|
| | A | U | θi | θe | Δθ | Qi | α |
| | m2 | W/m2.K | °C | °C | K | W | % |
| SO1 | 2153.888 | 0.144 | 20 | -13 | 33 | 10218.69 | 69% |
| PDL | 467.3016 | 0.191 | 20 | 3 | 17 | 1519.26 | 10% |
| S - with vegetation | 467.3016 | 0.2053 | 20 | -13 | 33 | 3165.44 | 21% |
| | | | | | | **14903.39** | |

**Table 10.** Total object heating balance - part A2.

| Part | S - without veg. | S - with veg | Saving | |
|---|---|---|---|---|
| | W | W | W | % |
| Construction | 146779.52 | 135477.50 | 11302.02 | 7.70% |
| Object | 240267.88 | 228965.86 | 11302.02 | 4.70% |

By transferring these values to the annual balance of energy needs for heating, we also get the value of savings in kWh/year or GJ/year. From this, when converted to current energy prices in this period, it is also possible to determine the total economic annual savings. Part A1 is captured by table 11-13 and part A2 table 14-16.

**Table 11.** Annual energy requirement for heating without a vegetated roof – part A1.

| Data of the object: | | |
|---|---|---|
| Town: | | Košice |
| External calculated temperature: | te | -13 |
| Length of heating period: | d | 223 |
| Average temperature in the heating season: | tes | 2.9 |
| Average indoor temperature: | tis | 20 |
| Heat loss of the object: | QHL | 240.2679 |

| Calculation of the expected heat demand for heating | | | | | | |
|---|---|---|---|---|---|---|
| d | D | ε | ηr | QHL | QÚK/rok | |
| days | K.days | - | - | kW | kWh/rok | GJ/rok |
| 223 | 3813.3 | 0.63 | 0.98 | 240.26788 | 428359.57 | 1542.09 |

**Table 12.** Annual energy requirement for heating with a vegetated roof – part A1.

| Data of the object: | | |
|---|---|---|
| Town: | | Košice |
| External calculated temperature: | te | -13 |
| Length of heating period: | d | 223 |
| Average temperature in the heating season: | tes | 2.9 |

| | | |
|---|---|---|
| Average indoor temperature: | tis | 20 |
| Heat loss of the object: | QHL | 228.9659 |

| Calculation of the expected heat demand for heating | | | | | | |
|---|---|---|---|---|---|---|
| d | D | ε | ηr | QHL | QÚK/rok | |
| days | K.days | - | - | kW | kWh/rok | GJ/rok |
| 223 | 3813.3 | 0.63 | 0.98 | 228.96586 | 408209.85 | 1469.56 |

**Table 13.** Total comparison of heating energy demand - part A1.

| The results of comparison | | | | | |
|---|---|---|---|---|---|
| | Heat energia demand | | Cost comparison | | |
| | | | Electricity | Gas | Heatpump |
| | kWh/rok | GJ/rok | €/rok | €/rok | €/rok |
| S - withou veg. | 428359.567 | 1542.094 | 68537.53 | 26772.47 | 17134.38 |
| S - with veg. | 408209.850 | 1469.555 | 65313.58 | 25513.12 | 16328.39 |
| Saving | 20149.72 | 72.54 | 3223.95 | 1259.36 | 805.99 |
| | **4.704%** | | | | |

**Table 14.** Annual energy requirement for heating without a vegetated roof – part A2.

Data of the object:

| | | |
|---|---|---|
| Town: | | Košice |
| External calculated temperature: | te | -13 |
| Length of heating period: | d | 223 |
| Average temperature in the heating season: | tes | 2.9 |
| Average indoor temperature: | tis | 20 |
| Heat loss of the object: | QHL | 15.1675 |

| Calculation of the expected heat demand for heating | | | | | | |
|---|---|---|---|---|---|---|
| d | D | ε | ηr | QHL | QÚK/rok | |
| days | K.days | - | - | kW | kWh/rok | GJ/rok |
| 223 | 3813.3 | 0.63 | 0.98 | 15.167465 | 27041.19 | 97.35 |

**Table 15.** Annual energy requirement for heating with a vegetated roof – part A2.

Data of the object:

| | | |
|---|---|---|
| Town: | | Košice |
| External calculated temperature: | te | -13 |
| Length of heating period: | d | 223 |
| Average temperature in the heating season: | tes | 2.9 |
| Average indoor temperature: | tis | 20 |
| Heat loss of the object: | QHL | 14.9034 |

| Calculation of the expected heat demand for heating | | | | | | |
|---|---|---|---|---|---|---|
| d | D | ε | ηr | QHL | QÚK/rok | |
| days | K.days | - | - | kW | kWh/rok | GJ/rok |
| 223 | 3813.3 | 0.63 | 0.98 | 14.903392 | 26570.39 | 95.65 |

**Table 16.** Total comparison of heating energy demand - part A2.

| The results of comparison | | | | | |
|---|---|---|---|---|---|
| | Heat energy demand | | Cost comparison | | |
| | | | Electricity | Gas | Heatpump |
| | kWh/rok | GJ/rok | €/rok | €/rok | €/rok |
| S - withou veg. | 27041.187 | 97.348 | 4326.59 | 1690.07 | 1081.65 |
| S - with veg. | 26570.387 | 95.653 | 4251.26 | 1660.65 | 1062.82 |
| Saving | 470.80 | 1.69 | 75.33 | 29.42 | 18.83 |
| | **1.741%** | | | | |

In the final table No. 3, you can see the overall comparison of parts A1 and A2, where you can see the difference in percentage savings with a vegetated roof, but also with different ratio of the area of the given roof. Therefore, the greater justification for the use of a vegetation roof, as far as heating is concerned, is for objects with a larger area of horizontal constructions compared to vertical ones.

**Table 17.** Final comparison of constructions with and without a vegetation roof.

| **Roof** | **Area ration roof/object** | **Heating energy demand** | | **Saving** | |
|---|---|---|---|---|---|
| | | **Without veg.** | **With veg.** | | |
| | | **kWh/a** | **kWh/a** | **kWh/a** | **%** |
| Part A1 | 43.48% | 428359.567 | 408209.72 | 20149.72 | 4.704 % |
| | 15.13% | 27041.187 | 2657.387 | 470.80 | 1.741 % |

## 4 Conclusion

Even the overall effect of the vegetation roof on the object depends on a number of factors, the result confirmed the effect on heating. At the same time, a secondary comparison based on the ratio proved its suitability for different area ratios. The subject of this article is not an economic principle, as it is not objective to compare one indicator and its financial side. On the other hand, this article is intended to enrich the view of vegetated roofs and to complement its advantages.

This work was supported by Development Agency under the contract (APVV APVV-18-0360 ACHIEve Active hybrid infrastructure for the sponge city).

## References

1. T. Susca, Green roofs to reduce building energy use? A review on key structural factors of green roofs and their effects on urban climate, (2019)
2. https://population.un.org/wup/Publications/Files/WUP2014-Highlights.pdf (na internete dňa: 10.03.2023)
3. https://data.worldbank.org/indicator/SP.URB.TOTL.IN.ZS?end=2021&name_desc=false&start=1960&view=chart (na internete dňa: 12.03.2023)
4. T. Takakura, S. Kitade, E. Goto, Cooling effect of greenery cover over a building, Energy and building, (2002)
5. T. G.Theodosiou, Summer period analysis of the performance of a planted roof as a passive cooling technique, Energy and building, (2003)

6. https://www.toronto.ca/legdocs/bylaws/2009/law0583.pdf (na internete dňa: 15.02.2023)
7. A. Niachou, K. Papakonstantinou, M. Santamouris, A. Tsangrassoulis, G. Mihalakakou, Analysis of the green roof thermal properties and investigation of its energy performance, Energy and building, (2001)

MATEC Web of Conferences **385**, 01018 (2023) https://doi.org/10.1051/matecconf/202338501018
*Young Scientist 2023*

# Biosolar roofs - The trend of the future

*Pavol* Knut[1*] *Maria* Kocurkova[1] and *Zuzana* Vranayova[1]

[1]Institute of Architectural Engineering, Faculty of Civil Engineering, Technical University of Kosice, Vysokoskolska 4, 042 00 Kosice, Slovakia

**Abstract.** This article shows the different types of biosolar roofs. An important factor that interests me is how a green roof can affect the efficiency of photovoltaics, how it can cool and how much more electricity it can produce than a classic roof. Localized energy generation through rooftop solar is gaining popularity in urban areas, and green roofs are often used for a range of services such as thermal insulation. In recent years, the use of biosolar green roofs to insulate heat and increase solar performance has increased. Two buildings observed by the research team are located in Sydney, Australia, of similar size, location and construction materials. One building has a biosolar roof and the other has a classic solar roof. Each solar array contains a series of environmental sensors including ambient temperature and global horizontal radiation. From the measurement, we can see the results that prove that the biosolar roof had a higher energy output by 4.5%, which proves that it produced 14.26MWh more electricity than the classic solar roof. Compared with previously reported studies and some simulation results, it is clear that the implementation of a bio-solar roof is beneficial for maximizing energy production and reducing greenhouse gas emissions.

## 1 Introduction

Climate change is exacerbating air pollution problems, which, combined with the increasing scale, duration and frequency of extreme climate events, poses a significant threat to cities [1]. Building rooftops represent a viable space for integrating new green infrastructure into high-density urban areas [2]. Urban rooftops also provide prime locations for photovoltaic (PV) systems. It is increasingly recognized that the two technologies can be combined to achieve mutual benefits in terms of energy efficiency and biodiversity goals [3]. Integrating plants into building structures through greening roofs has been suggested as a vital and important solution in building resilient cities [4]. Plants are able to improve air quality in several ways. They remove gaseous pollutants through their stomata, particles with their leaves, and organic compounds with plant tissues or through the soil [5]. Local climate is also an important factor affecting the rate at which plants can remove pollutants from the air, with warmer climates generally associated with increased outdoor phytoremediation potential of green spaces [6]. In recent years, a lot of research has been carried out on different ways of dealing with building roofs in order to improve thermal comfort, improve the energy efficiency of buildings and reduce the negative impact on the environment [7]. Several experimental and modeling studies have been conducted to assess the performance of biosolar roofs compared to conventional solar, with varying results. Pilot experiments conducted by Alshayeb and Chang [8] found an increase in energy production of 1.4% for model bio-solar roofs; where a study by Chemisana and Lamnatou [9] observed an increase in energy production of 1.29 and 3.33% for 5 day pilot scale experiments. Here I present a

* Corresponding author: pavol.knut@tuke.sk

unique opportunity to compare two roofs that are spatially opaque, with almost identical construction and dimensions, with similar age and roof infrastructure. In this study, they used a commercial biosolar roof as well as an independent control roof in Sydney, Australia to determine the independent effect of a biosolar installation through empirical observations and simulations.

### 1.1 Abbreviations

| | |
|---|---|
| BSGR – | Bio-solar green roof |
| CSR – | Conventional solar roof |
| GHI – | Global horizontal irradiance |

## 2 Methodology

Combining green roofs and solar photovoltaic systems can be an effective strategy for promoting sustainable development in buildings while reducing greenhouse gas emissions. By installing both technologies on a building's roof, it can enhance the building's thermal engineering and increase the efficiency of photovoltaic cells, thanks to the cooling effect produced by the green roof.

### 2.1 Description of the research

The purpose of this study was to compare the solar energy output of two nursing homes located in Barangaroo, Sydney, Australia. The first building, Daramu house, was constructed in 2019 and had a bio-solar green roof, while the second building, International House, was built in 2017 and had conventional solar panels installed on its roof, both buildings we can see in the Fig. 1. Both buildings were the first multi-story wooden office buildings in the country and had nearly identical roof infrastructure, with minor differences in the building maintenance unit model and design.

The research was conducted over a period of 237 days, starting from mid-spring (October) 2020 and ending in fall (May) 2021. During this time, the average daily sunshine in the Barangaroo district was 6.66 hours, while the average daily rainfall was 4.08 mm, and the average evaporation rate was 5.93 mm/day. The average daily temperatures in the region varied between 9 and 27.43 °C during the monitored period.

Both buildings had a total roof area of 1,863.35 $m^2$, with solar panel coverage of 593.96 $m^2$ and 567.44 $m^2$ for the biosolar and conventional roofs, respectively. The bio-solar roof used an extensive design with a variable substrate depth of 0.1–0.12 m and included a planted area of 1460.7 $m^2$, which accounted for 78.4% of the total roof area. The solar panels covered 40.66% of the planted area on the biosolar roof, as shown in Figure 1. The gray concrete slabs of 0.8 m thickness were used as a bio-solar roof foundation or roof surface for both buildings.

**Fig. 1** The study site is shown in the aerial photograph (center image), which includes Daramu House (bio-solar; left) and International House (conventional solar; right). These two buildings were almost identical, except for the roof covering, where the Daramu House had plant material while the International House had concrete. On the bio-solar roof, the plant material covered all areas below the panel surfaces.

## 2.2 Solar arrays

Since the biosolar and conventional roofs were constructed in different years (2019 and 2017, respectively), they used different types of solar panels. Furthermore, there were other differences between the buildings, which required the data to be adjusted to ensure accurate comparisons [10].

The bio-solar roof used 332 MAXEON 3 solar panels (SunPower, Australia; pNom 395W, efficiency 22.6%, Fig. 2), while the conventional roof used 346 NeON2 solar panels (LG, Australia; pNom 320W, efficiency 19.5%). The biosolar and conventional roofs had a total capacity of 131.14kWp and 110.72kWp, respectively. Both buildings used four three-phase inverters (27.6k-AU000NNU2, SolarEdge, USA) with an efficiency set at 98%. Given these differences, a series of corrections were made to the data to enable accurate comparisons between the buildings [10].

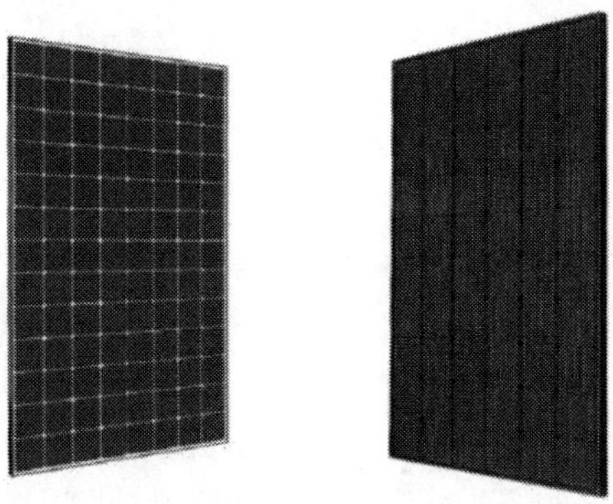

**Fig. 2** SunPower, Australia; pNom 395W, efficiency 22.6%, (left). LG, Australia; pNom 320W, efficiency 19.5%, (right).

The roofs were carefully designed and modeled before construction to optimize solar radiation and panel layout. Due to the presence of greenery on the biosolar roof, structural architects and solar engineers had to account for planting, maintenance, and plant survival in their designs. The solar panels on the biosolar roof were divided into different sections. Most of the panels (248) were placed above the main planted area, with an azimuth of 0° (to the north) and a tilt angle of 15°. The remaining panels (84) were arranged between the roof infrastructure, with an azimuth of 90° (east orientation) and an inclination angle of 2°. The center of the solar panels was positioned at an average height of 1 m above the substrate surface (0.8 m above the leaf zone) to support plant growth and facilitate maintenance.

In contrast, the conventional solar roof did not require any biological considerations, so the solar panels were arranged in an accordion style, with most of the panels (145) placed at 90° or 270° azimuth towards the center of the roof space (145 east and 145 west). This layout was chosen to maximize solar exposure, as determined by pre-installation modeling procedures. Additional panels (56) were placed between the building infrastructure, with an identical layout to the panels towards the center of the building. All panels had an inclination angle of 5° and were centered at an average height of 0.4m above the surface of the concrete slab [10].

## 2.3 Solar modelling

This statement is describing the methodology used to estimate the average annual incident solar radiation on each roof (Fig.3). A 3D model of the Barangaroo district was developed using Rhino 6 modeling software and DAYSIM in the Grasshopper's Honeybee plug-in. The solar radiation calculations were based on the Sydney CBD Representative Meteorological Year (RMY) file from EnergyPlus, Australia. The purpose of this modeling was to provide an estimate of the amount of solar radiation each roof would receive per year, which was an important factor in comparing the performance of the biosolar and conventional roofs [10].

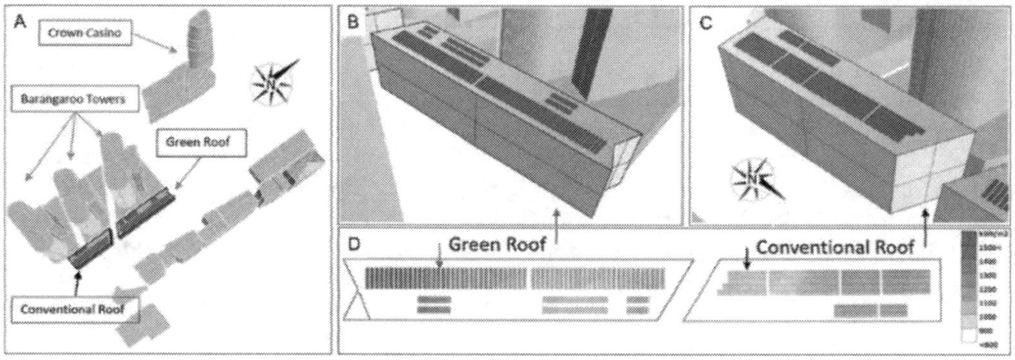


**Fig. 3** 3D model Rhino 6.

A) The Barangaroo district to determine the effect of urban geometries on reflectance and shading;
B) Biosolar field as built;
C) Conventional solar array;
D) Average annual solar radiation received for each roof.

## 2.4 Data collection and corrections

The solar arrays used in the study were equipped with various environmental sensors, including those for ambient temperature and global horizontal irradiance (GHI). Data on environmental and solar parameters were transmitted to the SolarEdge monitoring platform for management, which could also obtain time-matched temperature measurements from a nearby weather station. Solar output data were collected every two weeks during site inspections, and energy outputs were summed for both buildings after collecting data from each inverter. GHI was measured using rooftop pyranometers to record hourly average light intensity for each location.

To correct for differences between the two systems, four adjustments were made to the BSGR gross energy output, resulting in a reduction of 19.8% in the recorded energy output. The BSGR power was first reduced by 15.54% to account for differences in system capacity, as the BSGR had a capacity of 131.14 kWp compared to CSR's 110.72 kWp. The BSGR's outputs were then reduced by an additional 1.2% to account for age-related losses of panels on CSR and by 3.1% to account for differences in panel efficiency, with derating for efficiency and age based on the known differences or degradation rates from the two systems, according to the manufacturer's specifications. Finally, to account for the differences in temperature coefficients between the two systems, the energy outputs of the BSGR were reduced by 0.09% per °C for temperatures of each 1 °C panel above 25 °C.

Although this study analyzed a commercial BSGR with a spatially non-uniform CSR of nearly equal size, two considerations involving convective airflow could not be addressed. Convection above and below the panel could potentially affect the cooling potential of the system and thus the energy output. However, both roofs were modeled before construction, and optimal layouts were selected for each roof to maximize energy yield while accommodating the planting, survival, and maintenance of plant material on the BSGR. Thus, while this study was unable to quantify the impact of convective airflow, comparisons were made considering these differences in roof design. It is likely that other commercial-scale BSGRs would use panel designs based on these criteria, and most CSR solar arrays in the surrounding region are designed similarly to the CSR in this study, based on satellite imagery [10].

## 3 Results and discussion

In this paragraph, it is stated that the Rhino 3D solar model predicted a 6% annual difference in GHI exposure between BSGR and CSR due to surrounding buildings that both block and reflect light. However, during the 237 days of this study, the measured light exposure was greater than predicted by the Rhino model and varied significantly between seasons. The BSGR received more GHI than CSR in spring and autumn, while it received less GHI in summer. The observed solar exposure was more than double compared to the Rhino 3D predictive model. It is emphasized that monitoring environmental variables such as GHI and temperature is crucial for such assessments (Fig. 4) [10].

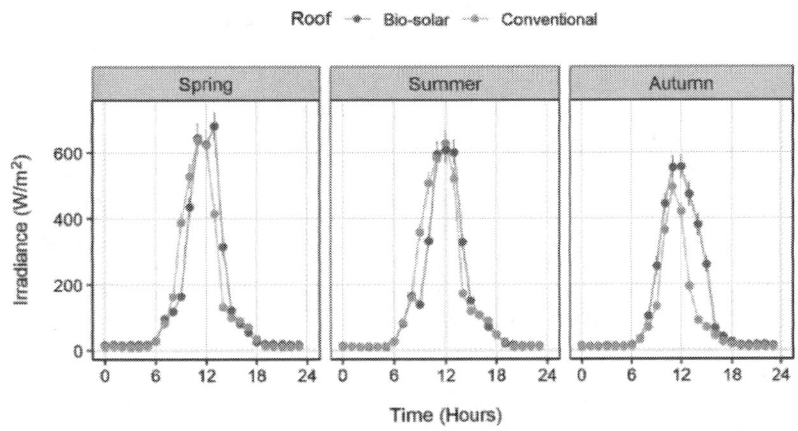


**Fig. 4** shows the mean hourly global horizontal irradiance (GHI) reported by on-site pyranometers during each season for both roofs. The variations in light availability within a season are largely attributed to urban geometries, while differences between seasons are due to the seasonal diurnal arc. During spring and fall, the BSGR received an average of 4.37% and 61.31% more GHI than the CSR, respectively. However, in summer, the CSR received an average of 5.67% more GHI than the BSGR. These differences in GHI exposure are likely due to the surrounding buildings that both block and reflect light, and highlight the importance of monitoring environmental variables such as GHI and temperature for accurate assessments of solar power outputs.

During the study period, BSGR maintained a lower average ambient temperature than CSR in all three seasons, with differences of 1.00, 1.12, and 0.72°C in spring, summer, and fall, respectively. However, during peak GHI hours (11:00-14:00), BSGR ambient temperatures were only slightly cooler than CSR, with differences of 0.44, 0.95, and 0.26°C in the respective seasons (Fig. 5). In a previous study conducted on the same roofs, researcher Fleck found a significant difference in temperature gradient between the two roofs at a distance of 1 meter from the roof surface/plant leaf, suggesting a potential cooling effect of BSGR on the solar array. Under-panel temperatures were found to be up to 6 and 11°C lower than CSR in spring and summer, respectively, but the effect was attributed to evapotranspiration and reduced latent heat reflectance/solar reflectance of the plant material rather than the orientation of the panel. It is possible that the effect of convection beneath the panel could also contribute to the cooling effect, although it was not specifically measured in this study [10].

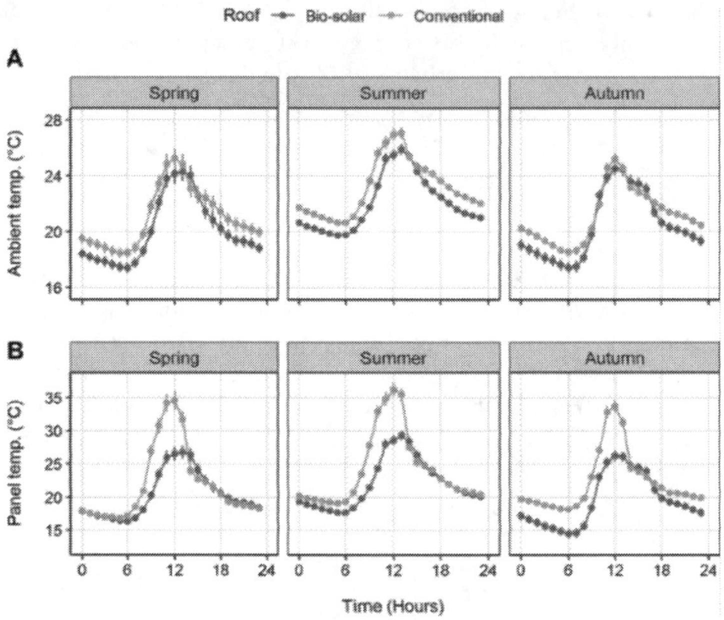


**Fig. 5** shows the mean hourly ambient temperature and panel temperature reported by on-site temperature sensors during each season for both roofs. Part A of the figure shows that the ambient temperatures of the BSGR roof and panels were on average 1.00, 1.12, and 0.72°C and 1.50, 2.10, and 2.88°C lower than CSR in spring, summer, and fall, respectively. During GHI peak hours (11:00 to 14:00 inclusive), ambient and panel temperatures were on average 0.44, 0.95, and 0.26°C and 4.68, 4.95, and 4.98°C lower than CSR in spring, summer, and autumn. Part B of the figure shows that the panel temperatures on BSGR were lower than those on CSR in all seasons, with the greatest difference observed during summer.

The study found that BSGR panel temperatures were significantly cooler than those at CSR, with reductions of 1.50, 2.10, and 2.88°C in spring, summer, and fall, respectively. During the highest GHI hours, BSGR was 4.68, 4.95, and 4.98°C cooler than CSR, which helped mitigate power loss due to temperature by 1.36, 1.44, and 1.44% in spring, summer, and autumn, respectively (Fig. 6.). The mean hourly energy output for BSGR was significantly higher than that of CSR, with BSGR producing on average 16.85 ± 0.76 (SEM), 15.66 ± 0.58, and 14.30 ± 0.74 kWh, while CSR produced 12.71 ± 0.58, 12.92 ± 0.92, and 2.3 ± 0.9 kWh in each season. The energy output differences were likely due to panel temperature differences caused by the evapotranspiration effect of the BSGR and the resulting cooler microclimate. The study also noted that differences in solar exposure made it difficult to isolate a primary causal effect on the increased energy output of BSGR. System performance for both systems was modeled under standardized lighting conditions to eliminate the effect of increased solar exposure noted on BSGR [10].

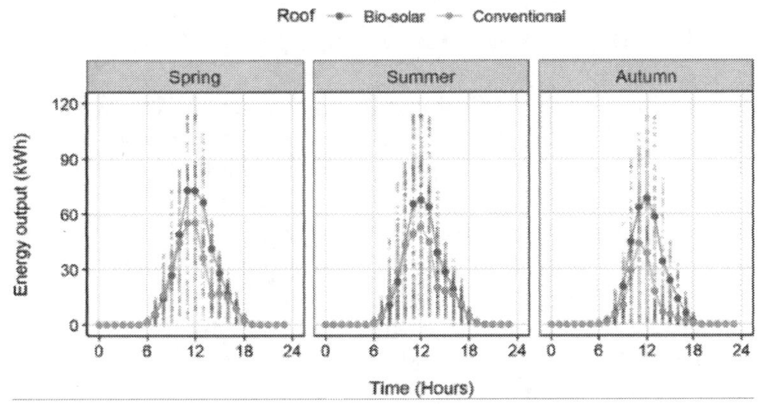


**Fig. 6** shows the average hourly energy output in kWh for both the BSGR and CSR roofs during each season. According to the figure, BSK produced significantly more energy output than CSR during all three seasons, with an average increase of 32.52%, 21.25%, and 107.29% in spring, summer, and autumn, respectively. The differences in maximum output between BSGR and CSR were also substantial, with BSGR producing 25.14 kWh, 20.35 kWh, and 29.8 kWh more energy in spring, summer, and autumn, respectively. These findings suggest that the BSGR was more effective at generating solar energy compared to the CSR, particularly during the autumn season where the energy output was substantially higher.

## 4 Conclusion

This study provides valuable insights into the energy output and greenhouse gas emissions of different types of roofs in a commercial setting. The use of a BSGR roof resulted in a higher energy output and lower greenhouse gas emissions compared to a conventional solar roof. The study also highlighted the potential of rooftop plants to further mitigate CO2 emissions. Cities are also responsible for significant greenhouse gas emissions related to energy consumption by burning fossil fuels for transport, energy use in buildings and appliances.

However, it is important to note that the study was conducted in a tropical humid climate zone and the effect of BSGR during winter months is unknown. Therefore, it is recommended that future studies be conducted in different climate zones to fully understand the effectiveness of BSGR in different environments.

For a better understanding of this technology such as the biosolar roof, I will continue to study information and new knowledge from various researches around the world.

This work was supported by the project run by the Slovak Research and Development Agency APVV-18-0360 "Active hybrid infrastructure towards to a sponge city" and by project SECOVE 101056201 "Sustainable energy centres for vocational excellence".

## References

1. Peter J. Irga, Robert Fleck, Ekaterina Arsenteva, Fraser R. Torpy, Biosolar green roofs and ambient air pollution in city centres: Mixed results, Building and Environment, Volume **226**, (2022), 109712, ISSN 0360-1323,

2. S. Yao, Q. Wang, J. Zhang, R. Zhang, Y. Gao, H. Zhang, J. Li, Z. Zhou, Ambient volatile organic compounds in a heavy industrial city: Concentration, ozone formation potential, sources, and health risk assessment, Atmospheric Pollution Research, Volume **12**, Issue 5, (2021), 101053, ISSN 1309-1042.
3. C. Nash, J. Clough, D. Gedge, R. Lindsay, D. Newport, Ciupala, MA, S. Connop. Initial insights on the biodiversity potential of biosolar roofs: a London Olympic Park green roof case study, Israel Journal of Ecology and Evolution, Volume **62**, (2016), 74-87. doi:
4. S. H. B. Motlagh, O. Pons, S. M. A. Hosseini, Sustainability model to assess the suitability of green roof alternatives for urban air pollution reduction applied in Tehran, Building and Environment, Volume **194**, (2021), 107683, ISSN 0360-1323.
5. P. Kumar, A. Druckman, J. Gallagher, B. Gatersleben, S. Allison, T. S. Eisenman, Uy Hoang, S. Hama, A. Tiwari, A. Sharma, K.V. Abhijith, D. Adlakha, A. McNabola, T. Astell-Burt, X. Feng, A. C. Skeldon, S. de Lusignan, L. Morawska, The nexus between air pollution, green infrastructure and human health, Environment International, Volume **133**, Part A, (2019), 105181, ISSN 0160-4120
6. A. Wania, M. Bruse, N. Blond, Ch. Weber. Analysing the influence of different street vegetation on traffic-induced particle dispersion using microscale simulations, Journal of Environmental Management, Volume **94**, Issue 1, (2012), Pages 91-101, ISSN 0301-4797.
7. A. Majed, Z. Gou. A Review of Roofing Methods: Construction Features, Heat Reduction, Payback Period and Climatic Responsiveness, Energies 11, Volume **11,** (2018) : 3196.
8. J. Alshayeb, J. D. Chang. Variations of PV Panel Performance Installed over a Vegetated Roof and a Conventional Black Roof. Energies, Volume **11**, (2018), 1110.
9. D. Chemisana, Chr. Lamnatou. Photovoltaic-green roofs: An experimental evaluation of system performance, Applied Energy, Volume **119**, (2014,) Pages 246-256, ISSN 0306-2619.
10. D. Fleck, R. Gill, T.J. Pettit, F.R. Torpy, P.J. Irga. Bio-solar green roofs increase solar energy output: The sunny side of integrating sustainable technologies, Building and Environment, Volume **226**, (2022), 109703, ISSN 0360-1323.

# Thermal energy with long-term accumulation

*Michal* Goras[1*], *Zuzana* Vranayova[1], and *Frantisek* Vranay[1]

[1]Technical University of Kosice, Faculty of Civil Engineering, Kosice, Slovakia

**Abstract.** Accumulation of thermal energy in an underground reinforced concrete thermal energy tank with a volume of 24m[3] of water, for 2 years. In these 2 years, 4 cycles are measured, namely 2 times charging and 2 times discharging. The measured data in these cycles indicate the charging and discharging time. By evaluating the data, we point out the losses of accumulated energy. Solar hot water collectors, which heat the medium in the thermal energy tank, serve as a source for recharging. The thermal energy tank has the potential to reduce the energy demand of buildings.

## 1 Introduction

In our European latitude and altitude, the demand for heat in winter is 10 times higher than in summer. At the same time, solar radiation reaches its peak in summer. The fifth to the eighth month is the most intense solar power in our country [1]. During this time, it is possible to heat or heat hot drinking water to the desired temperature directly with solar energy, without using other energy-intensive sources. The problem is that the demand for heating is minimal in this period, Figure 1. Therefore, it is necessary to use seasonal storage for surplus energy from the summer period for later use [2].

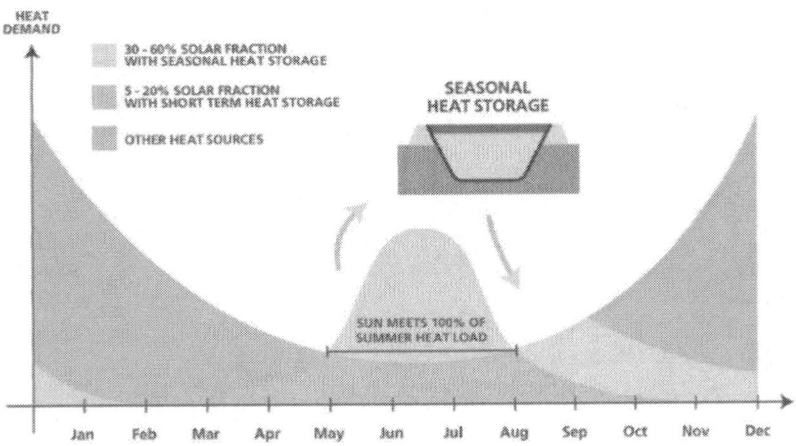


**Fig. 1.** Energy production and energy demand in year [2].

* Corresponding author: michal.goras@tuke.sk

# 2 Thermal storage tank

## 2.1 Thermal storage tank measured

Surface-selective hot water collectors with an area of 225 $m^2$ serve as a source of charging for our reservoirs. Water with glycol flows in these collectors. The heat is transferred to the water using exchangers. This water is stored in three thermal tank storage with a volume of 160$m^3$, Figure 2. Each thermal tank storage is of a different size for possible different uses. The thermal tanks storage are made of special concrete, insulated with thermal insulation and stored in the ground under the floor of the building.

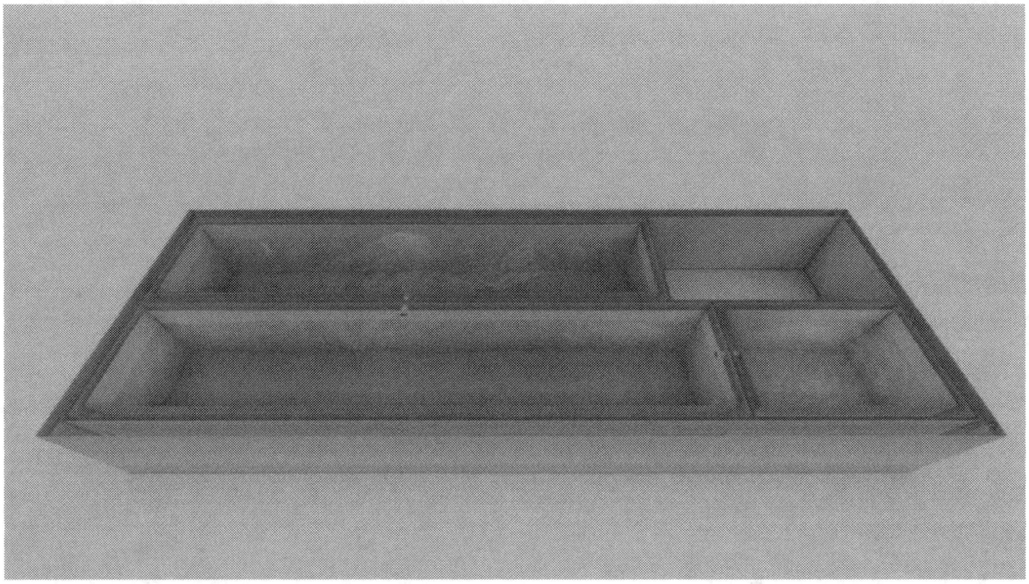

**Fig. 2.** 3- types of sizes Thermal energy storage.

## 2.2 Thermal storage tank in generally types

Thermal tanks storage are currently used, which are divided into sensible heat and latent heat. The storage capacity depends on the medium in which it is accumulated and also on the method of accumulation. Therefore, we divide it into chemical and thermal accumulation, Figure 3 [3].

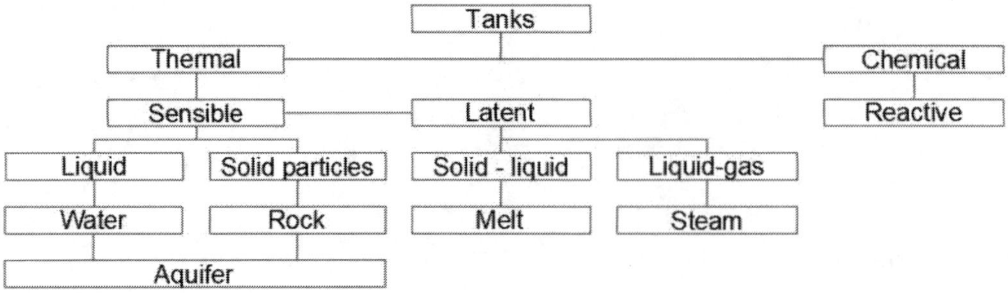


**Fig. 3.** Types of tanks [3].

Chemical energy is in the form of oil, gas or gasoline. There is bound energy released by combustion. However, so far it has not been possible to convert this energy into pirate conditions and efficiencies.Nowadays, only thermal storage tanks are used. They are divided into storage tanks for sensible and latent heat.

Latent heat is generated when the state changes. E.g. when switching from ice to water, just as much energy is needed as heating the water to 80 °C. Sensible heat represents the behavior of the temperature during the supply and consumption of energy.
Sensible heat is derived according to the equation. [4]

$$Q_{Sp} = m.c.(\theta_2 - \theta_1) \quad (1)$$

**Table 1.** Specific and volumetric heat capacity of materials for heat accumulation at 20 °C [4].

| Medium | Temperature range (°C) | Specific heat capacity (Wh / kg.K) | Volume heat capacity (Wh / m$^3$.K) | Density (kg/m$^3$) |
|---|---|---|---|---|
| water | 0-100 | 1.16 | 1160 | 998 |
| air | -273->1000 | 0.28 | 0.31 | 1.1 |
| oil | 0-400 | 0.44-0.5 | 350-450 | 800-900 |
| gravel, sand | 0-800 | 0.2 | 360-390 | 1800-2000 |
| granite | 0-800 | 0.21 | 570 | 2750 |
| concrete | 0-500 | 0.24 | 460-560 | 1900-2300 |
| brick | 0-1000 | 0.23 | 330-440 | 1400-1900 |
| Iron | 0-800 | 0.13 | 1000 | 7860 |
| gravel backfill (37% water) | 0-100 | 0.37 | 810 | 2200 |
| mix of salts (53KNO3 + 4NaNO2 + 7NaNO3) | 150-450 | 0.36 | 480-550 | 2561-2243 |

The heat in the $Q_{Sp}$ storage tank depends on the weight of the medium (m), the heat capacity (c) and the temperature difference. If we have a 200 liter tank and needed to heat it from 10 ° C to 60 ° C, the energy needed would be:

$$Q_{Sp} = 200 \cdot 1.16.(60 - 10) = 11600Wh = 11.6kWh \quad (2)$$

According to these values, water-price availability and specific heat were selected as media. [5]

## 3 Measuring

Data measurement took place over 2 years. During this period, measurements were also made where water and clay temperatures near the reservoirs were recorded, Figure 4.

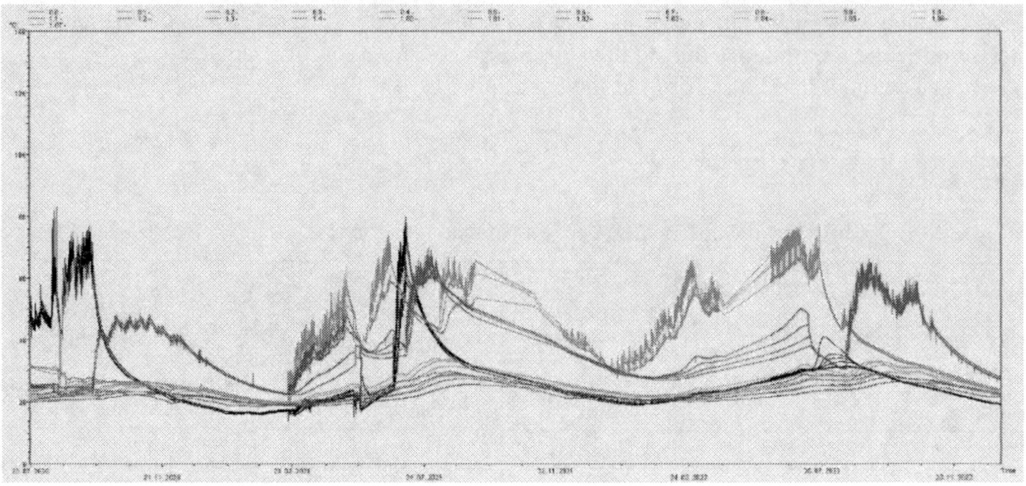

**Fig. 4.** 2-year measuring temperatures in tanks.

27 temperature sensors were placed in these thermal tanks storage and in the clay, which are shown in the following Figure 5, and the temperature course in the previous figure is adjusted according to the colors of the sensors.

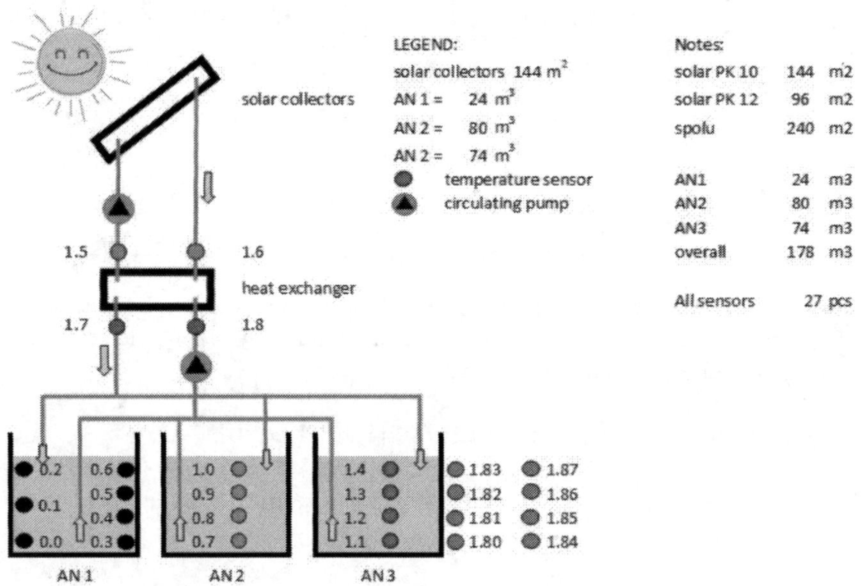


**Fig. 5.** Placement of temperature sensors.

# 4 Comparison AN1 tank

The analyzed tank AN1 with a volume of $24m^3$ is therefore the smallest thermal tank storage. In this thermal tank storage, we dealt with large heat losses caused by evaporation. Data measurement took place over 2 years, Figure 6.

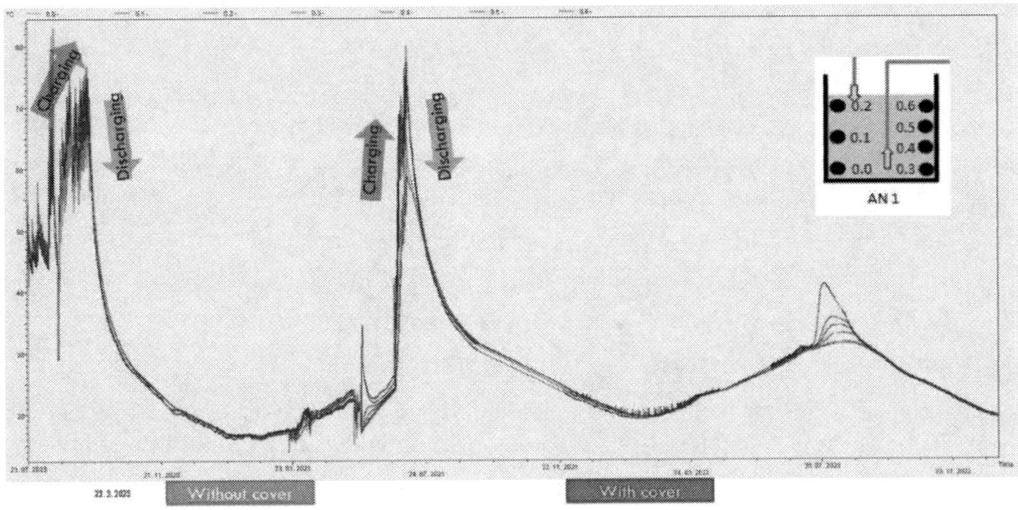


**Fig. 6.** 2-year measuring temperatures in AN1-tank.

## 4.1 Accumulation tank without cover

In the first year of measurement, large heat losses related to evaporation through the water surface were found.

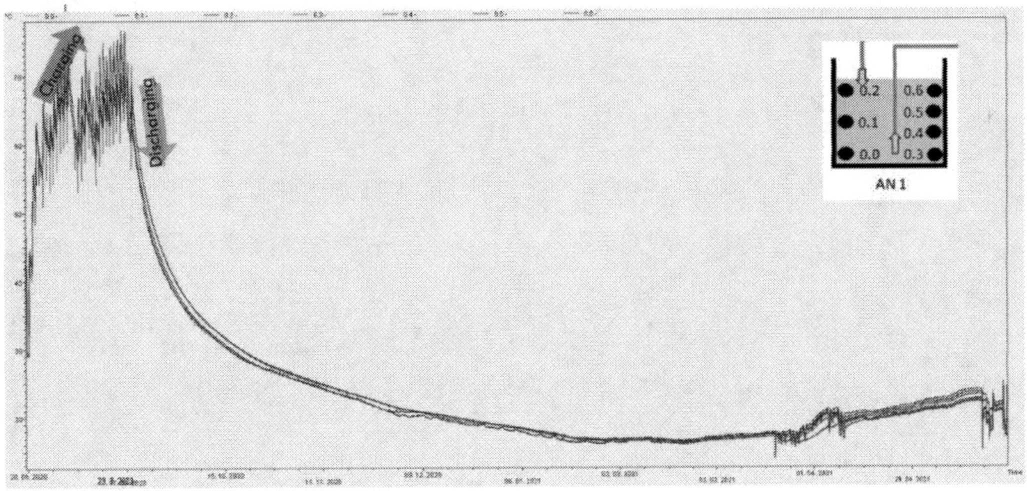


**Fig. 7.** First year measuring temperature without cover.

From the collected data, it can be seen that the discharge decreased rapidly quadratically without energy consumption, Figure 7.

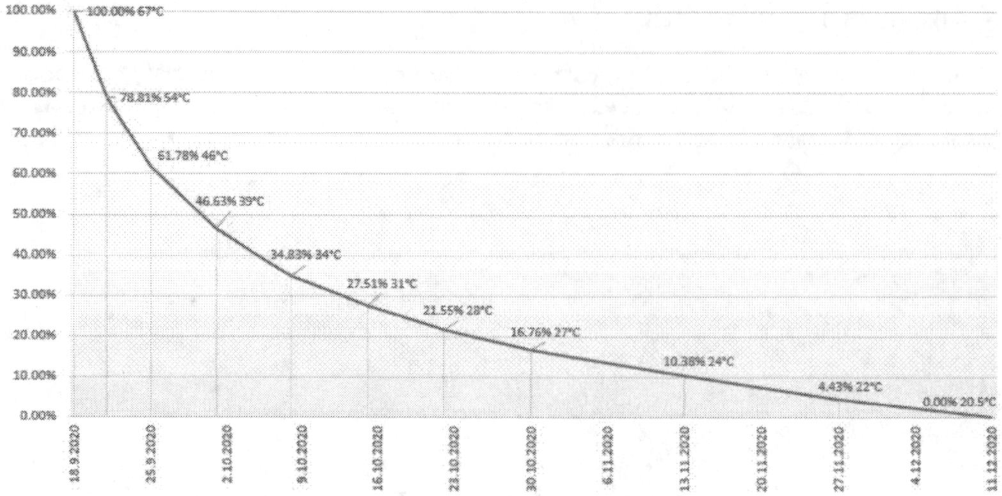


**Fig. 8.** Percentage graph without cover.

This graph is from the maximum medium temperature of 67°C to the original temperature where charging started, which is 18°C in 85 days. Such rapid discharge has little efficiency, Figure 8.

## 4.2 Accumulation tank with cover

Second year, we covered the water surface of the tank with thermal insulation of 20 mm thickness to minimize heat losses through the water surface.

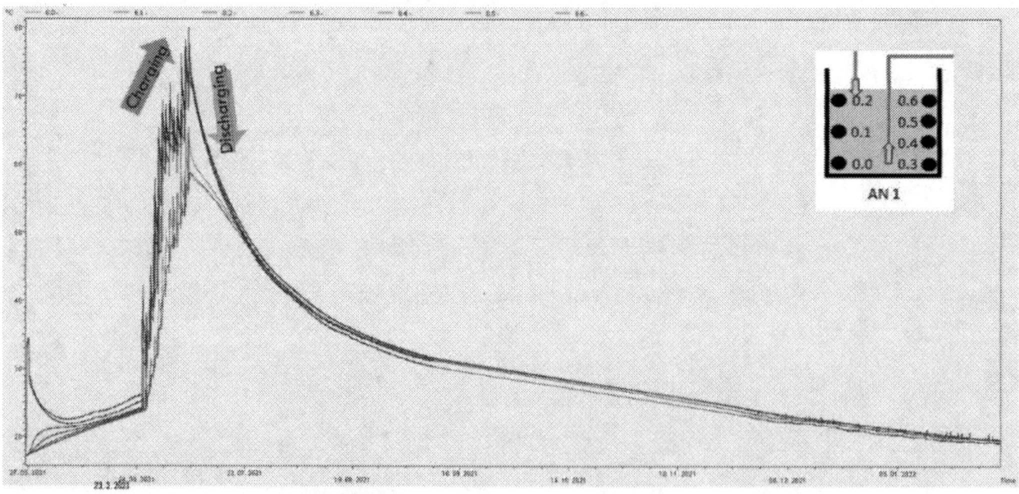


**Fig. 9.** Second year measuring temperature with cover

It can be seen from the ongoing temperatures that the discharge is not so fast and at a certain temperature it goes from a quadratic course to a linear course, Figure 9.

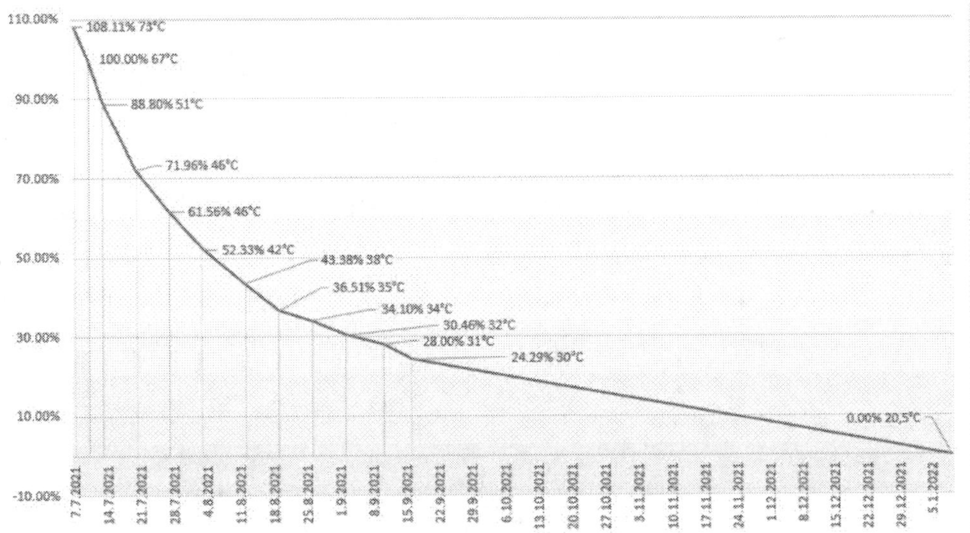


**Fig. 10.** Percentage graph with cover.

Although this year the tank was charged up to a temperature of 73°C, but for a better comparison, 100% is set to 67°C so that a comparison of the period is possible. This covered water surface proved to be an effective solution. For comparison 3 weeks of discharge the temperature before covering is 34°C and after covering the temperature is 46°C. The discharge of this tank to the original value of 18°C took 183 days, Figure 10.

## Conclusion

Such a solution helps with long-term accumulation and prevents heat loss through the water surface in the tank. Furthermore, physical processes such as heat losses, stratification, etc., will be investigated in the reservoirs.

The article was created with the support of the project Hungary-Slovakia-Romania-Ukraine ENI CBC Programme 2014-2020 HUSKROUA/1702 New Energy Solutions in Carpathian area.

## References

1. L. Riahi and C. Martinez · *District energy in cities* - UNEP (2019) http://www.districtenergyinitiative.org
2. D. Trier, F. Bava, C. Kok Skov, S. Stendorf Sorensen - *Solar District Heating Trends and Possibilities,* Helsinge, Denmark, (2018)
3. H. Ladener, F. Späte, *Solaranlagen* (Staufen bei Breisgau) ISBN: 80-247-0362-9, (2001)
4. Understanding Renewable Energy Systems *Carl Hanser Verlg Gmbh Co KG,* ISBN: 1 - 84407 -136 -7, (2005)
5. Technology Roadmap Energy-efficient Buildings: *Heating and Cooling Equipment.* International Energy Agency – IEA 9 rue de la Fédération, 75015 Paris, France, PRINTED IN FRANCE BY CORLET, (2011)

# Prototyping of an indoor mobile mapping system

*Jan* Brindza[1*], *Jan* Erdelyi[1] and *Peter* Kyrinovic[1]

[1] Department of Surveying, Faculty of Civil Engineering, Slovak University of Technology in Bratislava, Radlinskeho 11, 810 05 Bratislava, Slovakia

**Abstract.** The paper deals with prototyping a kinematic measurement system for digitizing indoor environments of buildings. The scanning system itself consists of several sensors for localization (inertial measurement unit, rotary encoders and a stereo camera), while its main part is a Velodyne lidar. The procedure of data processing is performed by an algorithm for simultaneous localization and mapping (SLAM). The paper describes the hardware of the measurement system. Part of the contribution is the proposal of a series of tests evaluating the quality of the prototype of the mobile mapping system. The tests verify the accuracy of length measurement, the accuracy of the determination of the inclination of the measuring system, and the accuracy of the lidar position determination.

## 1 Introduction

In today's era of digitization of processes and services, new methods and techniques must be implemented also in the construction industry. The recent trend in the construction industry is the use of Building Information Modeling (BIM) [1]. A BIM model collects available information about the building, which can be graphic, no-graphic (parameters) and in form of documentation. The use of the BIM also makes processes more efficient, e.g. the as-built documentation and the verification of the geometry of executed structural parts of a building is more complex. For this purpose, modern methods of spatial data collecting are needed, when detailed and accurate data must be collected in a relatively short time. These requirements are fulfilled by the technique of terrestrial laser scanning and photogrammetric methods, often in combination with conventional surveying methods (e.g., using a total station). To make data collecting more efficient often mobile mapping systems are deployed. Such a system consists of several sensor for determination of the spatial position and orientation of the measuring system itself, light detection and ranging sensor (Lidar) and cameras, etc. In fact, they create a kinematic laser scanning system that significantly contribute to the automation of the as-built documentation of buildings.

The paper deals with a proposal and prototyping of a mobile mapping system for digitizing the indoor environment of buildings. It consists of 5 subsystems, described in the following sections, and the entire mapping system is controlled using the robotic operating system ROS. As part of the data processing, simultaneous localization and mapping (SLAM) is used, within which the current position of the measuring system is simultaneously calculated and a model of the surrounding environment is created. SLAM can be divided

* Corresponding author: jan.brindza@stuba.sk

based on the sensors used into SLAM using lidar [2], [3], [4], SLAM using cameras [2], [3], [5], [6], [7] and SLAM using combination of several sensors [3]. There are many measuring systems on the market, but their price reaches several 10 000 €. Our goal is to design a prototype of a low-cost mapping system for the indoor environments of buildings. The paper mainly focusses on the components of the prototype of the mapping system. Part of the paper is devoted to operating system ROS. Also, a proposal of future testing of the entire systems is described.

## 2 Prototype of the mobile mapping system

The designed prototype of the mobile mapping system (Fig. 1) consists of several sensors, while the individual sensors compensate for the shortcomings of the others. The mapping system consists of lidars sensors (RPLIDAR, Velodyne), a stereo camera, an inertial measurement unit (IMU), rotary encoders, carrier and other electrical parts such as power supplies and others.

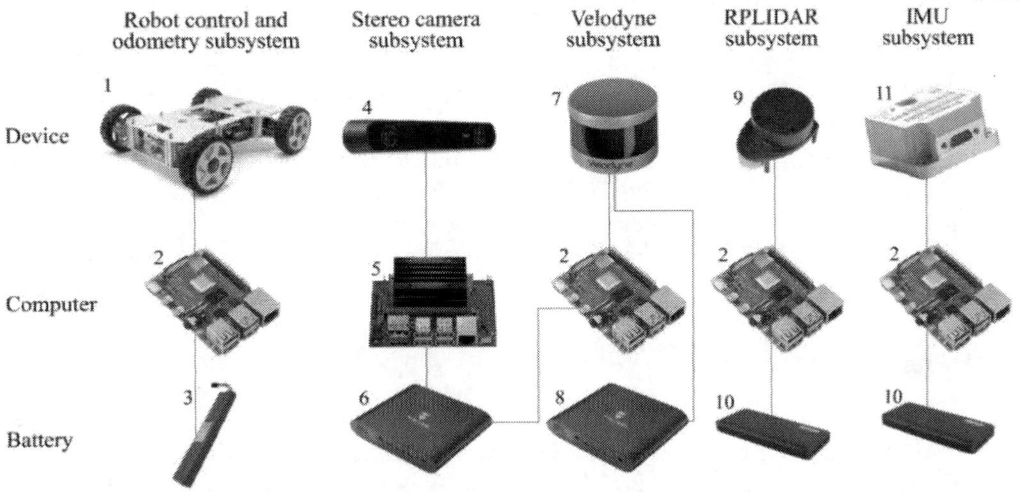


**Fig. 1.** Subsystem wiring diagram (1 - carrier, 2 - Raspberry PI 4 model B 8 GB, 3 - 22,2 V battery, 4 - ZED 2, 5 - NVIDIA Jetson Nano 4 GB, 6 - 5 V battery, 7 - Velodyne Puck, 8 - 12 V battery, 9 - RPLIDAR A1, 10 - 5 V battery, 11 - STIM 300).

The mobile mapping system is created of individual subsystems that are placed on a moving "Ackerman type" carrier. They are as follows:

- robot control and odometry subsystem,
- stereo camera subsystem,
- Velodyne subsystem,
- RPLIDAR subsystem,
- IMU subsystem.

### 2.1 Robot control and odometry subsystem

The main part of the subsystem is a pair of rotary encoders, which are located on the rear axle of the moving platform (Fig. 1(1)) [8]. The entire moving platform is controlled by the STM32F103RC Controller, which can be connected to a Raspberry Pi 4 8 GB (Fig. 1 (2)) computer. Measured data from the pair of the rotary encoders are forwarded via the STM32F103RC Controller to the Raspberry Pi 4 and from there to the NVIDIA Jetson Nano 4 GB (Fig. 1(5)) master computer, where they are stored. The entire subsystem is powered by a 22.2 V battery (Fig. 1 (3)), where the Raspberry Pi 4 is powered via the STM32F103RC Controller.

## 2.2 Stereo camera subsystem

This subsystem is created by a ZED 2 stereo camera (Fig. 1(4)) [9], which uses two high-resolution sensors to capture stereo images. From the images depth maps are generated consequently. The camera is also equipped with an IMU, barometer, magnetometer and two temperature sensors. The stereo camera's maximum field of view is 110° (H) x 70° (V) x 120° (D) with a depth range from 0.2 m to 20.0 m. The stereo camera is connected using a USB type A cable to the NVIDIA Jetson Nano 4 GB computer (Fig. 1(5)) and the entire system is powered by a 5 V battery (Fig. 1(6)). All data from the entire system is stored on the memory card of the NVIDIA Jetson Nano 4 GB computer, from which it is downloaded after the measurement is finished, and the data is post-processed. The results of the measurement are recorded is one file in *.svo format from the stereo camera.

## 2.3 Velodyne subsystem

The Velodyne Puck lidar (Fig. 1(7)) [10] is a 3D lidar that scans the surrounding environment in 16 channels and can measure up to 300 000 points per a second. Its maximum range is 100 m with a 360° horizontal field of view, a 30° vertical field of view (with +/- 15° up and down) and a rotation rate of 5 Hz – 20 Hz. The Velodyne Puck is powered by a 12 V battery (Fig. 1(8)) via interface box and is connected to the Raspberry Pi 4 8 GB computer (Fig. 1(2)) using an ethernet cable. The Raspberry Pi 4 8 GB computer is then powered by a 5 V battery (Fig. 1(6)).

## 2.4 RPLIDAR subsystem

The subsystem consists of three RPLIDARs A1 (Fig. 1(9)) [11] from Shanghai Slamtec Co., Ltd., which are located in three mutually perpendicular directions on a common platform. RPLIDAR A1 is a 2D lidar (measuring a single profile) with a maximum range of 12 m and scans in a 360 ° horizontal field of view. The scan rate is configurable between 2 Hz - 10 Hz, with a recommended scan rate of 5,5 Hz. The outputs from RPLIDAR A1 are scans in the form of a point cloud created by sets of 2D (X, Y) coordinates of the measured points. A trio of A1 RPLIDARs is connected to a Raspberry Pi 4 8 GB computer (Fig. 1(2)), which controls the initialization of the system and starts the measurement. The measured points (measurement records) are stored on the Nvidia Jetson Nano 4 GB master computer (Fig. 1(5)). The entire subsystem is powered by a 5 V battery (Fig. 1(10)).

## 2.5 IMU subsystem

The main part of the subsystem is the inertial measurement unit IMU STIM 300 (Fig. 1(11)) [12], which contains 3 accurate MEMS gyroscopes, 3 stable accelerometers and 3 stable inclinometers. The STIM 300 is connected to a Raspberry Pi 4 8 GB computer (2) using a RS422 to USB type A cable. The IMU is placed on a common platform with the RPLIDAR A1 (Fig. 1(9)) and the Velodyne Puck (Fig. 1(7)) lidars to ensure the measurement of the movement of the platform carrying the lidar sensors. The entire subsystem is powered by a 5 V battery (Fig. 1(10)).

### 2.6 System launch and control

The entire system is controlled by the operating system ROS [13], which is an open-source software development kit for robotics applications. ROS provides the tools and libraries needed for development of a robotic system. ROS does not replace, but instead works alongside a traditional operating systems [14]. Each computer of the mapping system is running Ubuntu 20.04 and using ROS1 Noetic Ninjemys, the latest version of ROS1.

Before launching the ROS system, it is necessary that all computers are connected to the same Wifi network. When starting the system, ROS must first be launched on the NVIDIA Jetson Nano 4 GB master computer, and then the ZED 2 stereo camera can also be launched. The Raspberry Pi 4 8 GB computers are connected to the NVIDIA master computer based on their IP addresses, and then all the sensors for the given computers are launched (using ROS). After connecting all the computers to ROS, which make up its nodes, the measurement recording to a *.bak file is started on the master computer. The recording from the stereo camera is saved in a file of type *.svo. Subsequently, the measurement of the geometry of the indoor environment of the building is performed. After the measurement, the data recording is finished and all the subsystems are turned off.

## 3 Proposal of testing

Set of tests needs to be done to verify the accuracy of lidar sensors and their characteristics as well as the accuracy of the measured data. Also, the verification of the data processing algorithm is significantly important. The lidar sensors (3 x RPLIDARs A1 and 1 x Velodyne Puck) will be tested in order to verify the accuracy of the measured lengths, to determine the inclination and the position of the lidars within the reference model. Reference model is a model of the indoor environment created, for example, by terrestrial laser scanning. The test will be based on the comparison of the reference values, from reference model obtained by terrestrial laser scanner, with the values measured by the proposed mobile mapping system.

When testing the accuracy of the measured lengths from the lidars, the lidars will be placed in front of a wall in equal stations. The position of the lidars will be measured using a total station by measuring their centres (centres of rotation). The wall's relative position will be determined by measuring at least 3 points on the wall's surface. For the lidars RPLIDAR A1, the distance measurement will be tested up to 15 m because the producer defines the distance measurement range as from 0.15 m to 12.00 m. For the lidar Velodyne Puck it will be up to 110 m, while its distance measurement range is maximum 100 m. The length measured by lidars will be compared by the results from the total station. The differences between lengths will determine the accuracy of the lidar's length measurement. Based on the results it will be determined the range interval from which the scanned points can be used for data processing.

The next series of tests have to be conducted are aimed at the quantification of the accuracy of determination of the measurement system's inclination (tilt). This kind of test are necessary because during the measurement using the mapping system, there may be vibrations caused by the acceleration of the carrier, which will cause inclination of the whole system. The inclination have to be removed from the measured data (the influence on the resulting point clouds) before further processing. The lidars of the corresponding subsystem will be placed on a positionable device. Due to the light weight of the lidars, for this can be also used the baseline of a historical tachymeter instrument (e.g. Carl Zeiss Jena BRT006). The positionable device will be used to set the reference values of inclination. The procedure can be divided into several steps: setting the reference inclination, measuring by the subsystem for a few seconds, setting a new value of inclination, and repetition of the whole procedure for regular intervals in the range 0° to 90°.

The accuracy of lidar positioning will be tested in order to verify the quality of the resulting point clouds, while the position of the scans within a reference model will be determined. During the testing, the lidars will be placed on reference points (e.g. observation pillar heads), on which a short measurement (approx.. 60 s) will be made. The scans will be transformed into the reference model (obtained by terrestrial laser scanning), thus obtaining the measured position of the pillar head. Deviations will be calculated as the difference between the measured and known coordinates of the reference points.

In addition to the above-mentioned analytical tests (testing the individual components and subsystems separately), we also suggest performing a global test, the results of which qualify the point clouds resulting from mobile mapping. The SLAM used will be tested in order to verify the quality of the measured scans, while the position of the scans within the reference model will be determined. The results will be evaluated as differences between the reference point cloud and the point cloud acquired by the proposed mobile mapping system.

## 4 Conclusion

The paper is devoted to development of a prototype of a low-cost mobile mapping system. The mapping system represents a kinematic laser scanning system for digitizing of indoor environments of buildings. It consists of a trio of 2D lidar, multichannel 3D lidar, an IMU, a stereo camera and rotary encoders all mounted on an "Ackerman type" carrier. The sensors used form subsystems: a robot control and odometry subsystem, an RPLIDAR subsystem, a Velodyne lidar subsystem, an IMU subsystem, and a stereo camera subsystem. All the subsystems are connected and controlled by the operating system ROS.

A proposal od series of test have to be done in the future is also described. The future work will focus on testing the suitability of all components in the mapping system. In order to find out whether all the components are necessary, test measurements will be performed. The results will be evaluated also in a global test of the mapping system. The aim of it is the comparison of the results from scanning with the point cloud acquired by a terrestrial laser scanner, which will be considered as a reference model of the scanned environment. In addition, improved SLAM algorithm with an improved scan matching algorithm [15] will be developed.

„This publication was created with the support of the Scientific Grant Agency of the Ministry of Education, science, research and sport of the Slovak Republic and the Slovak Academy of Sciences for the project VEGA-1/0272/22"

## References

1. Ch. Eastman, P. Teicholz, R. Sacks, K. Liston, *BIM handbook: A guide to building information modeling for owners, managers, designers, engineers and contractors*, John Wiley & Sons, 648, Hoboken, NJ, USA (2009)
2. S. Chen, B. Zhou, C. Jian, W. Xue, Q. Li, *A lidar/visual slam backend with loop closure detection and graph optimization*, Remote Sensing, **13**, 2720, Basel, Switzerland (2021)
3. J.S. Gutmann, K. Konolige, *Incremental mapping of large cyclic environments*, in Proceedings 1999 IEEE International Symposium on Computational Intelligence in Robotics and Automation, 318-325, Monterey, CA, USA (1999)
4. S. Thrun, D. Hahnel, D. Ferguson, M. Montemerlo, R. Triebel, W. Burgard, C. Baker, Z. Omohundro, S. Thayer, W. Whittaker, *A system for volumetric robotic mapping of*

*abandoned mines*, in 2003 IEEE International Conference on Robotics and Automation, **3**, 4270-4275, Taipei, Taiwan (2003)

5. M. Cummins, P. Newman, *Probabilistic appearance based navigation and loop closing*, Proceedings 2007 IEEE International Conference on Robotics and Automation, 2042-2048, Rome, Italy (2007)
6. C. Stachniss, *Robotic mapping and exploration*, Springer, **55**, Berlin, Germany (2009)
7. R. Valencia, J. Andrade-Cetto, *Mapping, planning and exploration with Pose SLAM*, Springer, **74**, 114, Berlin, Germany (2018)
8. Rosbot user manual https://static1.squarespace.com/static/5e76e0c52a318c0c1a850442/t/63f4408fe7287f4a231273ad/1676951781181/Rosbot+User+Manual.pdf
9. Stereo camera Zed 2 datasheet https://cdn2.stereolabs.com/assets/datasheets/zed2-camera-datasheet.pdf
10. Velodyne puck lidar datasheet https://www.amtechs.co.jp/product/VLP-16-Puck.pdf
11. RPLIDAR A1 datasheet https://bucket-download.slamtec.com/d1e428e7efbdcd65a8ea111061794fb8d4ccd3a0/LD108_SLAMTEC_rplidar_datasheet_A1M8_v3.0_en.pdf
12. IMU STIM 300 datasheet https://sensonor.azurewebsites.net/media/5z5lv25o/ts1524-r29-datasheet-stim300.pdf
13. https://www.ros.org/blog/why-ros/
14. J. M. O'Kane, *A gentle introduction to ROS*, 155 (2014)
15. P. Biber, W. Strasser, *The normal distributions transform: A new approach to laser scan matching,* in Proceedings 2003 IEEE/RSJ International Conference on Intelligent Robots and Systems, **3**, 2743-2748, Las Vegas, NV, USA (2003)

# Green infrastructure implementation and its influence on the hydrologic parameters of the urban catchment: Case study of the city of Trebišov, Slovakia.

*Kostiantyn* Sokolchuk[1]

[1]Institute of Hydrology, Slovak Academy of Sciences, Faculty of Horticulture and Landscape Engineering, Slovak University of Agriculture in Nitra

**Abstract.** This study presents changes in hydrologic characteristics of the urban catchment that are result of the implementation of the green infrastructure. Such data are used as a base for modeling of the urban drainage water flows. The research and its results are presented on the case study performed in the city of Trebišov, Slovakia. For the evaluation of hydrological parameters and their changes, available maps and a digital model of the terrain were used. Percentage of impervious, semi- and pervious surfaces, expected average curve number, travel time were determined, all in the particular sewer network subcatchments. Consequently, the areas of the city that have the highest development potential in the context of the implementation of green infrastructure were determined, were elaborated three different alternatives. The preliminary results of the current status research show, that the area of impervious areas is around 23% of the Trebišov inner city. Parks and squares account for about 10% of the city, and besides that, about a third of the area is occupied with concentrated private development. According to preliminary assessment, in the "maximal" GI implementation alternative the application of LID practices is appropriate for 12-15% of Trebišov area.

## 1. Introduction

Nowadays, the modelling of water flow in urban conditions is becoming an increasingly important task. The determination of the change in flow depending on the type of surface is especially relevant. This is important for the development of stormwater management plans, assessment of the risks of negative hydrological phenomena, assessment of the effectiveness of green infrastructure.

The purpose of this work is precisely a preliminary assessment of the prospects of various scenarios for the development of green infrastructure. The object of the study is the city of Trebišov, Slovakia. The city has a sufficiently extensive sewage network, different types of urban spaces, and there is a sufficiently large volume of data for it. The question of such assessment is relevant both in the context of the development of individual cities and as a check of individual steps of algorithms for modelling the efficiency of green infrastructure. The surface of the city, percentage of impermeable surface, curve number for individual sub-basins in the city, and changes of these parameters depending on several proposed scenarios of green infrastructure development were evaluated.

## 2. Materials, methods and methodology

Trebišov is a small industrial town in the easternmost part of Slovakia. The town is located on a plain, in the northern zone of the temperate climate. The lowland part is covered, in particular, by alluvial soils, civilized and muddy. Around Trebišov, most of the area is covered by chernozems and brown soils [1]. The water permeability of the soil is average or little, and in the flysch zone the soils are almost constantly wet, and the water-retaining function of mainly forest soils is performed even in the period without significant precipitation. The water regime of the soil is classified as wet; on slopes it is moderately wet [2]. For the purposes of this study, these soils were conditionally assigned to hydrological soil group C.

Open sources of information were used for data collection, namely ZBGIS, satellite images, data from Google Maps and OpenStreetMaps etc. The assessment was carried out using QGIS.

The territory of the inner city is approximately 7.6 km$^2$. The division of the city into partial watersheds was based on the available information about the sewage network and measurement points on it, the territory of the city was divided into 714 small watersheds. Due to the lack of data on measurements in the stormwater network in certain parts of the city, the analyzed watersheds cover only 4.8 km$^2$.

Curve number was used as an integral, digital indicator that allows evaluating the surface from a hydrological point of view. The curve number (CN) is a function of hydrologic soil group (HSG), cover type, treatment, hydrologic condition, antecedent runoff condition and impervious area in the catchment. Conceptually, CN can vary from 0 to 100, where a CN equal to 100 means that all the precipitation turned to surface runoff and no infiltration occurred, CN of 0 means that all precipitation infiltrated, so there was no runoff. Impervious surfaces were assigned a CN of 98 [3]. In order to identify this parameter, it is necessary to use special tables [4, 5]. and further refine these data using assessment of soil, land cover, etc.. Similar studies were also conducted for small natural catchments in Slovakia [6].

In the course of this study, curve numbers were determined using a combination of official tables, taking into account data from other similar cities, maps of land use categories and with clarification using satellite images and photos from the city. Averaging across catchments is done using QGIS. The classes according to which the territory was classified and the approximate determined values of cn are shown in Table 1.

**Table 1.** Cover types and their CN value for Trebišov.

| Cover type | Curve Number value |
|---|---|
| Roads | 98 |
| Buildings | 98 |
| Paved area | 98 |
| Ground | 90 |
| Green zones | 74 |
| Water surface | 100 |
| Cemetary | 86 |
| Crops | 80 |
| Housholds | 81 |
| Other | 83 |

In addition, the percentage of impermiable areas was estimated, where 100% impermeability was assigned to buildings, roads and paved areas, and a conditional 1% to other areas.

Current conditions and 3 proposed scenarios of green infrastructure development were evaluated. The assessment was carried out by changing cn and impermeability, the scenarios and their corresponding changes are shown in Table 2.

**Table 2.** Green infrastructure development scenarios.

| N of scenario | Description | Curve number transformation |
|---|---|---|
| 1 | Large building with areas larger than 1000 $m^2$ are covered with green roofs | Building cn changed from 98 to 84 [7]. |
| 2 | In addition to changes from scenario 1, identified paved areas transformed to permeable pavement. These areas became permeable | Paved areas cn changed from 98 to 82 |
| 3 | In addition to changes from scenario 2, identified areas with bare ground are transformed into open spaces, lawns with good grass condition (75% or more) | Ground cn changed from 90 to 74 |

## 3. Results

The preliminary results of the current town conditions show, that impervious area occupies around 23% of the Trebišov inner city. These areas include buildings, 12% of the total area of the city, road surface, about 6%, and identified separately large paved areas. Green zones, such as parks, squares, lawns with good grass conditions all account for about 10% of the city. However, it is worth noting that a significant share is concentrated in one large park. Besides that, about a third of the town area is occupied with concentrated private development, often occupied by gardens.

Figure 1 shows the detailed distribution of CN across catchments within Trebišov. The city center and industrial areas on the outskirts have the highest values. The average CN value is more than 86.5.

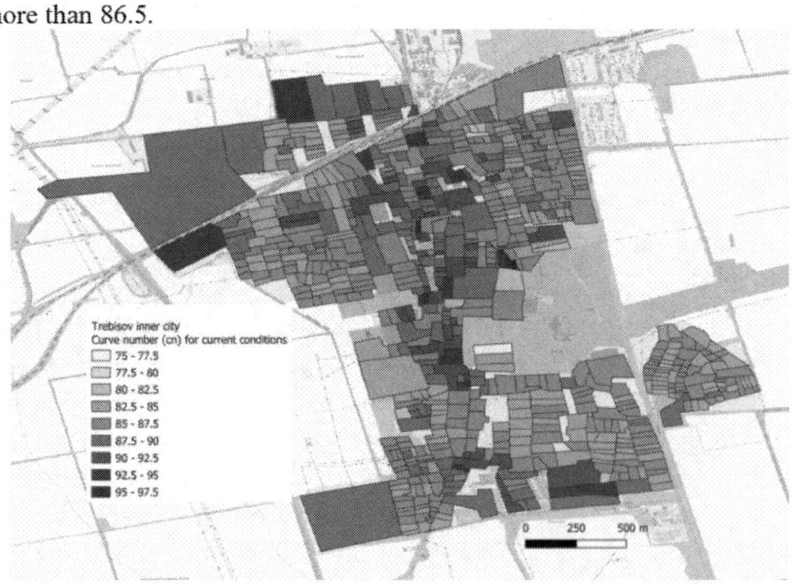


**Fig. 1.** Curve number (cn) for current condition, Trebišov inner city.

Figures 2.a and 2.b present the spatial distribution of cn according to scenarios 1 and 2, respectively. These changes cover, respectively, 6 and more than 13.3% of the total area of the received catchments. The main changes are taking place in the central parts of the city and on the outskirts. The total area proposed to be covered with green roofs is about 0.3 square kilometers. It is worth considering the significant cost of replacement, because the construction of green roofs of such an area would require at least 75 million dollars, with a price of 250 per square meter [8]- and projects of this scale cannot be completed quickly. However, the other benefits of green roofs are not discussed either. The values of CN consistently decrease to 86.2 and 85.8.

As part of this study, the technical conditions of specific buildings have not yet been evaluated, so there is a possibility that the installation of green roofs on the entire proposed area will be complicated.

a) b)

**Fig. 2.** Curve number (cn) provided by: a) scenario 1; b) scenario 2.

Also, upon completion of these two scenarios, we can estimate the change in water permeability of the studied area. Figure 3.a shows it for current conditions, Figure 3.b shows it after the changes proposed in scenario 3 are completed. Changes in this parameter are much more visible, with only one catchment having just over 60% impervious surface compared to 22 under current conditions.

3.a 3.b

**Fig. 3**. Percentage of the impervious area a) for current condition of Trebišov b) for scenario 2.

The last scenario (Figure 4) involves changes in large areas with bare soil or poor vegetation cover. The area of such plots within the investigated watersheds is 5.44%, accordingly, the total area of transformed sites is going to be almost 19%, or 12% of the area of the entire inner city of Trebišov. The average value of CN did not change significantly, decreasing to 85.68, however, the decrease is uniform, especially in the central and northern parts of the city.

**Fig. 4.** Curve number (cn), provided by scenario 3, Trebišov inner city.

## 4. Conclusions

The study shows that Trebišov has significant areas for the possible implementation of green infrastructure. At the same time, with a combination of several proposed scenarios, it is possible to achieve a fairly even decrease in the curve number and a significant decrease in the share of the impervious area.

Preliminary calculations show that the potential retention capacity in the scenario of transformation of large buildings, paved areas and areas with bare soil increases by 8%.

Research of this type is promising, in the future it is planned to compare potential retention capacity under different scenarios and conduct flow modeling for each of them. In addition, it is necessary to take into account the possibility that small private houses are disconnected from the storm water network.

## References

1. R. Šály, B. Šurina: Soils. Map on a scale of 1:500,000. In: Atlas of the Slovak Republic. Bratislava, Banská Bystrica: Ministry of the Environment of the Slovak Republic, SAŽP. Mr. 106–107. (2002)
2. Barka, T. Hlásny, A. Kičura, Et Al., J. Derbaľ, ed. Forest management and the threat of floods on the Slovak and Ukrainian sides of the Bodrog river basin. Zvolen: National Forestry Center, Forestry Research Institute (2015)
3. W. James, Editor: "Modeling the Management of Stormwater Impacts" Volume 6 Chapter 23, Authors: A. Pandit, J. Regan: What is the Impervious Area Curve Number? Publisher: CHI Guelph Ontario Canada. (1998)
4. A.D. Feldman, editor: Hydrologic Modeling System HEC-HMS: Technical Reference Manual. (2000). U.S. Army Corps of Engineers Hydrologic Engineering Center, HEC (2000)

5. R.H. McCuen. Hydrologic analysis and design. Upper Saddle River, NJ: Pearson prentice hall; (2005)
6. M.A. Vojtek, J.B. Vojteková, Land use change and its impact on surface runoff from small basins - a case of Radiša basin (Slovakia). Folia Geographica, **60(2)**, 104–125. (2019)
7. E. Fassman-Beck, W. Hunt, R. Berghage, D. Carpenter, T. Kurtz, V. Stovin, B. Wadzuk. . Curve number and runoff coefficients for extensive living roofs. J. Hydrol. Eng. **21(3)**:04015073.(2016)
8. P. Rosasco, K. Perini. Selection of (green) roof systems: A sustainability-based multi-criteria analysis. Buildings, **9(5)**, 134. (2019)

MATEC Web of Conferences **385**, 01022 (2023) https://doi.org/10.1051/matecconf/202338501022
*Young Scientist 2023*

# The assessment of the capacity of the existing detention reservoir – Lubina

*Jakub* Mydla[1*]

[1]Slovak University of Technology in Bratislava, Faculty of Civil Engineering, Department of Hydraulic Engineering, Radlinskeho 11, 810 05 Bratislava, Slovakia

**Abstract.** The presented article describes the assessment of the capacity of the existing detention reservoir with an alternative proposal for raising the dam and constructing an emergency spillway. The subject of the investigation was the capacity of the river bed under the polder too. The calculations were performed by numerical 2D modelling in the HEC-RAS program. Morphology data were obtained from own in-situ measurements and based on publically available data – DTM (digital terrain model).

## 1 Introduction

Flash floods can have devastating consequences, causing loss of life, damage to property and infrastructure, and severe economic impacts. They occur when precipitation falls rapidly and overwhelms the natural drainage capacity of the land, leading to a sudden and rapid rise in water levels. These floods can be particularly dangerous in mountainous or foothill areas where the topography is steep and the runoff can be intense. The occurrence of flash floods is influenced by a range of factors, including the intensity and duration of rainfall, the size and shape of the drainage basin, the soil type and vegetation cover, and the presence of urbanization or other human activity in the area. Climate change is also expected to increase the frequency and severity of extreme weather events, including flash floods [1].

Effective strategy for mitigating the impacts of flooding can be detention reservoir. A detention reservoir is a water management strategy that involves temporarily holding back a portion of floodwater in a designated area in order to reduce the peak flow downstream. The detention reservoir is characterized by a dam, a retention volume and control structures. The types of control structures that cause the transformation of the flood wave refer to the various elements of the detention reservoir that are designed to control the flow of water in and out of the basin. These may include control structures such as weirs, gates, and outlet pipes, as well as other features like spillways and emergency overflow channels that help to manage the flow of water during a flood event [2].

In Slovakia, the most investments were made in flood protection in the second half of the 20th century. The investigated detention reservoir was also built in this period.

* Corresponding author: jakub.mydla@stuba.sk

## 2 Description of the investigated detention reservoir

Above the village of Lubina in the district of Nové Mesto nad Váhom, there is a polder, which is created by an earth-fill dam on the stream of the same name (Lubinsky stream) in 0.930 rkm, with basic parameters:
- length in crest 94.0 m,
- the altitude of the crest is at the height 283.00 m a.s.l.,
- the height of the dam is approx. 7.6 m and
- the volume of the reservoir is about 20 thousand $m^3$.

Currently, the detention reservoir does not have emergency spillway, and the only structure that serves to transfer flows through the polder dam is a concrete culvert with a diameter of DN 800 mm (Fig. 1). Since such a condition is unsatisfactory from the point of view of the safety of the dam (danger of overflow), the capacity of the detention reservoir was verified and the design of the emergency spillway was processed.

**Fig. 1** Photo of culvert (a – intake, b – outlet).

Storage-Area-Elevation Curve for Lubina reservoir (Fig. 2) was determined based on data from DTM (digital terrain model – sources of LLS products: ÚGKK SR), which was supplemented by own geodetic measurements. Fig. 3 shows the DTM of the Lubina detention reservoir with a grid size of 1x1 m. Dots represent geodetic measurement locations. The dam itself, the riverbed above and below the dam, the location and size of the culvert were measured. The actual measurements were also important because there is no project documentation material about the structure. Current administrator - Slovak Water Management Enterprise, although it received the water construction in the administration, it did not receive any drawings or construction data for it.

The flooded area of the polder consists of pasture and scrub. The polder captures a watershed with a size of approximately 2.3 $km^2$, while this watershed is primarily used for agriculture. Table 1 shows basic hydrologic data of the flood wave on the Lubina stream in profile of detention reservoir.

**Table 1.** Hydrologic data of the flood wave (Source – Slovak Hydrometeorological Institute - SHMI)

| **Flood wave** | **$Q_{max}$ ($m^3.s^{-1}$)** | **Volume of the flood wave ($m^3$)** | **Duration of the flood (hour)** | **Time to peak (hour)** |
|---|---|---|---|---|
| $Q_{100}$ | 17.6 | 61 000.00 | 1.8 | 0.8 |

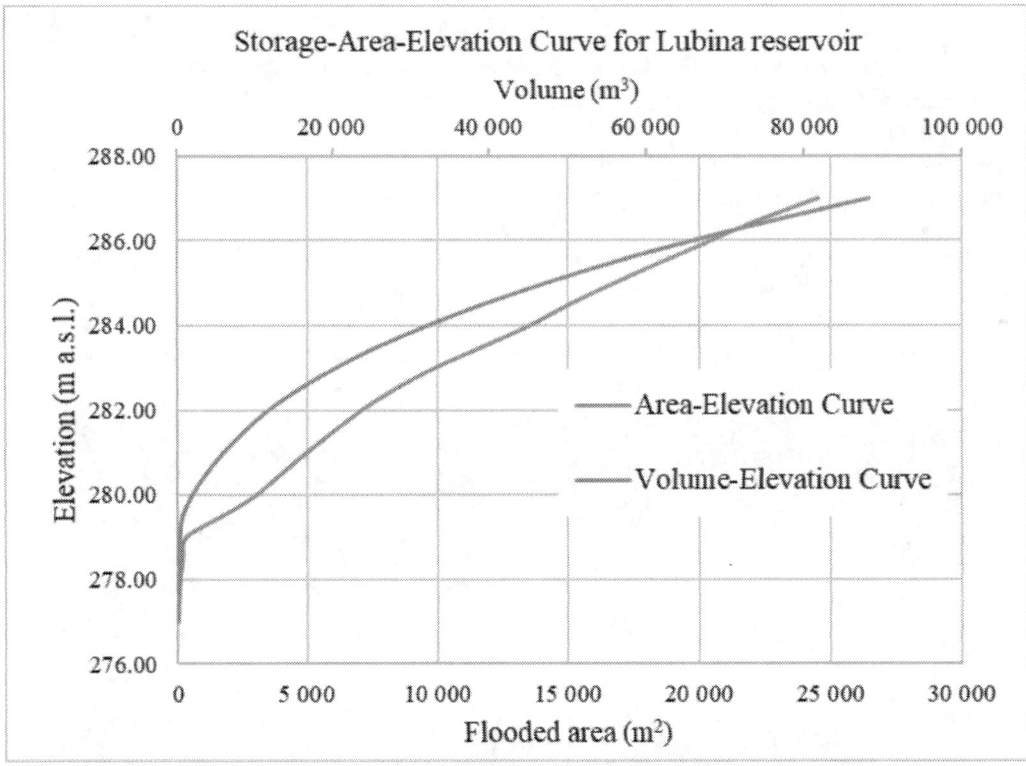


**Fig. 2** Storage-Area and Volume Elevation Curve for the Lubina reservoir.

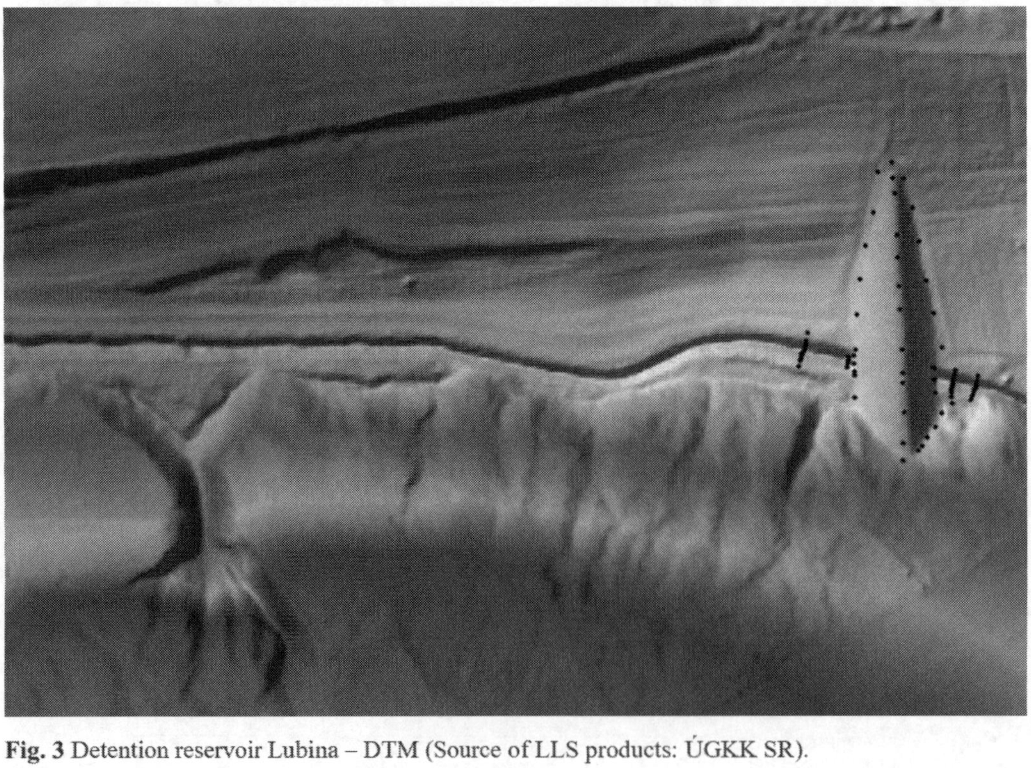

**Fig. 3** Detention reservoir Lubina – DTM (Source of LLS products: ÚGKK SR).

# 3 Mathematical modelling

## 3.1 Selected software – HEC-RAS

The HEC-RAS (Hydrologic Engineering Center´s – River Analysis System) software was used to build the 2D mathematical model. HEC-RAS has the ability to perform 2D unsteady flow routing using either the Shallow Water Equations (SWE) or the Diffusion Wave equations (DWE) numerical method. The SWE method is based on the principles of conservation of mass and momentum, and it can be used to simulate a wide range of unsteady hydraulic phenomena, including flood inundation, flow over complex topography and river channel morphodynamics. On the other hand, the DWE method is a simplified model that assumes that the lateral flow of water is negligible, and that the rate of change of water depth is proportional to the second derivative of water surface elevation. It is commonly used for simulating subcritical flows in relatively flat terrain, where the effects of lateral flow can be neglected. In general, the choice between the SWE and DWE methods will depend on the specific characteristics of the river system being modelled, as well as the modelling objectives and available data [3].

## 3.2 Procedure of creating model

In our case, the SWE method was used specifically SWE-ELM (Shallow Water Equations-Eulerian-Lagrangian Method). The geometry of the territory was created on the basis of data from the DTM and was refined mainly in the areas of the stream with geodetic survey. The buildings were modelled by raising terrain. The computational grid was square with a cell size of 2x2 m with local grid refinements. As upper boundary condition was used a flood wave $Q_{100}$ (source – SHMI) and normal depth was specified as the lower boundary condition. Manning´s roughness values were divided into three categories according to land cover (forest, meadow, stream).

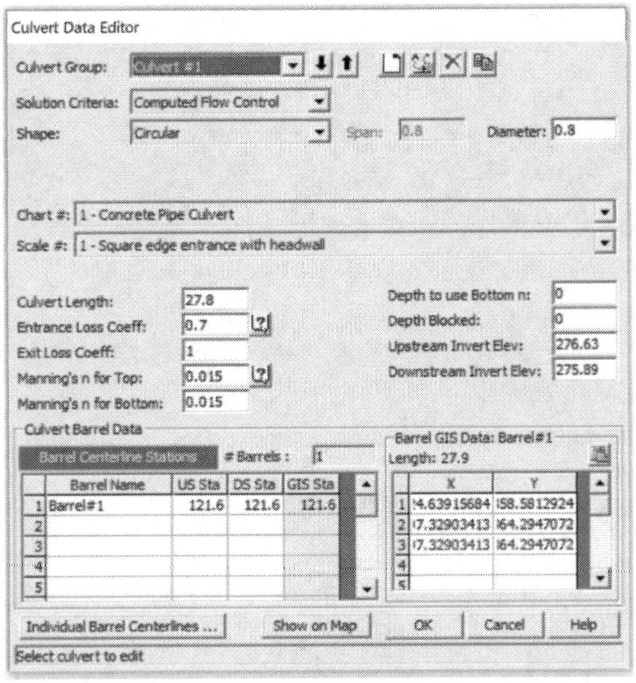


**Fig. 4** Illustration of the basic parameters of the culvert (Cut out from HEC-RAS).

Detention reservoir was modelled by connection between 2D flow areas and storage areas. The dam of detention reservoir was entered as weir/embankment and the existing culvert was also modelled. The length as well as the exact position of the inlet and outlet of the culvert was determined by geodetic survey. You can see other specified parameters for the culvert in the Fig. 4. Calibration of the model (Manning´s roughness or culvert parameters) was not possible due to lack of data.

## 3.3 Calculation scenarios

Three basic scenarios were calculated:
- current state (current dam height and without any emergency spillway),
- dam with emergency spillway and with current position of dam crest,
- dam with emergency spillway and increased dam crest.

### *3.3.1 Current state*

The simulation of a $Q_{100}$ flood wave showed that the current state of the detention reservoir is insufficient. During the flood, the dam would overflow and residential properties would be flooded, as you can see in the Fig. 5. In addition, there is a high probability of damage of the existing earth-fill dam when it overflows from the crest.

**Fig. 5** The maximum water level during the 100-year flood (current state).

### *3.3.2 Emergency spillway and current position of dam crest*

Different heights and lengths of spillways were modelled. The spillway was designed so that the level in the reservoir rose to a maximum elevation of 282.70 m a.s.l., i.e. 0.3 meters below the crest of the dam.

The length of the spillway of 10 m with the position of the spillway crest at an elevation of 282.00 m a.s.l. was chosen as the most suitable version. With the spillway designed in this way, the water level will reach a maximum height of 282.60 m a.s.l. This adjustment will prevent the dam from overflowing but will increase the flow beneath the dam compared to the current state.

### *3.3.3 Emergency spillway and raising dam crest*

Previous calculations have shown that the volume of the detention reservoir to flatten the flood wave is insufficient. After this finding, a variant with an increase of the dam was introduced. Changes of the size of the culvert were also investigated but these did not show desired results.

The design with a raised dam also includes a 10 m long emergency spillway with the spillway crest positioned at an elevation of 284.50 m a.s.l. With these parameters, the maximum water level would be 284.62 m a.s.l., and the flow beneath the dam would not flood the residential properties.

**Fig. 6** The maximum water level during the 100-year flood (Emergency spillway and increased dam crest).

## 4 Results

Mathematical modelling of the 100-year flow on the Lubinsky stream took place in the HEC-RAS software and was solved as 2-dimensional. The simulations showed that the detention reservoir with the current parameters has a minimum effect on the decrease of the flood wave. The design of the spillway without raising the dam would only protect the dam from overflowing, i.e. from possible damage of the dam, but the culminating flow would be even greater than in the current state. Raising the dam by approximately 2.0 m, the capacity of the detention reservoir would increase to such an extent that it would be able for safe passing of 100-year flood without major consequences on private properties. Fig. 7 shows the change in flow rates during individual simulations compared with the untransformed flood wave.

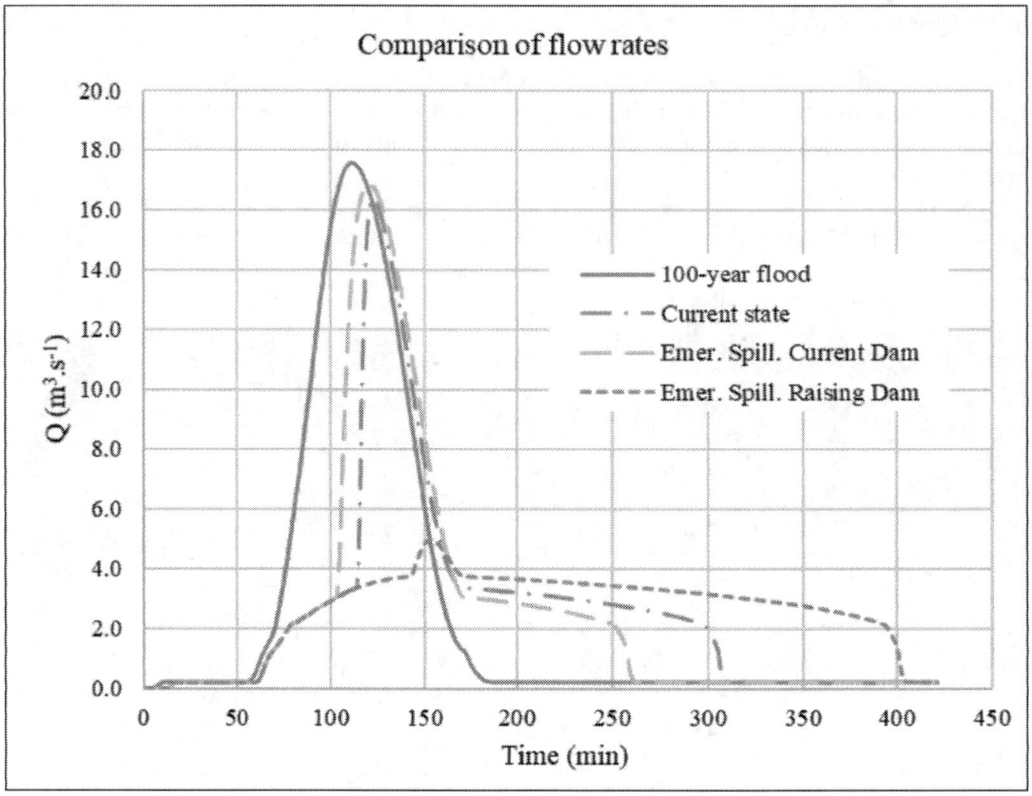


**Fig. 7** Comparison of flow rates for selected scenarios.

This article was supported by the project APVV-19-0383" Natural and technical measures oriented to water retention in sub-mountain watersheds of Slovakia" and VEGA Project No. 1/0728/21.

## References

1. A. Soltesz, L. Cubanova, A. Janik, D. Barokova, *Flood protection design for municipalities of Slovakia in various hydrological conditions,* In Proceedings of the 16th International Symposium on Water and Hydraulic Engineering, WMHE, 5-7 September 2019, Skopje, North Macedonia (2019)
2. M. Comaj, Smernica pre navrhovanie poldrov, VÚVH, Bratislava, 2004
3. G.W. Brunner, HEC-RAS River Analysis System, 2D Modelling User´s Manual, Version 6.0. Davis: US Army Corps of Engineers, Institute for Water Resourcces, 562 p. (2021)

# The use of mixed reality in construction

*Yaroslav* Nechyporchuk[1*], *Renata* Baskova[1], and *Jozef* Cabala[1]

[1]Faculty of Civil Engineering, Department of Construction Technology, Economy and Management, Technical university of Kosice, 042 00 Kosice, Slovakia

**Abstract.** This article describes Mixed Reality and its differences from Virtual and Augmented Reality. The article also deals with the concrete use of Mixed Reality in the construction industry, which is checking the conformity of the real building and the designed model. As part of the article, this method of use was actually used on the construction site. The work describes the procedure for using Mixed Reality to check the match between the real building and the model. Obstacles and problem areas that were discovered during use and will be subject to further research are also described in the article.

## 1 Introduction

"Construction 4.0" is a term used to describe the new generation of construction technologies and processes, also known as the fourth industrial revolution or Industry 4.0. It is a vision for a more efficient, sustainable, and innovative construction industry that enables the integration of advanced technologies [1]. Therefore, digitization plays a significant role in the construction industry, which helps with identifying, assessing, and mitigating risks while also improving project certainty [2]. Technologies that fall under Construction 4.0 include Building Information Modeling (BIM), 3D Printing, Internet of Things (IoT), Robotics and Automation, Virtual, Augmented and Mixed Reality, and Artificial Intelligence (AI).

Virtual, Augmented and Mixed Reality are elements of the construction 4.0 platform that help increase the level of digitization of the construction sector. VR, AR and MR offer excellent opportunities for architecture and construction through new approaches to Building Information Modeling (BIM). These technologies offer architecture and construction professionals the opportunity to personally experience the built environment in an immersive atmosphere and enable professionals to perform on-site visualization for construction planning and as-is verification [3]. Virtual reality is an interactive, computer-generated representation of a real or artificial world or activity [4]. A computer, smartphone, special headset or other display device can be used to view VR. Augmented reality essentially creates a digital window in which digital content is overlaid on the physical environment you're standing in. It can be said that Mixed Reality combines virtual and augmented reality. Only the headset is used for MR. In this case, the user is immersed in digital content but still aware of his physical surroundings, so he can interact with both physical and digital objects at the same time. VR, AR and MR can be use in stakeholder engagement, design support, design

* Corresponding author: yaroslav.nechyporchuk@tuke.sk

review, construction support, operations and management, training, but today estimated levels of adoption this technologies in construction in stage of testing or basic implementation [5].

## 2 Methodology

The use of mixed reality (MR) in construction has great potential and provides many opportunities for improving individual procedures and construction as a whole. In the framework of the article, the procedure of one of the ways of using MR was experimentally verified. Mixed reality was used to detect deviations of the BIM model from the real building and incorporate them into the model (transformation of the BIM model into a digital twin) for the purpose of its further use for facility management or other uses. The procedure of the experiment is shown in Fig. 1.

In the case study, building documentation in the form of a BIM model was used (1). The next step was pairing the model with the real building and looking for divisions between them (2). When a specific discrepancy is detected, a task must be created with a description of the discrepancy, photo documentation, etc. When creating a task, it is necessary to enter the responsible person or responsible persons. This procedure is repeated after each detected discrepancy (3). Next, the model is edited by the project participant who is responsible for the part in which the error was found (4). If necessary, the already repaired model is re-checked for compliance with the real building. The result is a digital twin of the building that can be used for facility management or other uses.

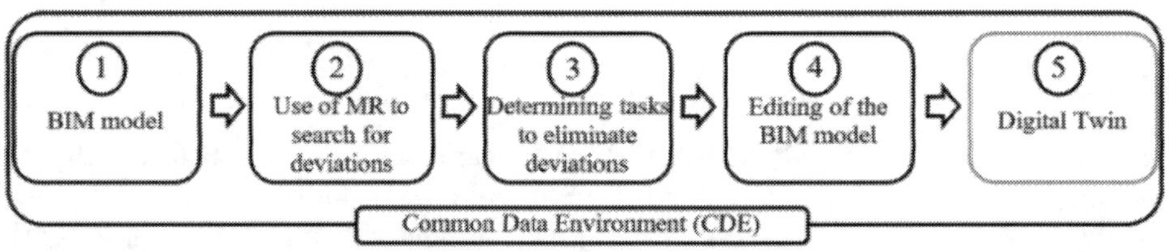


**Fig. 1.** Procedure of the experiment.

## 3 Using mixed reality to create a digital twin

The proposed procedure was used on the "C" building, which is part of the multifunctional complex "Popradská". An apartment building designed as a ten-story building with two underground and eight above-ground floors. The built-up area of the house is 1441,74 $m^2$. The procedure was repeated on a smaller part of the building - an apartment unit, the area of which is 87,44 $m^2$. BIM model of building was created in ArchiCAD.

### 3.1 Creating a project in a common data environment

A Common Data Environment (CDE) is a platform that allows project teams in the construction industry to share and manage project information from a central location. The CDE is used to store, manage, and share project data and documents throughout the project lifecycle. The purpose of a CDE is to provide a secure and structured platform for all project stakeholders to share, manage and collaborate on project information. The benefits of using a CDE include improved communication, reduced errors, faster decision making, and better project outcomes [6].

CDE from Trimble - Trimble Connect was used in this work. This software makes it possible to create projects and add to them various project documentation in the form of BIM models, drawings. text documents or other formats. Trimble allows you to add all project participants to one workspace and divide them into groups. Groups are created by the user (CDE manager) himself, so there can be both standard construction participants and specific participants. As part of the experiment, a full-fledged working group was not created, because it did not affect the result of the experiment. The project group consisted of two participants. The first participant uses mixed reality in construction site and adds tasks when finding inaccuracies. The second participant represents all construction participants who can be represented on the project and to whom the first participant gives tasks, if inaccuracies are found. After adding participants, 3D models of the building and an apartment unit were uploaded to the project. The files were uploaded in IFC format. Any computer, phone or tablet can be used to reproduce the scope of the project. To display the volume in mixed reality, you need to use a headset. Trimble XR10 with HoloLens 2 was used as part of the experiment. They are distinguished from the usual model of HoloLens 2 by the presence of a helmet that ensures safety on the construction site. HoloLens 2 must have the Trimble MR app loaded to display mixed reality. Pairing the model with the building and searching for deviations

### 3.2 Alignment of the model and the real building

In order to compare the model with the real structure, they need to be combined. There are two ways to do this automatically. The first method is pairing using a QR code. For this, it is necessary to create a QR code in the Trimble Connect program and place it in the selected place on the model. The QR code on the model must be placed in such a place that can be accurately determined on real buildings, for example, the corner of a doorway. The QR code must be printed at the appropriate scale. Then it must be placed in the appropriate place on the building. After completing the preparatory steps, the model is imaged using Microsoft hololens 2 in the Trimble MR program. To combine the model and the real building, select the Align tab from the main menu and select the Marker tool. The program will open a tool for reading QR codes. It is necessary to scan the QR code that was previously placed in the appropriate place of the building. After that, the building is automatically aligned with the model. Next, it is necessary to check whether the comparison was accurate and if it was inaccurate, then manually correct it. To do this, select the Fine Tune tool in the Align tab. After that, a cube will appear with which you can move or rotate the model.

The second method of automatic matching does not require preliminary preparation. The Plan tool located on the Align tab is responsible for the second method. After activating which method, must be to selected two vertical structures that are not parallel on the real building and their corresponding structures on the model. The floor level can also be matched automatically or it is possible to match it in the same way (select a horizontal structure in the real building and the corresponding structure on the model). Next, it is necessary to check whether the comparison was accurate and if it was inaccurate, then manually correct it as described above.

Then there is a comparison of the model and the real building. Mixed reality allows to simultaneously see structures in the form in which they were designed and in the form in which they actually exist.

### 3.3 Adding a task to remove the deviation

After all preparatory processes are completed, a visual comparison is made to control accuracy and find inaccuracies. If these inaccuracies exist, the software enables their measurement. An example of measuring inaccuracy can be seen in Figure 2.

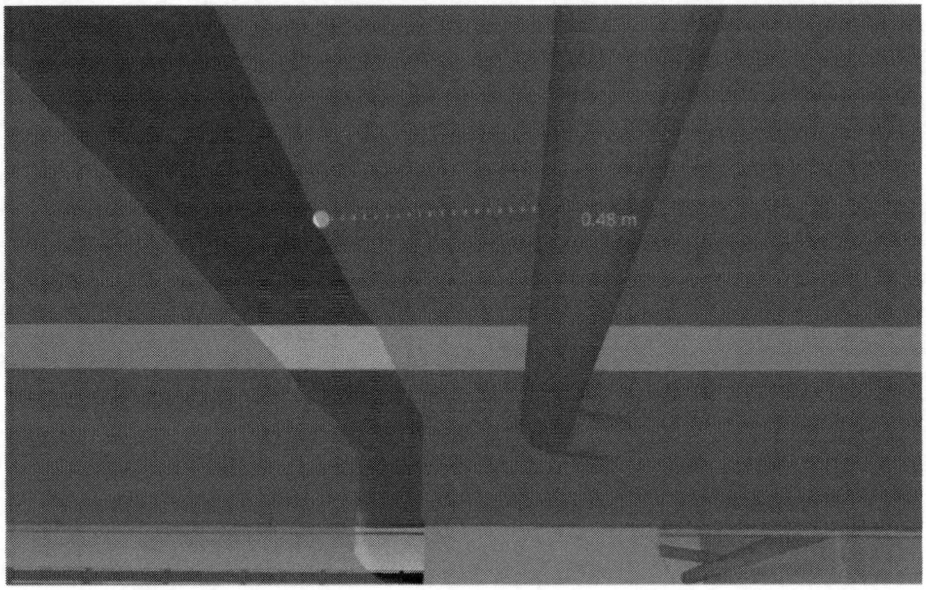


**Fig. 2.** Procedure of the experiment.

The next step is to create a task to eliminate the problem. To do this, select the "To do" tab and the "Create To do" function in the main menu. This tool allows you to create a task that will be placed in a specific location of the model. In addition, the task can be supplemented with a description of the problem, a mixed reality image that shows the real building and the model (with dimensions between them if necessary), and the responsible participant or participants of the project. Also, each task has its own priority (low, normal, high, critical). In Figure 3, you can see an example of creating a task.

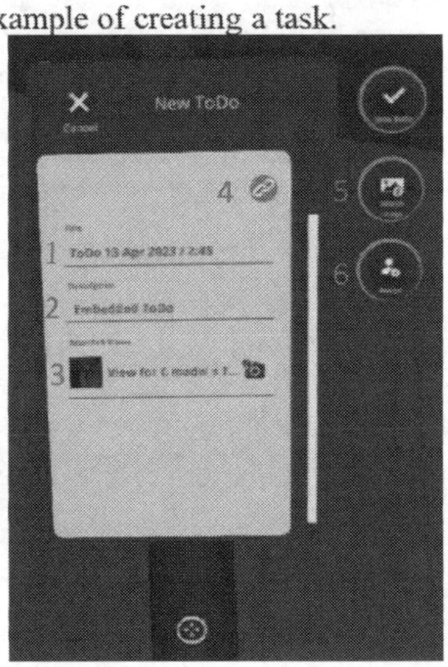


**Fig. 3.** Task creation window (1- task name, 2- description of the task, 3- added mixed reality photo, 4-task priority selection, 5-adding other photos, 6- adding a person responsible for the task).

### 3.4 Editing and other operations with the model

After creating a task, the person responsible for its execution receives a notification about the creation of the task. In the application environment, the responsible person can see all the properties and attached files of this task, and its specific placement in the model. A comment can be added to the task, and the percentage of its completion can be marked. After making changes (if needed), it is marked as closed.

**Fig. 4.** Display of the task in Trimble Connect. (1- task list, 2- display of the task in model, 3- added mixed reality photo, 4-close task button, 5-task details).

### 3.5 Summary of results and future research

The experiment described in this study was conducted twice. In the first case, the model of the entire building was used, and in the second, the model of a residential unit (apartment), which is part of this building. This was done to better define the problem areas and shortcomings of the described method. During the course of the study, several shortcomings were identified, including:

- insufficient accuracy of alignment of the model and the real building,
- too large a step of manually adjusting the placement of the model,
- inaccuracies in displaying a larger model.

During the research, methods of connecting the model and the real structure were used. Each method was used on both models. In each of the cases, the combination of the model and the building was not perfect and sufficiently accurate. To solve this problem, it was necessary to manually adjust the position of the model in real space. The following problem was detected when manually moving the model: the step of manually adjusting the placement of the model is too large. The minimum step by which it is possible to move the model in real space is 5 cm. In some cases, this distance is too large to precisely adjust the position of the model in space. This makes it impossible to accurately compare the real building and the model.

The next drawback was found when comparing the work with the larger and smaller model. When working with a larger model, it was found that the model is not displayed correctly in each part of the structure. After the model was automatically matched to the building, the position of the model was manually adjusted. For manual adjustment of the position, a point on the border of the door opening was selected. It was found that the position of structures located further from this point is not displayed correctly. To solve the problem, the automatic combination was performed several more times. For each of them, different points were selected for manual adjustment of the model's position. The model is still displayed incorrectly. Next, the actual dimensions of the structures were checked for compliance with the projected dimensions. The check showed that the projected dimensions fully correspond to the real ones. No similar problems were found when displaying the smaller model. From this, it was concluded that the size of the model affects the correctness of its display. Another reason for incorrect display of the model can be problems when converting the model from one format to another. The model created in Archicad was converted to IFC format for playback in Trimble Connect. Incorrect display of the model could be caused by conversion to the IFC format.

Further research will be aimed at identifying the reason for the incorrect display of the model in real space. It is necessary to determine whether the reason for the incorrect display is the problem during conversion, or whether there is a limitation in the size of the model that can be displayed correctly.

## Conclusion

Construction 4.0 brought new digital technologies, including mixed reality. This technology has many advantages. With the use of mixed reality, customers can be involved in the design process of projects and thus can visualize the result earlier. Mixed reality makes it possible to test different design options and their effects on a future project. The use of mixed reality in construction helps to minimize errors and disagreements during planning and construction. As part of the work, the possibility of using mixed reality to check the accuracy of the construction was tested. This option can be used either to modify the model after the construction is finished for the purpose of its further use, for example in Facility Management, or to check the accuracy of the structure during construction. When using mixed reality, it was found that behind this theory there are shortcomings that must be resolved before using this technology in practice. Future research will be connected with solving these problems.

Paper presents a partial research results of project KEGA 009TUKE-4/2022 "An interactive tool for designing a safe construction site in an immersive virtual reality" and VEGA research grant No. 1/0336/22 "Research on the effects of Lean Production/Lean Construction methods on increasing the efficiency of on-site and off-site construction technologies".

## References

1. E. Forcael, I. Ferrari, A. Opazo-Vega, J. A. Pulido-Arcas, *Construction 4.0: A Literature Review,* Sustainability **12**(22), 9755 (2020)
2. H. Begic, M. Galic. Z. Dolaček-Alduk, *Digitalization and automation in construction project's life-cycle: a review*, ITcon **27**, 441-460 (2022)
3. Y. Huang, S. Shakya, T. Odeleye, *Comparing the Functionality between Virtual Reality and Mixed Reality for Architecture and Construction Uses*, JCEA **13**, 409-414 (2019)
4. H. F. Moore, M. Gheisari, *A Review of Virtual and Mixed Reality Applications in Construction Safety Literature*, Safety **51**, 1-16 (2019)
5. J. M. D. Delgado, L. Oyedele, P. Demian, T. Beach, *A research agenda for augmented and virtual reality in architecture, engineering and construction*, AEI **45**, 1-21 (2020)
6. A. Sawhney, M. Riley, J. Irizarry, CONSTRUCTION 4.0: An Innovation Platform for the Built Environment(Routledge, New York, 2020)

MATEC Web of Conferences **385**, 01024 (2023) https://doi.org/10.1051/matecconf/202338501024
*Young Scientist 2023*

# Green facades - Their use in the sponge city

*Maria* Kocurkova[1*], *Pavol* Knut[1], and *Zuzana* Vranayova[1]

[1]Institute of Architectural Engineering, Faculty of Civil Engineering, Technical University of Kosice, Vysokoskolska 4, 042 00 Kosice, Slovakia

**Abstract.** Greenery in the urban environment is slowly disappearing and is being replaced by new construction. This trend has several consequences. One of them is the overheating of the urban climate and the associated increased need for energy for cooling. We are trying to eliminate this negative impact by creating green structures on buildings. As such, greenery brings a number of benefits. It helps to cool buildings, but also the urban climate, through the shading effect of leaves and the evaporation of trapped water. There are a number of studies around the world that investigate green roofs, or green facades. The effect of green leaf area index on cooling by shading and the effect of the distance of the façade from the envelope are investigated. Another influence is the orientation to the cardinal directions and the careful selection of plants. One of the main factors influencing the functionality of green facades is the choice of plants in relation to the climatic conditions in the country. In this article I discuss the adaptation of green facades to the climatic conditions in Slovakia. In Slovakia this issue is minimally studied and green facades occur only sporadically or only seasonally.

## Abbreviations

ET - Evapotranspiration(mm/time)

## 1 Introduction

The surface materials of cities have a higher heat capacity than greenery, which causes negative impacts on the urban environment [1]. This phenomenon is called the urban heat island effect and causes overheating of cities and the associated increase in energy demand. Green areas are limited in urban environments and therefore structures such as green roofs and walls are a suitable solution to heat islands, restoring the hydrological cycle and many other environmental and social benefits. In [2, 3] While studying experimental studies, we learned about the effect of green facades on passive cooling by evapotranspiration [4]. In addition to water evaporation, shading also contributes to cooling, which depends on the type of plants used and the area of the green wall. The cooling effect is estimated to be 60-80% from shading and 20-40% from evapotranspiration. [5] An important element in

* Corresponding author: maria.kocurkova@tuke.sk

evapotranspiration is the substrate used, often lightweight substrates with few fine particles and a higher mineral content are used in green facades [6].

In the design of green structures, it is important to understand the relationships between the different input media and the results in other experimental green structures. These green facades that we can base our designs on may be in other climates, the harmony between the facade design and a well set up irrigation system is important. Using continuous irrigation for these facades generates a daily water consumption of 0.5-20 l/m2 This volume of water needed for irrigation is unsustainable in arid areas and therefore the use of other sources of fresh water of sufficient quality for irrigation should be considered. For example, grey water, which we produce on average 120 litres per person, meets this condition [7, 8].

Surveys designed to detect water flow are quite challenging. There are methods such as gas exchange chambers on the scale of a single leaf or lysimeters on the scale of the canopy [9]. Although these devices are considered relatively accurate, they are expensive and have significant maintenance requirements . ET can also be estimated empirically using a full water balance approach or alternatively based on temperature data or using energy balance equations . The Penmann-Monteith energy balance model is widely used on daily, weekly or monthly time scales [10, 11, 13].

This paper focuses on the comparison of green façades in different climates and their detailed designs and differences. The design of green facades is a complex process depending on the appropriate choice of structure, substrate and plants. Green facades are one of the elements that have the potential to improve the environment in the future and thus increase the quality of life in densely built-up areas. The aim of this paper is to deepen the knowledge of this issue and to apply new findings to research. Our task is to use the findings to define the issue to our climatic conditions and create research that will bring new knowledge about green facades.

## 2 Material and methods

### 2.1 Study area

The first of the facades is located in Central Europe, specifically in the Austrian capital of Vienna. The altitude at which this city is located is 171m above sea level and the average temperatures in summer are 26°C during the day and 14°C at night. Temperatures in winter are around 5°C during the day and -1 at night. In the months of August and September when the research was conducted the temperatures were 25°C and 20°C. The second city, Perth, is located in Western Australia in a Mediterranean climate. Temperatures average over 30°C in summer and 18°C at night. In the cooler months temperatures reach around 20°C during the day and 9°C at night. At this site research was conducted in March and April when average temperatures were 29.6°C and 26°C.

### 2.2 Experimental setup

Research in Vienna Fig.1 worked on two green facades installed on a school building. The first wall was made of aluminium gutters and a coffered system was used on the second. These façades face the south side of the world and are located in a dense urban area. The plants on these facades were selected from native plants [12].

**Fig. 1** Green facade design in Vienna.

Compared to the facades in Perth, Australia, they use only one cardinal direction. Research in Australia Fig. 2 has focused on the north and west sides of the world. For the experimental green façades, both a façade shielded by a tarp and a façade without shading were placed. Climbing plants were planted in corten pots and aluminium trellises were applied for support. So-called groundcover plants were used and were selected from non-deciduous Australian and non-Australian species [13].

During the research in Vienna, the facades were watered with clean drinking water. In research in Australia, green facades were watered with fresh water on days off and with grey water on school days. The grey water came from handwashing sinks and from the drainage of drinking water fountains. The grey water was retained in a central storage pit. Three irrigation scenarios were identified: well irrigated (more than two irrigations per day) normally irrigated (two irrigations per day) and limited irrigation (no irrigation per day) [13].

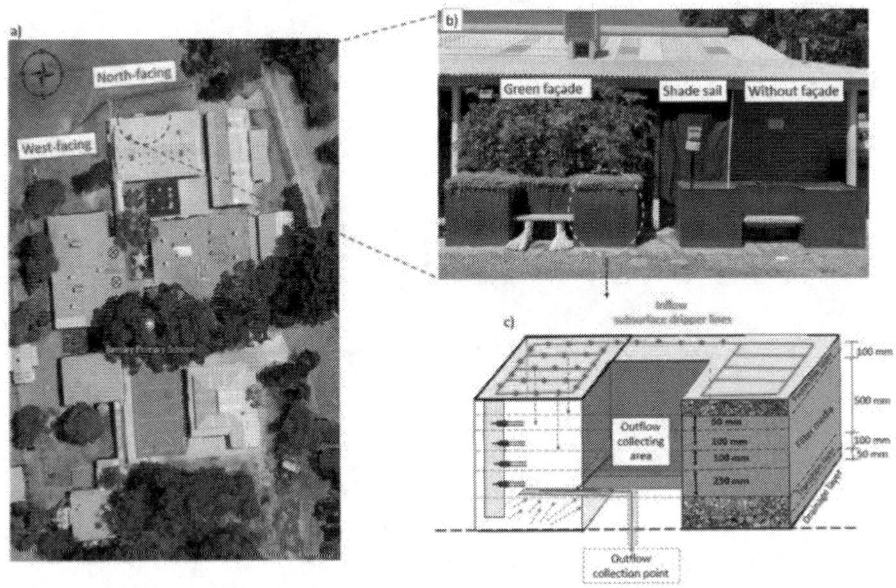


**Fig. 2** Green facade design in Perth.

### 2.3 Collection and analysis of data

The research in Vienna took place in the summer of 1.8.-31.9.2017. The authors of the research primarily selected the warm summer months. The amount of water that evaporated was calculated with the difference of inflow and outflow. To obtain the energy of evaporation, the value was divided by the mass of one mole of water (18.02 g/mol) and multiplied by the standard enthalpy of evaporation (43.99 kJ/mol), which is given at 25 °C [12]. PT 1000 sensors behind green structures were used to determine the temperature differences between shielded and unshielded facades.

Measurements were conducted on green walls in Australia from November 2020 to November 2021. Sensors for soil volumetric water content, temperature, solar radiation, wind speed and relative humidity were deployed in different areas of the façade. Together with the monitoring of inflows and outflow drains from the planters, a full water balance can be carried out [13]. Throughout the experiment, data were measured at 5 minute intervals. Air temperature, relative humidity, wind direction and wind speed, and precipitation were measured. Temperature sensors were installed at a height of 150 cm above the ground at a spacing of 20 cm. Sensors were placed before and after the green layer to measure wind speed [13].

## 3 Results

### 3.1 Results of research in Vienna

The evaporation value was not constant when measured, it depended on the amount of water received from rainfall or irrigation. The evaporation energy reached its maximum on 11 August at 121.80 kWh. Other results can be seen in the graphs.

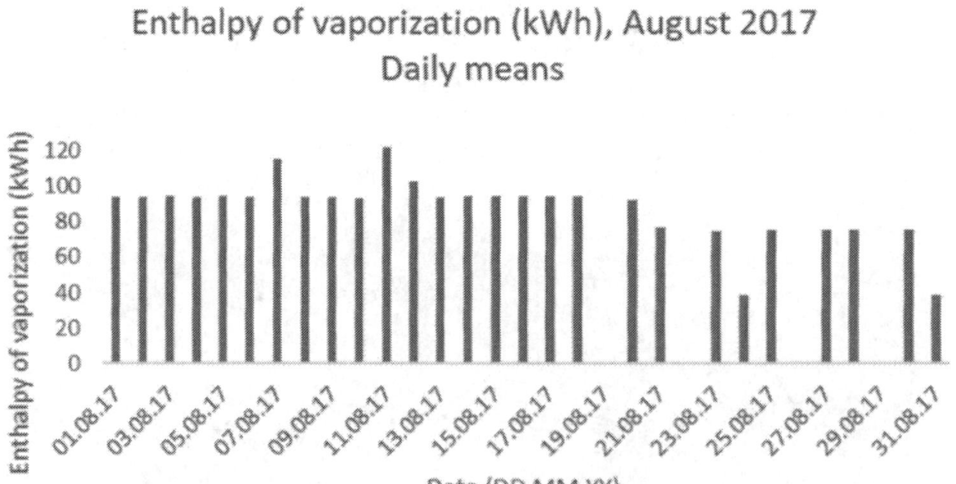


**Fig. 3** Daily average evaporation energy values (kWh) in August 2017.

These diagramsFig.3 show the daily amount of energy released by evapotranspiration. In order to regulate the irrigation of the plants, it was no longer applied daily from mid-August onwards and was further reduced in September. This can be observed on the evaporative energy balance, as the values are very close to zero. It can be assumed that with higher evapotranspiration and therefore higher evaporation energy, the environment and the building become even cooler.

According to the graph for September Fig.4 , the following results can be summarized. The lowest amount of evaporated water was -5.28 L. The highest amount of evaporated water was 170.26 L. The total amount of evaporated water was 2410 L. The lowest amount of evaporation energy was -3.31 kWh. The highest amount of evaporation energy was 106.83 kWh.

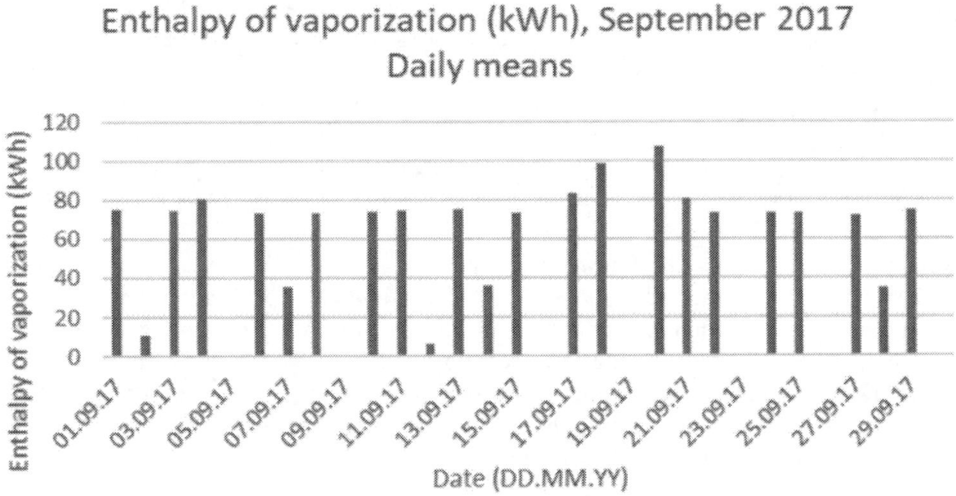


**Fig. 4** Daily average evaporation energy values (kWh) in September 2017.

A summary of the results from the chart for August is as follows. The minimum amount of evaporated water was -0.49 L. The maximum amount of evaporated water was 194.12 L. The total amount was 3774 L. The minimum amount of evaporation energy was -0.30 kWh. The maximum amount of evaporation energy was 121.80 kWh.A summary of the results from the graphs is as follows. The minimum amount of water evaporated was -5.28 L. The maximum amount of water evaporated was 194.12L. The total amount was 6184 L. The minimum amount of evaporation energy was -3.31 kWh. The maximum amount of evaporation energy was 121.80 kWh. This evaporation process removes heat energy from the environment, so higher values are better for the cooling effect of the environment. [12]

## 3.2 Results of research in Perth

After evaluating the collected data, it was found that the volumetric water content decreased in drier environments, either with less irrigation or in drier periods. When irrigation was sufficient, the soil water content was close to the maximum values and there was more runoff. Measurements also showed that the water content increased with greater depth. Using the water balance equation, changes in water storage at different layers were applied. According to the evaportranspiration values, the day times from 7:00 am to 6:00 pm and the night times from 6:00 pm to 7:00 am were determined. Evapotranspiration values were lower at night times and reached the highest values between 12:00 and 16:00 when solar radiation was highest.

The lowest measured volumetric water content was 8 and 10%, which is 5% lower than the wilting point. This means that if the lowest value of volumetric water content of the substrate was measured, the ET was zero. These results are consistent with those of previous studies. If there is little water in the substrate ET is slow, if there is enough water in the substrate ET is fast. The ET phenomenon takes place in soil, on vegetation and on other surfaces. This process is expected to reduce facade surface temperatures. In this study, the temperature difference between the area behind the façade and the area behind the shading sheet will be representative of ET. . In this experiment, a significant negative correlation was obtained for relative humidity and ET rates of green facades (Pearson's coefficient -0.91) Strong positive correlations were found between solar radiation, ET rates and the resulting evapotranspiration cooling. All differences of significant statistical nature ($p < 0.005$, paired t-test) were found in the ambient air temperature range. [13] Using correlation analysis, it was shown that the coupling between volumetric water content and ET was less than the effect of climatic conditions. The water balance results showed significant differences for individual facades. The northern façade with non-native greenery and limited irrigation had a low ET, which gradually increased with normal and later good irrigation. For the west façade with native foliage, the lowest results were obtained with normal watering, then with limited watering, and the highest results were obtained with good watering. With good watering, increasing the amount of water did not work to increase evapotranspiration, only the volume of runoff increased. When water was available to the plants, that is, when it was between the wilting point and the point of full capacity, that is, between 10-20% for the north and 8-14% for the west facades, ET adjusted. For pots without ground cover plants, we reach the desired temperature 100 cm above the soil. The green facades limited the wind speed by approximately 50%. This shows that ground transpiration did not affect evapotranspiration cooling. The green facade was shown to cool the space behind the facade. The air was 10.6 and 7.9°C cooler on the north façade and 6.1 and 4.7°C cooler on the west façade. On average, 35 and 15% of the cooling was due to ET cooling.

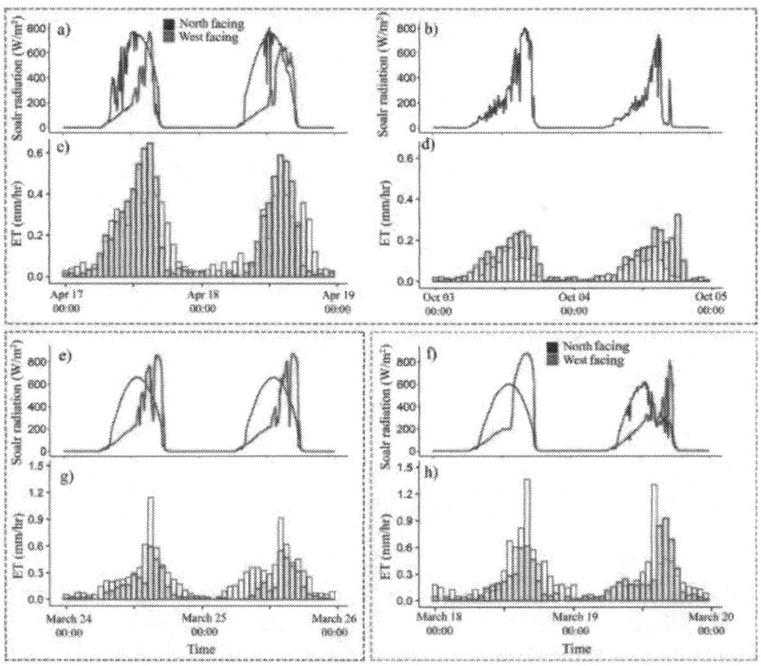


**Fig. 5** Solar radiation and hourly ET rates of the façades, under the two days of limited-watered (red box; a-d); normal-watered (blue box; e, g); and well-watered scenario (green box; f, h).

Native plants on the west facade achieved more evapotranspiration cooling than non-native plants. The green facade is also largely influenced by climatic conditions such as air temperature, wind speed, humidity, and solar radiation and is most affected by any irrigation method. The highest ET cooling was achieved at temperatures above 30°C. The expected effect of volumetric humidity on evapotranspiration was not as strong as expected. This could have been caused by frequent irrigation, high humidity averaging above 50%. The results of the measurements confirmed the results of previous studies, in which it was shown that the ET rate was better under good irrigation than under a condition with limited irrigation.

## 4 Conclusion

The study in the city of Vienna, Austria, was conducted inland. The main priority of the authors was to prove that a façade with greenery on it reaches lower temperatures than a façade without greenery. The temperature reduction was 9.3°C for the trough system and 11.6°C for the coffered system. The research demonstrates different results for the different façade systems. In terms of night-time cooling, the cassette system is more effective.

Research in the city of Perth , which is located in a Mediterranean climate, three irrigation regimes were monitored. Soil water and temperatures were investigated at four depths. During the study the following was found, limited irrigation and normal irrigation had constant moisture content, in contrast to good irrigation where runoff was variable and water drained away due to gravity drainage. The average evapotranspiration under limited, normal and good irrigation was 1.8, 6.9, 7.9mm per day on the north facing façade where non-native deciduous species were used and 2.4, 3.8, 4.0mm per day on the west facing façade where native non-native deciduous species were used. When comparing watering with normal water and good water, no positive results were found for evapotranspiration rates, but runoff was accelerated, up to 10 l/day. Evapotranspiration was closely related to solar radiation, air temperature and wind speed. Also, the rate of evapotranspiration and regeneration of green vegetation that was stressed by adverse conditions depended on the type of plants that were used.

Research shows that green facades can help combat heat islands. By designing green structures more frequently on buildings, we could achieve a noticeable cooling of the urban environment through the use of greenery. However, the issue of irrigation is still not solved. Not every city has enough drinking water, so the use of grey water is preferable. As the above-mentioned studies also show, the choice of greenery influences not only the durability but also the effectiveness of green walls. When designing and constructing these facades, it is necessary to observe all these aspects for the green facade to work properly. As is evident when comparing the two studies, exterior green facades cannot be generalised globally, but need an individual approach to their design according to climatic conditions. By applying the findings of these two studies, we intend to create an example of a green façade in our climatic conditions using plants typical for the Slovak Republic. And to propose the type of irrigation that will achieve the highest evapotranspiration rate.

This work was supported by the project run by the Slovak Research and Development Agency APVV-18-0360 “Active hybrid infrastructure towards to a sponge city” and by project SECOVE 101056201 “Sustainable energy centres for vocational excellence“.

## References

1. Ch. Butters ;A. Cheshmehzangi ;P. Sassi. Cities, energy and climate: Seven reasons to question the dense high-rise city, Journal of Green Building (2020); 15 (3): **197-214**.
2. Ch. Butters, A. Nordin, Khai,. District Cooling: A Key Solution for Hot Climate Cities, Palgrave Macmillan, Volume **11**, (2017), ISBN 978-981-10-6637-5
3. G. Pérez, J. Coma, L.F. Cabeza. Vertical greening systems to enhance the thermal performance of buildings and outdoor comfort, Building and Environment, Volume **2**, (2018), Pages 99-108, ISBN 9780128121504.

4. F. Convertino, G. Vox, E. Schettini. Evaluation of the cooling effect provided by a green façade as nature-based system for buildings, Building and Environment, Volume **203**, (2021), 108099, ISSN 0360-1323.
5. R. Bakhshoodeh , C. Ocampo , C. Oldham. Exploring the evapotranspirative cooling effect of a green façade. Building and Environment, Volume **226**, (2022), 103822, ISSN 0360-1323.
6. Ch. Berretta, S. Poë, V. Stovin. Moisture content behaviour in extensive green roofs during dry periods: The influence of vegetation and substrate characteristics, Journal of Hydrology, Volume **511**, 2014, Pages 374-386, ISSN 0022-1694.
7. L. Shashua-Bar, D. Pearlmutter, E. Erell. The cooling efficiency of urban landscape strategies in a hot dry climate, Landscape and Urban Planning, Volume **92**, (2009), Pages 179-186, ISSN 0169-2046
8. Department of Health, Code of Practice for the Reuse of Greywater in Western Australia. (2010).
9. T. Klein, E. Rotenberg, F. Tatarinov, D. Yakir. Association between sap flow-derived and eddy covariance-derived measurements of forest canopy CO2 uptake, Building and Environment
10. S. Flores Larsen, C. Filippín, G. Lesino. Modeling double skin green façades with traditional thermal simulation software, Solar Energy, Volume **121**, (2015), Pages 56-67, ISSN 0038-092X.
11. I. Susorova, M. Angulo, P. Bahrami, B. Stephens. A model of vegetated exterior facades for evaluation of wall thermal performance, Building and Environment, Volume **67**, (2013), Pages 1-13, ISSN 0360-1323.
12. T. Salonen et. al . Thermal Effects of Vertical Greening in Summer: An Investigation on Evapotranspiration and Shading of Façade Greening in Vienna. Buildings (2022), 12, 1705.
13. R. Bakhshoodeh, C. Ocampo, C. Oldham. Evapotranspiration rates and evapotranspirative cooling of green façades under different irrigation scenarios, Energy and Buildings, Volume **270**, (2022), 112223, ISSN 0378-7788.

MATEC Web of Conferences **385**, 01025 (2023) https://doi.org/10.1051/matecconf/202338501025
*Young Scientist 2023*

# Perception of ecological aspects as part of flood protection

*Marian* Dobransky[1*], *Peter* Bujansky[1], and *Gao* Zhenjun[2]

[1]Department of Environmental Engineering, Faculty of Civil Engineering, Technical University of Kosice, Vysokoskolska 4, 042 00 Kosice, Slovak Republic
[2]Three Gorges University, No. 8, Daxue Road, Yichang City, Hubei Province

**Abstract.** Modern trends in water management are more than ever inclined towards ecological solutions to avoid technical stream-channel regulation of riverbeds in the design of flood defence measures. Modifications to straighten the riverbeds that were considered in the past, removed the natural meanders of the river, increased the longitudinal slope of the riverbed, and accelerated the outflow from the threatened area. The goals of building protective dams and controlled floodplains are also limited by the negative perception of conservationists and ecologists, which leads to stagnation and discouragement from the proposals. The aim of this article is to compare the given perception of ecological aspects in different countries, how they approach them and how they are applied in studies of flood solutions. Among the ecological aspects, the following points of view will be observed: (1) options for regulation and design of flood areas, (2) duration of flooding time for the ecosystem or the landscape, (3) redefinition of riverfront in cities and (4) concept of resilience.

## 1. Introduction

According to the terminology of the EEA (European Environment Agency) glossary, flood is defined as an unusual accumulation of water above the ground caused by high tide, heavy rain, melting snow or rapid runoff from paved areas [1]. In Slovakia, the Flood Protection Act defines a flood as follows: a flood is a natural phenomenon in which water temporarily inundates an area that is not normally covered by water [2].

A flood occurs as a result

a) increasing the flow of water in the watercourse,

b) the creation of an obstacle or the creation of an obstacle in a watercourse, on the bank of a watercourse or on a construction, object or device crossing a watercourse, which caused the water to swell and spill over into the adjacent territory,

c) long-lasting precipitation or intense precipitation, snow melting or the simultaneous occurrence of these phenomena,

d) inflow of water from precipitation or inflow of water from melting snow on the surface from the adjacent area,

* Corresponding author: marian.dobransky.2@tuke.sk

e) rise of the groundwater level above the surface because of long-lasting high-water level in the adjacent watercourse or because of long-lasting precipitation.

As in other countries, the literature also describes historical floods. Historical flood marks on the territory of Slovakia date back to 1775 and 1813 on a linden (marks on a tree) in Drahovce, or evidence of the verse chronicle of the village of Cífer from 1813. [3]. Uncertainties in the determination of historical extreme design flows contributed to methods for estimating design values based on measured series of maximum flows. Currently, most of the water gauge stations are managed by the Slovak Hydrometeorological Institute, which provides information, monitoring and warnings related to the water level regime of our rivers.

The aim of the article is to examine the attitude towards ecological aspects in different countries and to describe ecological factors that apply to a given location. However, the narrow specification is the focus on flood protection and its elements precisely in connection with the effects on ecology and the environment.

## 2. Material and methods

In various countries, the needs, and requirements for the design of flood protection measures differ according to natural aspects such as geology, hydrology, climate, ecology, and so forth, but also due to the influence of the human factor, which is often defined by political agendas, laws, and awareness of the threat of floods as such. For this reason, I will point out different solutions and reasons for the effects of ecological aspects from other countries and how they dealt with them.

### 2.1 Dike displacement

The Ministry of the Environment, Baden-Württemberg, supported an extensive program called the 'Integrated Rhine Programme' that was ratified in November of 1988. The word "integrated" reflects an interdisciplinary co-operation in which ecological aspects are supposed to be ranked equally with those of water management, agriculture, and settlement needs. The Programme takes the viewpoint that the Upper Rhine valley must be seen as a geological, hydrological, and ecological unit. The reason for the creation of the given program was the public opinion favoured ecological aspects of water and nature management, and that remedial measures, based on flood management criteria alone and with little concern for environmental consequences, did not find general acceptance.

An environmental impact assessment, that was undertaken in this context, found that the previously recommended storage volume was not sufficient to regain the demanded state of flood protection. These findings were based on ecological considerations such as moderate inundations heights and dynamic flooding of the polders ('ecological flooding'). Therefore, the Integrated Rhine Programme launched a survey covering the entire Upper Rhine to identify and evaluate possible retention areas. All potential sites were investigated, essentially following three options: (1) Storage in regulated polders, (2) extending areas of natural flooding by displacing existing dikes away from the Rhine, or (3) 'ecological flooding' [4].

### 2.2 Duration of the flood

Heavy rain falling in Far East monsoon areas of Russia, can create floods with runoff about 12 – 20 $m^3.s^{-1}$ per $km^2$ in small river basins. For comparison, the runoff of melting of snow during the spring flood does not exceed 4 $m^3.s^{-1}$ per $km^2$ in the same river basins. An increasing of volume and frequency of extreme precipitation, particularly in areas with a monsoon climate in the Far East of Russia are forecasted by experts [5]. Preliminary evaluation gave the following results: the frequency of floods caused by heavy rains in the Far East will increase by 1.2 - 1.5 times. So, if now they are repeated 1 time to 10 - 15 years in the future, they will be repeated with time intervals to 7 - 12 years. The number of floods will increase by 45 - 60% in rivers of Far East region of Russia. and that is a big challenge for sustainable urban development in the region [6].

In current studies, the regulation of rivers as a measure to mitigate flash floods flows, proposes a system of self-regulating flood dams (SRFD). These dams are distributed on the river network and can save additional water volume from flash floods in temporary detention reservoirs created by SRFD. The main objective is to develop methods for analysis how the site location of SRFD meets the environmental criterium. Because ecological criterions were in the focus of the research, one of the tasks was to analyse and evaluate duration of flooding time for each ecosystem or landscape for the SRFD parameters during flash floods. Results of modelling in GIS environment were interpreted in raster GIS database which combined the landscape layer and critical time interval formula. This allowed a comparison of locations and possibility to chooses a location that has a minimal impact on the environment [7].

### 2.3 Waterfront design and redevelopment

When designing flood protection measures for built-up areas of cities, we come across the term riverfront. The "riverfront" stated here represents the water edge area, where the city meets the river. Many developed countries such as the U.S.A., U.K., Japan and have started to ecologically redevelop the riverfront in order to ensure that the rivers are restored to their natural condition [8, 9]. The ecological improvement of riverfronts can be divided into two types: (1) river restoration and (2) river rehabilitation. Restoration has been referred to projects that have a high potential of meeting the natural condition of the river, while river rehabilitation was specific to the river that had been urbanized to the extent that there was limited land available to achieve ecological improvements. This indicates that the implementation of river restoration and rehabilitation depends on the current condition of the riverfront. Riverfront development does not necessarily aim for restoring the natural setting of the riverbank. Improving its ecological setting would be sufficient if processes for full restoration cannot be conducted [10].

### 2.4 Concept of resilience

The "4Rs" of resilience (i.e., robustness, rapidity, redundancy, and resourcefulness) were adopted as a theoretical basis for the flood-adaptive green infrastructure planning. The 4Rs represent the four attributes included in the concept of engineering resilience, which is the ability to return to the pre-existing state after a disturbance and to maintain a single equilibrium state. (1) Robustness refers to the ability of a system to withstand the impact of floods, while (2) rapidity refers to the time it takes for a city system to be restored to its original stable state after a flood has occurred. (3) Redundancy involves introducing alternative elements or systems to reduce the probability of failure and to better adapt to floods. Finally, (4) resourcefulness refers to the facilities and systems of disaster prevention, such as disaster notifications and shelters, which can identify problems and mobilizing resources.

One way of interpreting the 4Rs is Flood-adaptive green infrastructure design matrix. A design matrix was constructed by classifying the types of flood-adaptive green infrastructure according to the strategy and the scale of the target area. According to the integrated design matrix, small, medium, and large areas were classified based on the size of the target area. Small areas were spaces in residential and commercial areas, and areas around footpaths with green infrastructure types as rainwater infiltration facilities, detention ponds, permeable pavements, trees, bioswales, artificial waterways, green roofs, green walls, and rain gardens. Medium areas were residential and commercial areas in flood-sensitive areas and from the point of infrastructure were defined as green roofs, green walls, rain gardens, urban gardens, parks, wetlands, and green streets (alleys). Large spaces where residential, commercial, and green areas that encounter water or are encircled.

These elements have been applied in several studies and models in South Korea, where the reduction of the rainwater runoff through the expansion of the green infrastructure can minimize the damage due to the urban flood by reducing the peak discharge as Gangnam- u´s park area (6,210,000 $m^2$) in Seoul [11].

## 3. Conclusions

In general, the design of flood protection levees (dikes) is one of the most used solutions for the protection of municipalities in Slovak republic. Proposition of the idea about displacing the existing dams, even after proving the improvement of the outflow from the study area, is not taken into consideration and further design stages. Dealing with property rights is probably the most complicated obstacle for offsetting large structures as levees. New proposals try to widen the width of the cross-section profile between protective levees primarily because of the drop in water surface elevation and not because of the creation of a potentially environmental and ecological point of view.

Similar studies of runoff models for sub-basins, with the aim of finding ideal cross-sectional profiles, are also being carried out in Slovakia. Looking at the time intervals and duration of floods in the catchment area does not place much emphasis in the design of dams (like system of self-regulating flood dams). Still the primary parameter in the design of dams with the aim of flood protection is the maximum possible regulated outflow from the dam. The regulated flow must be designed in such a way that the profiles of the riverbed of the area below the dam can safely transfer the mentioned flow. However, this theory could be used in the restoration and design of wetlands, with the goal of regulated floodwater capture. The restoration and design of artificial wetlands is a current topic that deals with the ways of regulating water levels in the catchment areas of wetlands. To what extent it is necessary and possible to divide the given volumes of flood waves so that flood protection measures are met, and at the same time the conditions for the functionality of wetlands are met is questionable, but it is necessary to point out the possibility of creating new valuable habitats.

A partial concept of resilience could be applied to the redefinition of the waterfront. The zone between the river and the inner city is indefinable or even non-existent in most places in Slovakia. The inundation and protected area of the river primarily fulfils the function of water discharge during a flood, but during average and minimum flows, the river corridor is used minimally. Another negative point of view is that functional objects are not proposed in the flow profile of the river, whether they are recreational, sports or even natural elements. Robustness could be the factor, which should be used in the design of stream modifications. The goal should be to redefine the riverfront to fulfil flood protection measures, but restore access to water, whether for people or animals in way, that it is durable and long-lasting.

This work has been supported by project of the Ministry of Education of the Slovak Republic APVV-20-0281 Mitigation of hydrological hazards - floods and droughts – by exploring extreme hydroclimatic phenomena and project SK-CN-21-0043 Study on the technology of attracting fish at the entrance of dam fishway.

## References

1. Glossary of European Environment Agency https://www.eea.europa.eu/help/glossary/eea-glossary/flood
2. Act no. 7/2010 Coll. Flood Protection Act – in Slovak (2002)
3. P. Pekárová, V. Bačová Mitková, J. Pekár, P. Miklánek, D. Halmová, S. Liová, *Historical floods in Slovakia and their significance in hydrology*; *Historické povodne na území Slovenska a ich význam v hydrológii* – In Slovak (2018)
4. M. Disse, Flood Events in the Rhine Basin: Genesis, Influences and Mitigation, *Natural Hazards*, **23**(2), 271-290 (2001)
5. P.C.B. Milly, R.T. Wetherald, K.A. Dunne, T.L. Delworth, *Nature*, **415**, 514-517 (2002)
6. S.G. Dobrovolskii, M.N. Istomina, *Water Resources*, **36**, 491-506 (2009)
7. V. Badenko, N. Badenko, A. Nikonorov, D. Molodtsov, V. Terleev, J. Lednova, V. Maslikov, *Ecological Aspect of Dam Design for Flood Regulation and Sustainable Urban Development*, MATEC Web of Conferences, **73,** 03003 7, 03003 (2016)
8. D. Miller, *The Hudson River Estuary Habitat Restoration Plan,* New York State Department of Environmental Conservation (2013)
9. Benoy co., *INCITY MEGA for SCPG in Wenzhou*, China (2018)
10. N. Redzuan, N.S. Abdul Latip, *Principles of Ecological Riverfront Design Redefined* DOI: 10.15415/cs.2016.41002 (2016)
11. H. Lee, K. Song, G. Kim, J. Chon, *Flood-adaptive green infrastructure planning for urban resilience*, Landscape and Ecological Engineering, **17**, 427–437 (2021)

# Theoretical research of the layout of shear studs on steel-concrete composite beams

*Kristyna* Hrabovska[1*]

[1]Brno University of Technology, Faculty of Civil Engineering, Institute of Metal and Timber Structures, Veveri 95, 602 00 Brno, Czech Republic

**Abstract.**This research is focused on the distribution of shear studs on steel-concrete composite beams. In practise, two different centre to centre spacings of coupling elements are commonly used, depending on the shear force. The main aim of this work is to find the best position on the beam to change the axial distance of the coupling elements, so that the smallest possible number of them can be used. Making a change at this point on the beam will ensure the most economical solution for the amount of the shear studs used as well as their appropriate utilization. In this work, equations for the design of the amount and the placement of the coupling elements on the beam, which can speed up the design in practice, were derived. In addition, the influence of the individual input parameters of the calculation on the distribution of shear studs was also investigated.

## 1 Introduction

Steel-concrete constructions use the favourable properties of the two materials: the high tensile strength of steel and conversely the high compressive strength of concrete. For the active interaction of the steel profile and the concrete slab, it is necessary to ensure their effective connection. Coupling elements, which can have different shapes, are used for this purpose. The Eurocode 4 [1] for the design of composite structures only contains an assessment procedure for welded shear studs with head, which is the most often used type of shear couplings. Their advantage over other types of coupling elements is their easy and quick connection and identical properties in all direction due their symmetrical shape.

Research focused on the optimization of coupling elements and development of new types is in progress in many countries. Linear shear couplings are often used in bridge structures. The linear type is e.g., a puzzle-like cut of web of the steel profile, which ensures the connection with the concrete slab. Research [2] dealing with the shape of the individual dowels, specifically, the radius of their rounding, was conducted at the Czenstochowa University of Technology in Poland. Experiments focused on the fatigue of the rib shear connectors were carried out by a united team from Germany and Belgium at the university in Aachen [3]. The researchers examined rib connectors with clothoid shaped dowels and exposed them to cyclic loading. Data from the experiment should be used to predict the service life of the structures. Another research effort is to achieve the

* Corresponding author:hrabovska.k@fce.vutbr.cz

demountability of composite structures and the possibility of reusing its parts after the end of their service life. This is related to the development of removable coupling elements, e.g., blind bolts. Experimental measurements of the load-bearing capacity of various blind bolts and their comparison with shear studs were carried out at the University of Isfahan in Iran [4]. This research tested bolts of different diameters and strength classes. The results show that the initial stiffness of the beam with demountable coupling elements is higher than the beam with welded shear studs. On the other hand, a beam with blind bolts exhibits an earlier slip and nonlinear behaviour. A similar experiment was performed at the Guangzhou University in China [5]. Here, the researchers investigated the influence of the individual input parameters on the load-bearing capacity of blind bolts. Specifically, the parameters were the size of the preload, the diameter of the bolt, the strength class of bolts and the size of pre-drilled hole. The obtained data served to create a model for a more detailed parametric study. Joint research by a team from Iran and Australia [6] was focused on the cyclic behaviour of bolted shear connectors. Retrofitted coupling elements were examined in a united work of Australian universities [7]. This work compared classic shear studs and blind bolts. Retrofitted shear couplings can be used in reconstruction of construction. Most research efforts deal with the optimization of the shape of the coupling element and do not pay the attention to their distribution on the beam. The layout of shear couplings was examined on Kumming University of Science and Technology in China [8]. The researchers compared the magnitude of force, slip and deformation on a two-span continuous beam with two different centre to centre spacings of shear studs. The results were compared with a beam with an equal distribution of the coupling elements.

The distance between the shear connectors is directly dependent on the shear force on the beam. The denser location is therefore situated near the supports, where the greatest shear force occurs. This force decreases towards the centre of the span, and thus the centre to centre spacing of the coupling elements gradually increase. If all the distances were different from each other, the assembly would be very time-consuming. Thus, only two different spacings of shear studs are used in practise. The aim of this research is to follow up on the previous works [9] and [10] and to find out how which parameters affect the distribution of coupling elements. At the same time, we also try to find the most suitable position on the beam to change the centre to centre spacing of shear studs, so that shear coupling are used effectively and the smallest possible number of then can be used.

## 2 Methods

There are two types of the coupling calculation – elastic and plastic. Plastic calculation is possible for the 1$^{st}$ and 2$^{nd}$ cross-section class and allows to take in account the redistribution of the shear force to all coupling elements. This allows the shear studs to be placed equally along the length of the beam. Specifically, it means, that the calculated amount of shear studs is equally distributed between the support and the point on beam with the maximum bending moment (e.g., for a simply supported beam loaded with uniformly distributed load, this is exactly half the span). As was already said in the previous work [10], the actual value of the load is not included in the plastic calculation. Instead, the maximum load-bearing capacity is considered, which leads to frequent oversizing of the beam. This can be prevented by using the partial coupling where the number of the shear couplings can be reduced. By modifying the relationship for the calculation of the partial coupling by substituting the actual value of the bending moment on the beam, the calculation gets closer to reality. On the other hand, the actual load enters the elastic calculation in the form of a shear force. Although the elastic calculation can be used for the 1$^{st}$ and 2$^{nd}$ cross-section class, it produces very conservative values. In fact, the coupling elements near the supports are used the most, because this is where the greatest shear force occurs. Their use towards the center of the span gradually decreases along with the shear force. For this reason, the coupling elements are placed more densely near the supports. The spacing of the shear studs is proportional to the shear force on the beam. However, since the shear force does not figure in the plastic calculation, the elastic calculation must be used for the placement of the coupling elements.

When calculating in the elastic region, it is assumed that the shear force acting on one shear stud corresponds to the increase of the normal force in the concrete, which can be expressed as

$$V_{L,\mathrm{el}} = \frac{V_{\mathrm{Ed}} \cdot S_{\mathrm{c}} \cdot s_{\mathrm{l}}}{n \cdot I_{\mathrm{i}}} \tag{1}$$

where $V_{\mathrm{Ed}}$ is the acting shear force, $S_{\mathrm{c}}$ is the first moment of concrete slab, $s_{\mathrm{l}}$ is the centre spacing of shear studs, $n = E_{\mathrm{a}} / E_{\mathrm{c,eff}}$ is the working factor and $I_{\mathrm{i}}$ is the second moment of ideal cross-section.

According to the reliability condition, the ratio of the shear force and the number of the coupling elements in the transverse direction must be smaller than the design load-bearing capacity of one shear stud $P_{\mathrm{Rd}}$. From this condition and from equation (1), a relationship can be derived for the calculation of the maximum possible centre to centre spacing of the shear couplings for the given force. The maximum shear force on the beam corresponds to the axial distance of the shear studs, which will be marked as $s_{\mathrm{l,min}}$. As the shear force decreases, the centre to centre spacing of coupling elements increases further. Therefore, the axial distance of the shear studs corresponding to the maximum shear force on the beam is calculated as

$$s_{l,min} = \frac{P_{\mathrm{Rd}} \cdot n_{\mathrm{r}} \cdot n \cdot I_{\mathrm{i}}}{V_{max} \cdot S_{\mathrm{c}}} \tag{2}$$

where $P_{\mathrm{Rd}}$ is the design load-bearing capacity of the shear stud, $n_{\mathrm{r}}$ is the number of shear studs in transverse direction, $n$ is the working factor, $I_{\mathrm{i}}$ is the second moment of ideal cross-section, $V_{\mathrm{max}}$ is the maximum shear force on the beam and $S_{\mathrm{c}}$ is the first moment of the concrete slab.

If the maximum shear force $V_{\mathrm{max}}$ corresponds to the centre to centre spacing of coupling elements $s_{\mathrm{l,min}}$, then the force $V_{(\mathrm{x})}$ at any point $x$ on the beam corresponds to the axial distance $s_{\mathrm{l.(x)}}$.

The above equations apply in general to all beams. For a simply supported beam with a uniformly distributed load, the maximum shear force is calculated as

$$V_{\mathrm{max}} = \frac{1}{2} qL \tag{3}$$

where $q$ is the value of the uniformly distributed load on the beam and $L$ is the span of the simply supported beam.

Similarly, the force V(x) at any point of the beam can be expressed as

$$V_{(\mathrm{x})} = \frac{1}{2} qL - qx \tag{4}$$

where $q$ is the value of the uniformly distributed load on the beam, $L$ is the span of the simply supported beam and $x$ is the distance from the support to the point on the beam where the centre to centre spacing of the coupling elements changes.

The number of shear studs on a section of a beam can be found as the length of a given section divided by the size of the centre to centre spacing used on that section. If two different sizes of the axial distance of the coupling elements are used on the beam, one corresponding to the maximum shear force and the other for the mid-span section corresponding to the force $V_{(\mathrm{x})}$ at the point $x$, where the axial distance of studs has changed, then the total number of shear couplings is calculated as

$$n = \frac{2x}{s_{\mathrm{l,min}}} + \frac{L-2x}{s_{\mathrm{l,(x)}}} \tag{5}$$

where $L$ is the span of the simply supported beam, $x$ is the distance from the support to the point on the beam where the centre to centre spacing of the coupling elements changes,

$s_{l,min}$ is smallest axial distance of the coupling elements on the beam corresponding to the force $V_{max}$ and $s_{l,(x)}$ is the maximum possible axial distance of the coupling elements corresponding to the force $V_{(x)}$.

By substituting equations (3) and (4) into relationship (5) and other adjustments, the equation for calculating the number of shear studs is created depending on the distance from the support to the place where the centre to centre spacing of shear coupling will change. It can be seen from the equation that it is a quadratic function in the form

$$n = \frac{4}{s_{l,min} \cdot L} \cdot x^2 - \frac{2}{s_{l,min}} \cdot x + \frac{L}{s_{l,min}} \tag{6}$$

where $L$ is the span of a simply supported beam, $x$ is the distance from the support to the point on the beam where the centre to centre spacing of coupling elements changes and $s_{l,min}$ is the smallest axial distance of coupling elements on the beam corresponding to the force $V_{max}$.

The graph of a quadratic function is a parabola. Equation (6) can be easily converted into vertex form to obtain the coordinates of the vertex

$$V\left[\frac{L}{4}; \frac{3L}{4 \cdot s_{l,min}}\right] \tag{7}$$

where the first coordinate indicates the position on the beam where it is most appropriate to change the axial distance of shear studs, so that the smallest possible number of the studs can be used. The second coordinate indicates the specific minimum number of shear studs, when the centre to centre spacing of coupling elements is changed at the desired point.

It follows from the first coordinate of relation (7) that for a simply supported beam of any length loaded with a uniformly distributed load, it is always most advantageous to change the axial distance of the studs in the quarter of the span. This would of course be true under ideal conditions, where the maximum centre to centre spacing of the coupling elements can theoretically go to infinity when considering the shear force near the centre of the span. Normally, the axial distance of studs is limited by design principles. The maximum centre to centre spacing should not exceed the smaller of the values $6 \cdot h_c$ or 800 mm, where $h_c$ is the height of the concrete slab. By combining equations (2) and (4), where the axial distance is replaced by the maximum permissible value, we create the relationship

$$x = \frac{1}{2} L - \frac{P_{Rd} \cdot n_r \cdot n \cdot I_i}{S_c \cdot s_{l,max} \cdot q} \tag{8}$$

where $L$ is the span of the simply supported beam, $P_{Rd}$ is the design load-bearing capacity of the shear stud, $n_r$ is the number of the shear studs in transverse direction, $n$ is the working factor, $I_i$ is the second moment of the ideal cross-section, $S_c$ is the first moment of the concrete slab, $s_{l,max}$ is the maximum allowable centre to centre spacing of shear studs according the design principles and $q$ is the value of the uniformly distributed load on the beam.

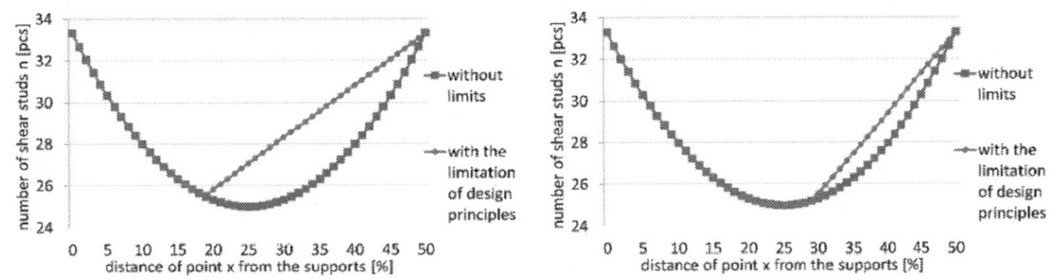


**Fig. 1.** Graphs showing the dependence of the number of shear studs on the change of their centre to centre spacing at the distance $x$ from the edge of the simply supported beam without and with the limitation of design principles. Left for $x<L/4$ and right for $x>L/4$.

By modifying the equation (8), a form can be obtained in which it is evident that it is a linear function whose graph is a straight line. Limiting the centre to centre spacing of studs means that from a certain point on the beam, the axial distance can not be increased anymore. Graphically, it looks like a straight line that intersects the parabola at a certain point, as can be seen in Fig. 1. To find the minimum number of coupling elements on the beam, it is essential whether the straight line crosses the parabola before or after the vertex. If the intersection of the parabola and the straight line is beyond the vertex, the limit of design principles is irrelevant and the best position to change axial distance of studs is in the quarter of the beam. On the other hand, if the intersection is before the vertex, it means the limit of design principles have to be considered. Then the ideal position for the change of axial distance of studs is precisely at the intersection of these function, which is the value calculated by equation (8). This means that the best place on the beam to change the centre to centre spacing of the coupling elements is the smaller of values of $L / 4$ and the result of equation (8).

The calculation is a bit more complicated for continuous beams. Again, equations (1) and (2) apply. For simplicity, only beams with equally length of span are considered. Specifically, beams with 2 and 3 spans are further considered. The calculation here is complicated by the different magnitude of the shear force in the outer and inner supports, as well as by more options for arranging of the uniformly distributed load on the spans. For now, a uniformly distributed load along the entire length of continuous beams is considered.

For a beam with two equally sized spans, the maximum shear force is on the internal support with the value $V_{max} = (5 / 8)\ qL$. The shear force at any point $x$ on the beam is therefore calculated as $V_{max} - qx$, where $x$ is the distance from the inner support. There is a different value of the shear force above the outer and inner support. Therefore, it is not possible to consider changing the centre to centre spacing of studs at the same distance from outer and inner supports. A place must be found on the beam with the same value of the shear force. So, the distance $x$ is measured from the inner support and the distance from outer supports for the change of the centre to centre spacing of the coupling elements (point on the beam with shear force of value $V_{(x)}$) is at a distance of $x - (1 / 4)\ L$, as shown in Fig. 2. From Fig. 2, the equation for the number of the shear studs on a beam with two equal spans loaded with a uniformly distributed load can be derived. Its form is similar to equation (5) for a simply supported beam, only the lengths of the individual sections are different. The equation is therefore as follows:

$$n = \frac{2x - \frac{1}{4}L}{s_{l,min}} + \frac{\frac{5}{4}L - 2x}{s_{l,(x)}} \tag{9}$$

where $L$ is the length of the span of the two-span beam, $x$ is the distance from the inner support to the point on the beam where the centre to centre spacing of coupling elements changes, $s_{l,min}$ is the smallest axial distance of the coupling elements on the beam corresponding to the force $V_{max}$ and $s_{l,(x)}$ is the maximum possible axial distance of the coupling elements corresponding to the force $V_{(x)}$.

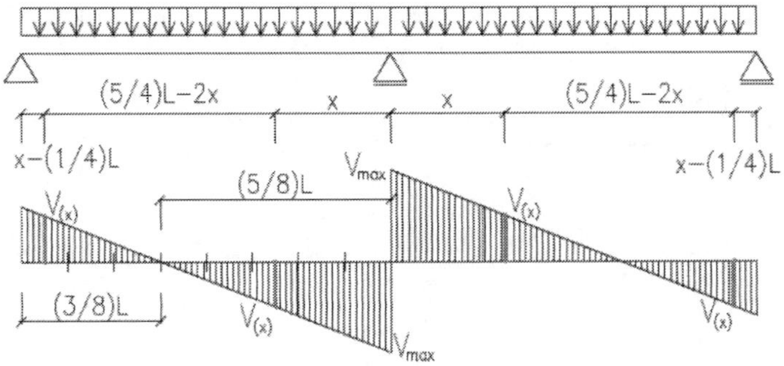


**Fig. 2.** A continuous beam with two spans of the same length loaded with a uniformly distributed load with a marked distance from supports to points on the beam with shear force of value V(x).

By substituting of calculations for $V_{(\mathrm{max})}$ and $V_{(\mathrm{x})}$, equation (9) can be modified to the form

$$n = \frac{32}{5 \cdot L \cdot s_{\mathrm{l,min}}} \cdot x^2 - \frac{4}{s_{\mathrm{l,min}}} \cdot x + \frac{2L}{s_{\mathrm{l,min}}} \tag{10}$$

where $L$ is the length of the span of the two-span beam, $x$ is the distance from the inner support to the point on the beam where the centre to centre spacing of the coupling elements changes, $s_{\mathrm{l,min}}$ is the smallest axial distance of the coupling elements on the beam corresponding to the force $V_{\mathrm{max}}$.

It is again a quadratic function, only with different parameters. Equation (10) can be converted into vertex form, where the first coordinate indicates the ideal position on the beam for the change of the axial distance of the studs under ideal conditions. For a beam with two spans, it is (5 / 16) $L$ from the inner support (i.e. (1 / 16) $L$ from the outer support). By introducing the limit of the design principles into equation (2) with the shear force $V_{(\mathrm{x})}$ applied, the equation can be obtained for determining the ideal position for centre to centre spacing of studs in the form

$$x = \frac{5}{8} L - \frac{P_{\mathrm{Rd}} \cdot n_{\mathrm{r}} \cdot n \cdot I_{\mathrm{i}}}{s_{\mathrm{l,max}} \cdot S_{\mathrm{c}} \cdot q} \tag{11}$$

where $L$ is the length of the span of the two-span beam, $P_{\mathrm{Rd}}$ is the design load-bearing capacity of the shear stud, $n_{\mathrm{r}}$ is the number of the shear studs in transverse direction, $n$ is the working factor, $I_{\mathrm{i}}$ is the second moment of the ideal cross-section, $S_{\mathrm{c}}$ is the first moment of the concrete slab, $s_{\mathrm{l,max}}$ is the maximum allowable centre to centre spacing of the shear studs according the design principles and $q$ is the value of the uniformly distributed load on the beam.

As with a simply supported beam, this is a linear function, graphically a straight line. And again, it is decisive whether it intersects the parabola before or after the vertex. The ideal position for the change of the axial distance of the coupling elements can then be determined as the smaller of the values of (5 / 16) $L$ and the result of relation (11), where $x$ is the distance measured from the inner support.

The same method can be used for a continuous beam with three equal spans loaded with a uniformly distributed load. In this case, the maximal shear force is on the inner support too. Its value is $V_{\mathrm{max}} = 3 / 5\, qL$. Then the shear force $V_{(\mathrm{x})}$ at any point x of the beam is equal to $V_{\mathrm{max}} - qx$, where $x$ is the distance from the inner support measured towards the edge of the beam. Fig. 3 shows the length of the individual sections of the beam to places with the same value of the shear force $V_{(\mathrm{x})}$. From Fig. 3 we derived the equation for the number of the studs on the beam with three spans

$$n = \frac{10}{L \cdot s_{\mathrm{l,min}}} \cdot x^2 - \frac{6}{s_{\mathrm{l,min}}} \cdot x + \frac{3L}{s_{\mathrm{l,min}}} \tag{12}$$

where $L$ is the length of the span of the three-span beam, $x$ is the distance from the inner support measured towards the edge of the beam to the point on the beam where the centre to centre spacing of the coupling elements changes, $s_{l,min}$ is the smallest axial distance of the coupling elements on the beam corresponding to the force $V_{max}$.

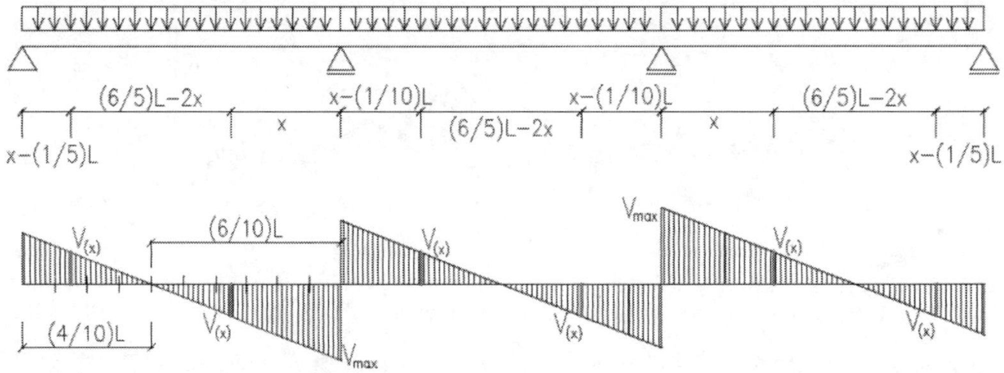


**Fig. 3.** A continuous beam with three spans of the same length loaded with a uniformly distributed load with a marked distance from supports to points on the beam with shear force of value V(x).

As in the previous case, relation (12) is a quadratic function, where first coordinate of the parabola's vertex gives the ideal point for the changing of the centre to centre spacing of the shear studs without affecting the design principles. Specifically, it is a value of (3 / 10) $L$ from the inner support towards to edge of the beam. The distance from the outer support and in the middle span can be read from Fig. 3. Again, by introducing the limit of the design principles into equation (2) with the shear force $V_{(x)}$ for the three-span beam applied, the equation can be obtained for determining the ideal position for the centre to centre spacing of the studs in the form

$$x = \frac{3}{5}L - \frac{P_{Rd} \cdot n_r \cdot n \cdot I_i}{s_{l,max} \cdot S_c \cdot q} \tag{13}$$

where $L$ is the length of the span of the three-span beam, $P_{Rd}$ is the design load-bearing capacity of the shear stud, $n_r$ is the number of the shear studs in transverse direction, $n$ is the working factor, $I_i$ is the second moment of the ideal cross-section, $S_c$ is the first moment of the concrete slab, $s_{l,max}$ is the maximum allowable centre to centre spacing of the shear studs according to the design principles and $q$ is the value of uniformly distributed load on the beam.

So, the ideal place on the beam to change the axial distance of coupling elements is found as the smaller of the values of (3 / 10) $L$ and result of relation (13), where $x$ is the distance from the inner support towards to edge of the beam.

As mentioned above, this is not always applied for continuous beam, as it depends on the distribution of the uniformly distributed load on individual spans of the beam. For a two-span beam, if only one span is loaded, this position obtained from the vertex of parabola can shift by (1 / 16) $L$, i.e., 6.25 % of the span length. For a three-span beam considering different placement options for uniformly distributed loads, the position of the ideal point on the beam can shift by a maximum of (1 / 20) $L$, i.e., 5 % of span length. The distance of places on the beam with marked the same value of $V_{(x)}$ for different configurations of loads can be seen in Fig. 4.

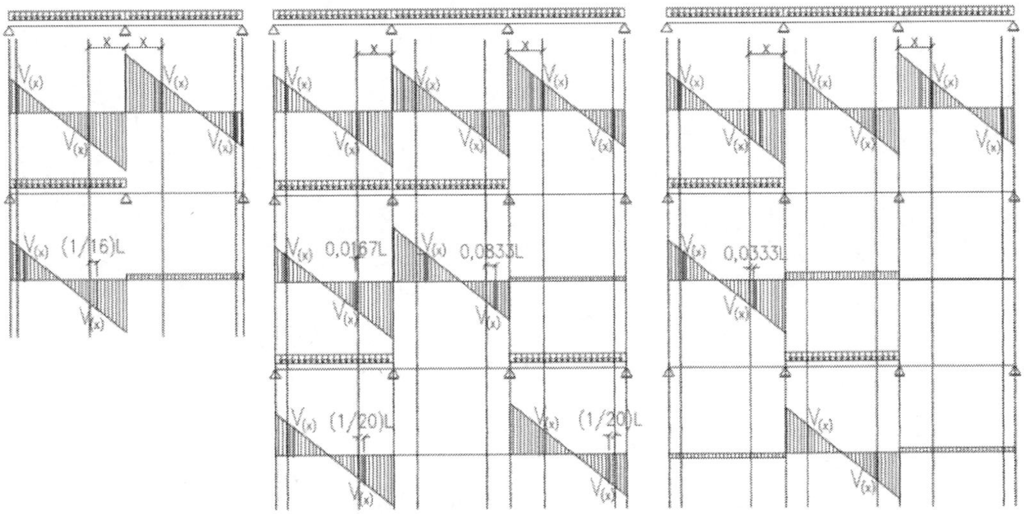


**Fig. 4.** Options for the distributing of the uniformly distributed load on individual spans of a continuous beam with two and three spans of same length.

## 3 Results

The derived equations were inserted into a spreadsheet and compared with the result of a parametric study, where the number of the coupling elements on the beam was calculated. A parametric study was carried out for a simply supported beam and for continuous beams with two and three spans of the same length. The distance $x$ was varied, which indicates the position on the beam, where the centre to centre spacing of the shear studs was changed. Specifically, the distance $x$ was increased in units of percentages of span length.

As basic samples for the parametric study, beams from IPE 300 profile made of steel S235 and concrete slabs of height 80 mm and effective width 2200 mm, made of concrete C25/30 were considered. The coupling was ensured by shear studs of length of 50 mm, shank diameter of 16 mm made from strength class 4.8. The beams were assumed with a length of spans of 10 m, loaded with a uniformly distributed load of 15 kN/m.

For a simply supported beam, the individual input parameters were further changed and their effect on the position of point $x$ was observed. The length of the span and the value of the load have a direct influence on the ideal position for changing the axial distance of the coupling elements. However, other parameters can also affect the position of point $x$. These are the variables that will affect the load-bearing capacity of the shear stud or the value of the maximum centre to centre spacing of the coupling elements, e.g., the strength of the concrete, the height of the concrete slab, the size of the steel profile or the strength class and the dimensions of shear studs.

Finally, the minimum amount of the coupling elements obtained by the elastic calculation using the point $x$ on the beam to change of centre to centre spacing of shear studs and the number of coupling elements from the plastic calculation considering the partial coupling were compared.

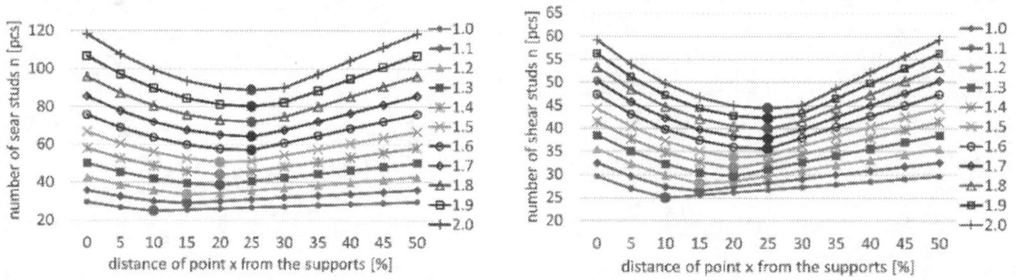


**Fig. 5.** Graphs of the dependence of the amount of coupling elements on the distance of point *x* from the edge of the simply supported beam for the percentage increase in the span length (left) and uniformly distributed load (right). Minimum values are marked in the graphs.

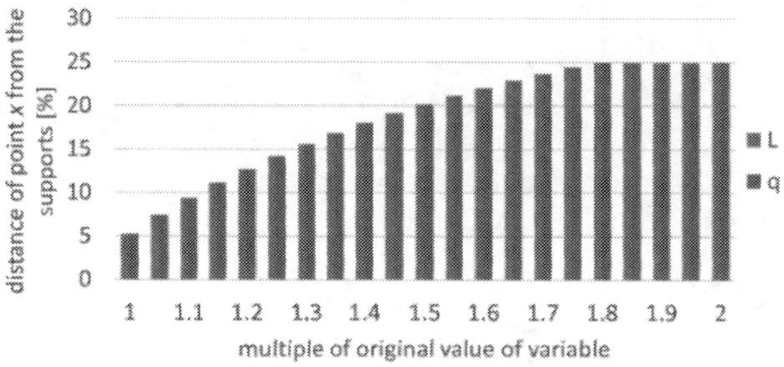


**Fig. 6.** Graph comparing the position of point *x* depending on the percentage increase in the span length and the value of the uniformly distributed load.

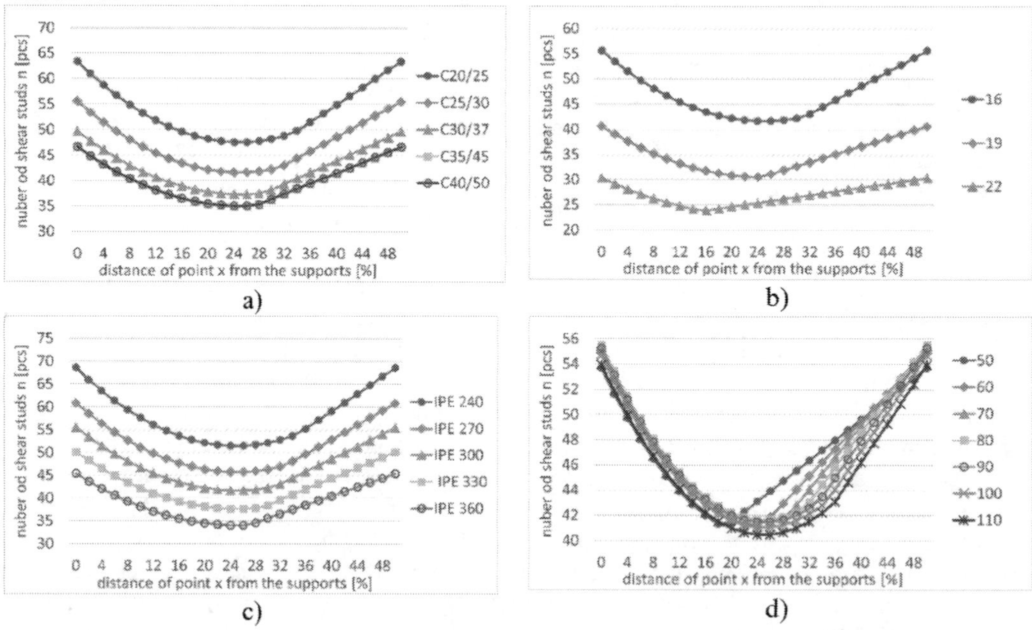


**Fig. 7.** Graphs of dependence of the number of the coupling elements on the position of point *x* for changing individual input parameters. Specifically, graph a) for the change in the strength of concrete, graph b) for the change in the diameter of shank of shear stud, graph c) for the change in the IPE profile and graph d) for the change in the height of the concrete slab.

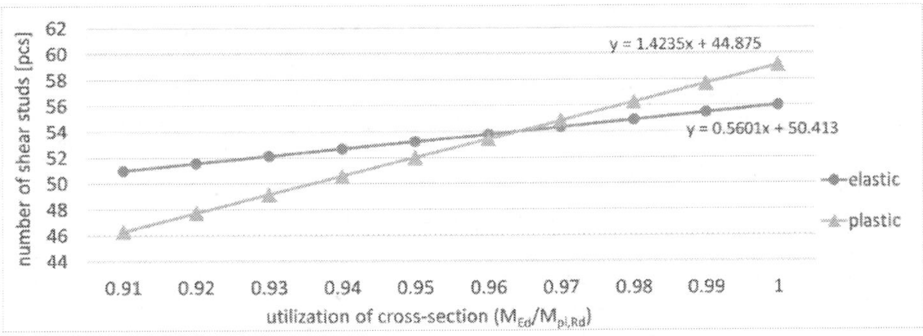


**Fig. 8.** Graph comparing the number of shear studs depending on the utilization of the cross section for plastic and elastic calculation.

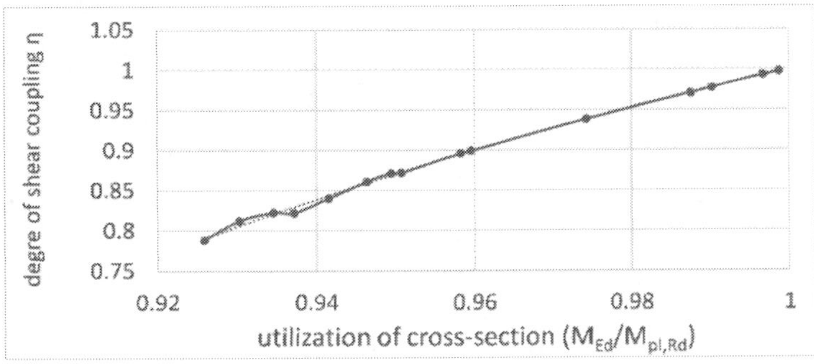


**Fig. 9.** Graph showing the dependence of the degree of shear coupling on the utilization of the cross section of the composite beam (the ratio of the bending moment to the plastic load-bearing capacity of the cross-section).

## 4 Discussion

The results of the parametric study prove the correctness of the derived relations for a simply supported beam as well as for the continuous beams with two and three spans. In the parametric study, the number of the coupling elements on the beam was calculated with different position of point *x*, where the axial distance of the coupling elements was changed. The position of point *x* was changed by units of percent of length of the beam span. The minimum amount of the shear studs was achieved by changing the centre to centre spacing of the coupling elements at the point of the beam, which is equal to the value from the above relationships, specifically, the value from relation (8) for the simply supported beam, from relation (11) for the two-spans beam and the relation (13) for the three-span beam. The correctness of derived equation can also be seen from the graphs in Fig. 5. The dependence of the amount of the shear studs on the position of point *x* for a simply supported beam is shown there. It is clear from the shape of the curves, that there are parabolas, but only up to the certain point, where the maximum axial distance of the shear studs according to the design principles is reached. From this point the parabola becomes a straight line. The shapes of the function are also similar in the case of the two-span and three-span continuous beam.

In Fig. 5 we can also see that the percentage increase in the length of the beam and the value of the uniformly distributed load causes a different increase in the number of the coupling elements. When the span of the beam changes the amount of the shear studs increases faster than when the load changes. However, regarding the ideal position for changing the centre to centre spacing of the coupling elements, for the same percentage increase in both quantities, the position on the beam is the same. (Locations with the minimum number of the coupling elements are marked in the graphs.) This dependence is better visible in the graph in Fig. 6, where the position of point $x$ is shown at the percentage increase in the length of the beam and the value of load. The same dependence applies to both parameters. At the same time, the graph shows that increasing the distance $x$ from the supports stabilizes at a value of 25 % of the span and does not increase further. This is due to the fact that here is an extremum of the function for a simply supported beam and there is no more advantageous position for changing the axial distance of the coupling elements further up the beam.

The values of the input parameters were changing in the parametric study and their influence on the position of point $x$ on the beam was investigated. The height of the concrete slab, the strength of the concrete, the strength of the steel, the size of the steel profile and the dimensions and strength of the shear studs were considered. These parameters have only a limited effect on the position of point $x$ because they interact with each other. E.g., the load bearing capacity of shear studs $P_{Rd}$ (which occurs in the calculation of the position of $x$) is calculated as the smaller of two values and depends on which one enters the calculation. In Fig. 7 there are graphs for the position of point $x$ on the beam depending on the selected parameters. Moving the position for the ideal change of the axial distance of the coupling elements closer to the support will cause an increase in the strength of the concrete (Fig. 7a), the height or diameter of the shank of the shear studs (Fig. 7b) or an increase in the steel profile (Fig. 7c). A smaller value of $x$ is also achieved by reducing the span of the beam, the value of the uniformly distributed load, the height of the concrete slab (Fig. 7d) and the effective width of the slab.

At the end, a comparison of the minimum amount of the coupling elements in the elastic and plastic calculations was made. For the elastic calculations, the number of the shear studs was calculated using point $x$ on the beam for the ideal the centre to centre spacing change. For plastic calculations, the partial coupling was considered. The number of the shear studs was calculated for the values of the bending moment on the beam corresponding to the percentage quotient of the plastic load-bearing capacity $M_{pl,Rd}$ from 50 % utilization to the full capacity. This means that in some cases the elastic calculation would no longer comply here. A parametric study was carried out for different beams, where the height of the concrete slab, the strength class of concrete and steel and the size of the IPE profile was varied. For each configuration, the size of the bending moment (equal to the percentage value of the plastic load-bearing capacity) was determined, when the same number of shear studs is used in both calculations, as can be seen in the graph in Fig. 8. These intersections of plastic and elastic calculations were found to be dependent on the degree of shear coupling. This dependence is shown in the graph in Fig. 9. In the graph we can see that with a degree of shear coupling from 0.79 to 0.99, the match of the number of shear studs from both calculations is achieved when the value of the bending moment is approx. 92-99 % of $M_{pl,Rd}$. The beam no longer complies in the elastic calculations for this value of the bending moment. On the other hand, the elastic calculations to cross-section classes 1 and 2 are quite conservative. This dependence proves that even in plastic calculations, the coupling elements can be placed using the elastic area, because in both cases, similar numbers of the shear studs are used. This means that there is a more dense placement of the shear studs at the supports. By redistributing the shear studs, a better utilization is achieved than with an even distribution.

## 5 Conclusion

The results of the research proved the correctness of the derived relations for the ideal position on the beam of the changing the centre to centre spacing of the coupling elements. This was done for the simply supported beam as well as for the two-span and three-span continuous beam with spans of the same length. It is always the smaller of the two values, where the first is given by the extremum of the function, i.e., the vertex of the parabola, and the second follows from the result of the equation, according to the type of beam. The span of the beam and the value of uniformly distributed load have a direct effect on the determination of the position $x$. The location of load on the individual spans of a continuous beam can also influence the ideal position for the change of the axial distance of the shear studs. In this case, the point $x$ on the beam moves by the maximum of 6.25 % of the span for the two-span beam and by the maximum of 5 % of the length of the span for the three-span beam. The ideal position for changing the centre to centre spacing of the coupling elements can also be affected to a limited extent by other parameters such as the strengths of the materials or the dimensions of the individual parts of the composite beam or the shear stud.

Furthermore, the research shows that even in the case of cross-section classes 1 and 2, where the beam is commonly calculated by plastic calculations (the elastic calculation would not be comply), the coupling elements can be redistributed using the elastic calculation. This is possible due to the approximate matching of the number of the shear studs in the elastic and plastic calculation using partial coupling. The shear studs can be located more densely near the supports and with a greater axial distance in the middle of the span. In this way, we achieve a better use of the individual coupling elements than in the case of their equal distribution on the beam, when the force redistribution is assumed.

This research was only concerned with beams loading with uniformly distributed load. The same principle can be probably also used for other types of beams with different type of load.

The article was prepared as a part of the Specific University Research project at the Faculty of Civil Engineering of the Brno University of Technology No. FAST-S-22-8006 and No. FAST-S-23-8317.

## References

1. EN 1994-1-1: Eurocode 4: Design of composite steel and concrete structures – Part 1-1: General rules for building, Brussels, CEN, 2004
2. P. Lacki, P. Kasza, K. Adamus, Optimization of composite dowels shape in steel-concrete composite floor, Comp. Struct., Vol. **222**(2019), 110902, Elsevier Ltd., ISSN: 0263-8223, https://doi.org/10.1016/j.compstruct.2019.110902
3. G. Christou, J.Hegger, M. Classen, Fatigue of clothoid shaped rib shear connectors, J. of Const. St. Rese., Vol. **171**(2020), 106133, Elsevier Ltd., ISSN: 0143-974X, https://doi.org/10.1016/j.jcsr.2020.106133
4. A. Ataei, M. Zeynalian, A study on structural performance of deconstuctable bolted shear connectors in composite beams, Struct. [online], **148** (2018), 508-254, ISSN: 23520124, doi:10.1016/j.struc.2020.11.065
5. Y. Zhang, B. Chen, A. Liu, Y. Pi, J. Zhang, Y. Wang, L. Zhong, Experimental study on shear behaviour of high strength bolt connection in prefabricated steel-concrete composite beam, Compos. Part B: Engineering [online], **159** (2019), 481-489, ISSN: 13598368, doi:10.1016/j.comositesb.2018.10.007
6. A. Ataei, M. Zeynalian, Y. Yazdi, Cyclic behaviour of bolted shear connectors in steel-concrete composite beams, Eng.Struc., Vol. **198**(2019), 109455, Elsevier Ltd., ISSN: 0141-0296, https://doi.org/10.1016/j.engstruct.2019.109455
7. I.E.J. Handerson, X.Q. Zhu, B. Uy, O. Mirza, Dynamic behaviour of steel-concrete composite beams retrofitted with various bolted shear connectors, Engineering

Structures [online], **131** (2017), 115-135, ISSN: 01410296, doi:10.1016/j.engstruct.2016.10.021

8. Y. Luo, D.H. Zhou, B. Duan, Study on the influence of step changed spacing of shear stud connector on the behaviour of continuous composite beam, IOP Conference Series: Earth and Environmental Science [online], **153** (2018), ISSN: 1755-1307, doi:10.1088/1755-1315/153/5/052060
9. K. Hrabovská, J. Břečka, Theoretical research of dimensional optimization and choice of material strength in steel-concrete composite beams, IOP Conf. Ser.: Mater. Sci. Eng., **1209**(2021), 012057 doi:10.1088/1757-899x/12091/1012057
10. K. Hrabovská, Theoretical research of the influence of changes in input parameters on load-bearing capacity and quantity of shear coupling, Conference Young Scientist 2022, 27-29 June, Čingov, Slovakia (2022) (to be published)

# Flame spreading across the green roof

*Klara* Hobzova[1*], *Martin* Pilinszki[1], *Daniel* Skrek[1], *Petr* Benes[1] and *Romana* Benesova[1]

[1]Brno University of Technology, Faculty of Civil Engineering, Veveri 331/95, 602 00 Brno

**Abstract.** The main request for roof exposed to fire is the classification Broof (t3), which prevents the transfer and spread of flame on the surface, this is not required for green roof according to Czech fire standards and regulations. A fire can spread in three ways: direct contact, flying burning particles and thermal radiation. In most cases, the spread of fire between objects occurs with the help of radiation. The article deals with the difference in flame spread over the surface of a green roof during a normal fire without the influence of radiant heat and the Broof (t3) test, where the surface is exposed to radiant heat. The aim was to determine the dependence of flame spread on a flat green roof due to the effect of radiation during a fire and the possibility of endangering neighboring buildings.

## 1 Introduction

The requirements for a roof covering with vegetation, from the point of view of fire safety, are not currently in the Czech Republic directly defined. In 2019, one composition of the green extensive roof was tested by the company ISOVER [1], and this issue is further discussed in the Czech Republic in articles [2], [3]. Abroad, the Canadian Université Laval deals with this topic, with publications [4], [5], [6].
Due to the susceptibility of plants to ignition, green roofs may raise concerns about fire hazards.

When designing roof formations of green roofs, we must take care to assess the spread of flame across the roof shell. Due to the different seasons, there are significant changes in the moisture and consistency of green plants, shrubs, grasses, etc. The most unfavourable is the summer season, when the so-called "dry season" occurs. Very often during this period there are a lot of fires caused by ignition (human factor or high temperatures), etc.

Another important factor is the location of various technological elements on the roof cladding (HVAC, chimney vents, ventilation shafts, FVE, etc.), which by their operation can contribute to the occurrence of a fire. Green roofs in a modern concept may not only be located on the top floor of the building, but also as a roof structure above the garage stall.

A fire can spread in three ways: direct contact with the flame, flying burning particles, and thermal radiation (radiation). The spread of a fire between objects in most cases occurs through radiation because heat can be transmitted over long distances, even without any solid or liquid matter, between the focus and the surrounding environment.

The experiment was carried out on two samples of the same dimensions and composition, the only difference being the way the fire was applied to the roof cladding. The aim was

* Corresponding author: 157465@vutbr.cz

to find out the differences between the fire under normal conditions and in the fire danger area. The results from both tests were compared and thus the influence of the fire danger area on the spread of flame across the green extensive roof cladding was determined. These results were discussed with the relevant Czech fire standards.

## 2 Description of the current state

Requirements for roof cladding from the point of view of fire safety are given in CSN 73 0810 - Fire protection of buildings – General Requirements [7]. In the fire danger area, the classification Broof (t3) is required, which prevents the transmission and spread of flame over the surface. If the roof is outside the fire danger area and its area is up to 1500 $m^2$, it does not need to have any classification, if the area exceeds 1500 $m^2$, it must have the classification Broof (t1).

This aspect is not required for green roofs according to Czech technical standards and regulations. According to recommendations and experience, fire strips are carried out using washed river aggregates called "pebble".

Due to fire regulations and standards [8], [9] a fire strip with a width of min. 2000 mm or a construction of type DP1 exceeding the outer surface of the roof cladding by 300 mm or other equivalent treatment preventing the spread of fire is required for neighbouring buildings. When applied to green flat roofs with an area of more than 1500 $m^2$, a strip of so called "pebble" would have to be applied from the attic structures.

## 3 Experimental tests

Measurement conditions:

- Air Pressure: 1020 hPa,
- Wind speed: 2.1 m/s,
- Exterior temperature: 13.0 °C.

Roof composition from the top layer:

- Growing mat of sedum pre-grown on a supporting surface (fresh weight 24.9 kg/$m^2$, weight after drying 16.2 kg/$m^2$ - values determined by the laboratory),
- organic roof substrate extensive, thickness 240 mm,
- filtration geotextile, surface weight 300 g/$m^2$,
- nop foil with geotextile, nop foil height 8 mm,
- waterproofing membrane foil, thickness 1.5 mm,
- standard particleboard base.

The roof layer base was made in accordance with [10] Art. 6.4.2.2 b of 16 mm thick particle boards cut into boards 250 mm wide with straight edges. The joints between the boards were 5,0 ± 0,5 mm.

The structure of the roof was not anchored to the base. All layers of the roof were laid on the test base, which was bounded around the perimeter by OSB boards with a height of 300 mm.

For the experimental test, a model of an extensive green flat roof with a size of 1200 x 3000 mm was used, which was stored at an inclination of 5°. On this surface were stored 2 burning pyres according to standard regulations [10] shown in Fig. 1.

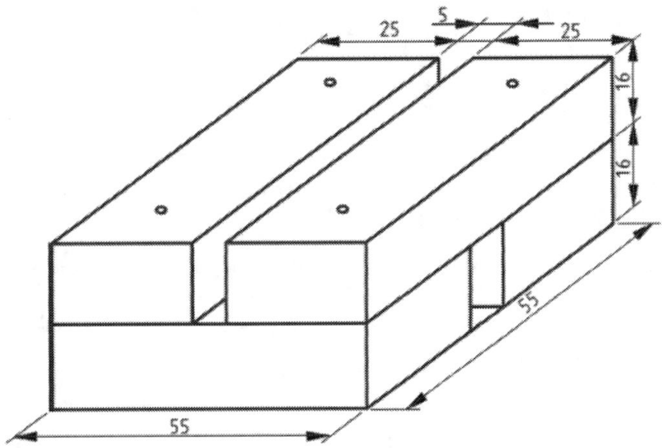


**Fig. 1** Pyres [10].

The test took place in the fire testing laboratory of PAVUS a.s. in Veselí nad Lužnicí, the progress of the fire is shown in Fig. 2. In total there were two measurements. The first test simulated a normal fire, the second one was Broof (t3), i.e. with an additional radiant panel.

The test took place over a period of 30 minutes, where at the end the spread of the flame was measured on the surface of the green roof.

**Fig. 2.** The progress of the fire test in the fire testing laboratory of PAVUS a.s..

# 4 Measurement results

## 4.1 Test specimen No 1 (normal fire – without radiant panel)

The measurement results of sample No 1 are recorded in Table 1, Table 2, Table 3. Fig. 3 shows the affected size surface by fire during the test.

**Table 1.** Behaviour of the test sample during the test / slope 5°.

| Time (minutes) | Observations |
|---|---|
| 0. – 2. | There was no substantial change in the sample surface |
| 3. – 8. | On the surface the pyres burn, due to the air flow from the fan the flame is directed upwards, in the place of the flame and around the borders of the plant brown, does not burn, there is a slight development of smoke |
| 9. – 14. | In the immediate vicinity of the borders the plants are burnt, there is a slight development of smoke |
| 15. – 23. | The boundaries have burned out and are spontaneously extinguishing, there is no flame on the surface, there is a weak smoke development |
| 24. – 30. | There is no flame on the surface, there is a slight smoke development |
| 31. | End of test, sample opening |

**Table 2.** External spread of fire.

| Spread upwards | | Spread downwards | |
|---|---|---|---|
| **Distance (mm)** | **Reaching Time (min:s)** | **Distance (mm)** | **Reaching Time (min:s)** |
| 100 | - | 100 | - |
| 300 | - | 300 | - |
| 500 | - | 500 | - |
| 700 | - | - | - |
| 900 | - | - | - |
| 1100 | - | - | - |
| 1300 | - | - | - |
| 2000 | - | - | - |

**Table 3.** Damage to sample.

| Description of damage | Value |
|---|---|
| Maximum length of burnt material upwards (mm) | 70 |
| Maximum length of burnt material downwards (mm) | 45 |
| Extent of internal damage upwards – substrate under the vegetation carpet without damage (mm) | 0 |
| Extent of internal damage downwards – substrate under the vegetation carpet without damage (mm) | 0 |
| Damaged external area – scorched plants due to heat (m2) | 0.010 |
| Flame-burning material falling off the roof surface | No |
| Flame burning material falling off the underside | No |
| Burning through, creating holes | No |

**Fig. 3.** Sample No 1 after completion of the test. [3]

## 4.2 Test specimen No 2 (exposed to radiant panel (Broof(t3))

The measurement results of sample No 2 are recorded in Table 4, Table 5, Table 6. In Fig. 4 shows the size affected surface by fire during the test.

**Table 4.** Behaviour of the test sample during the test / slope 5°.

| Time (minutes) | Observations |
|---|---|
| 0. – 3. | The sample is stressed by radiant heat from the radiation panel, there has been no substantial change in the surface |
| 4. – 5. | On the surface the pyres burn, due to the air flow from the fan the flame is directed upwards, in the place of the flame and around the pyres the plants turn brown, they do not burn, under the radiation panel the plants begin to wither, there is a slight development of smoke |
| 6. – 10. | In the immediate vicinity of the pyres the plants are burnt, under the radiation panel the plants dry and brown, there is a slight development of smoke |
| 11. – 16. | The pyres have burned down and are spontaneously extinguishing, there is no flame on the area, the area of dead and brown plants under the radiation panel is growing, there is a slight build-up of smoke |
| 17. – 30. | There is no flame on the surface, under the radiation panel the plants are brown, sometimes blackened, on the other surface the state without substantial changes, there is a weak development of smoke |
| 31. | End of test, sample opening |

**Table 5.** External spread of fire.

| Spread upwards | | Spread downwards | |
|---|---|---|---|
| Distance (mm) | Reaching Time (min:s) | Distance (mm) | Reaching Time (min:s) |
| 100 | 6:30 | 100 | - |
| 300 | - | 300 | - |
| 500 | - | 500 | - |

| | | | |
|---|---|---|---|
| 700 | - | - | - |
| 900 | - | - | - |
| 1100 | - | - | - |
| 1300 | - | - | - |
| 2000 | - | - | - |

**Table 6.** Damage to sample.

| Description of damage | Value |
|---|---|
| Maximum length of burnt material upwards (mm) | 120 |
| Maximum length of burnt material downwards (mm) | 70 |
| Extent of internal damage upwards – substrate under the vegetation carpet without damage (mm) | 0 |
| Extent of internal damage downwards – substrate under the vegetation carpet without damage (mm) | 0 |
| Damaged external area – scorched plants due to heat (m2) | 0.023 |
| Flame-burning material falling off the roof surface | No |
| Flame burning material falling off the underside | No |
| Burning through, creating holes | No |

**Fig. 4.** Sample No 2 after completion of the test. [3]

## 5 Discussion

Given that the spread of fire between objects in most cases occurs through radiation, and this can increase the fire area by more than 2x according to the results of the experiment, it is possible to discuss the requirement of standards, namely the design of type DP1 exceeding the outer surface of the roof cladding by 300 mm [8] and [9]. An experiment would need to be conducted, for example, on semi-intense greenery, where the elevation of adjacent roofs by only 300 mm is debatable not only in terms of the effect of significant radiation, but also the potential impact of burning particles.

## 6 Conclusion

Experimental tests show that under the action of a radiant panel, the area affected by fire will increase by up to 2.3 times. This value was determined by the proportion of areas damaged to external surfaces, in normal fire and under the action of a radiant panel, for vegetation carpets from sedum.

Fire strip requirement with min width. 2000 mm [7] is sufficient, according to experimental research, for extensive and semi-intense greenery up to a maximum height of about 1.5 m, however with intense greenery it would be advisable to choose a fire strip with a width greater than the proposed maximum height of greenery. The recommended requirement around enclosing building structures of 500 mm width is sufficient for extensive green roofs given the experimental tests carried out. For semi-intense and intense green roofs that extend over the edge of the building structures may compromise the rim shell structures in a fire, this requirement is insufficient.

## References

1. Saint-Gobain Construction Products CZ a.s., Divize ISOVER. Vegetation extensive roof with certification BROOF (t3). TZB-info [online]. [cit. 2023-03-03]. Available from: https://stavba.tzb-info.cz/strechy/22382-vegetacni-extenzivni-strecha-s-certifikaci-broof-t3 (2021)
2. K. Hobzová; M. Pilinszki, Fire resistance testing of green flat roofs. Juniorstav 2022. JUNIORSTAV. Pod Nemocnicí 590/23, 625 00 Brno: ECON publishing, s.r.o. page 49-54. ISBN: 978-80-86433-76-9. ISSN: 1190-1535 (2022)
3. D. Skřek; M. Pilinszki; K.Hobzová, Flame spreading across the green roof. JUNIORSTAV 2023 - 25. international doctoral conference of civil engineering. 1. ECON publishing, s.r.o. Pod Nemocnicí 590/23, 625 00 Brno: page 52-58. ISBN: 978-80-86433-80-6 (2023)
4. N. Gerzhova, P. Blanchet, CH. Dagenais, S. Ménard, J. Côté. Flammability Characteristics of Green Roofs., 13 (2020)
5. N. Gerzhova, P. Blanchet, CH. Dagenais, S. Ménard, J. Côté A Conceptual Framework for Modelling the Thermal Conductivity of Dry Green Roof Substrates (2019)
6. N. Gerzhova, P. Blanchet, CH. Dagenais, S. Ménard, J. Côté. Heat Transfer Behavior of Green Roof Systems under Fire Condition: A Numerical Study (2019)
7. ČSN 730810 - Fire protection of buildings - General requirements. online: Office for Technical Standardization, Metrology and State Testing (2016)
8. ČSN 730802 ed.2 - Fire protection of buildings - Non-industrial buildings. online: Office for Technical Standardization, Metrology and State Testing (2020)
9. ČSN 730804 ed.2 - Fire protection of buildings - Industrial buildings. online: Office for Technical Standardization, Metrology and State Testing (2020)
10. ČSN P CEN/TS 1187:2012 – Test methods for external fire exposure to roofs (2012)

# Application referring to the effect of continuous deck occurring in the joint used in the ceiling structure

*Klara Freudenberger*[1*]

[1]Department of Steel and Timber Structures, Faculty of Civil Engineering, Slovak University of Technology in Bratislava, Radlinskeho 11, 810 05 Bratislava, Slovak Republic

**Abstract.** The usage of hybrid structures brings advantages in their application. Structures with appropriately combined materials within the supporting structure are able to withstand higher loads and overcome larger spans. Effective structural solutions can also be achieved by combining materials such as cross laminated timber, concrete, and steel. The result is the achievement of a more efficient construction and at the same time achieve an aesthetic, but also an ecological structure, furthermore, creating a slimmer ceiling structure can ensure financial savings. In order to achieve the maximum use of the potential of the structure, it is necessary to create suitable structural connections. By creating a suitable joint among the ceiling decks and the beam, it is possible to create a load-bearing ceiling structure that behaves like a continuous deck. In this paper, numerical analysis of selected types of joints are performed and compared. During the examination, an adjustment has been made for alternative joint variation. After updating the proportions of components in this specimen, alternative sample of joint was able to withstand loads applied during analysis. Examined joint connection offers alternative to original joint connection.

## 1 Introduction

Due to emphasis of using sustainable materials witch are also able to withstand high construction standards is hybrid construction appropriate option. Main idea of designing hybrid structures is using combination of materials in their best possible way of their application [1]. Combining steel with concrete or even also with timber are mainly used combinations in this type of constructions. Effective combination of materials allows to obtain a structure that is able to accept higher loads and also withstand longer spans of the ceiling structure [2].

The effective design of the ceiling load-bearing structure allows to reduce the height in the ceiling, the easiest way to get the best solution is creating a slim-floor [3]. The limitation of the usage of continuous deck is a frequent problem that arises when applying a slim-floor system. Appropriate detail of ceiling connections is the key to a troubleshooting.

* Corresponding author: klara.freudenberger@stuba.sk

# 2 Hybrid structure

To achieve more and more efficiency of hybrid ceiling systems, it is necessary to achieve a result with the most capable modules of each component within the system. By using suitable combination of materials is possible to get more efficient structure, especially with ceiling constructions, the height of the floor construction also plays a significant role [4].

Based on this reason, there is a tendency to create the thinnest possible load-bearing structure, the so-called Slim-Floor, this can be achieved by placing the ceiling panels on the lower flange of the beam. This principle creates reduction in height by the height of the ceiling beam, considering that the ceiling slab are no longer placed on the upper edge of the steel beams (Figure 1). A steel beam combined with massive timber ceiling panels, mostly used cross-laminated timber (CLT), is common combination mostly used in slim floor construction. However, the use of a simple solution consisting of the use of a hot formed H-shaped cross-section has significant disadvantages, especially in a fire situation [5].

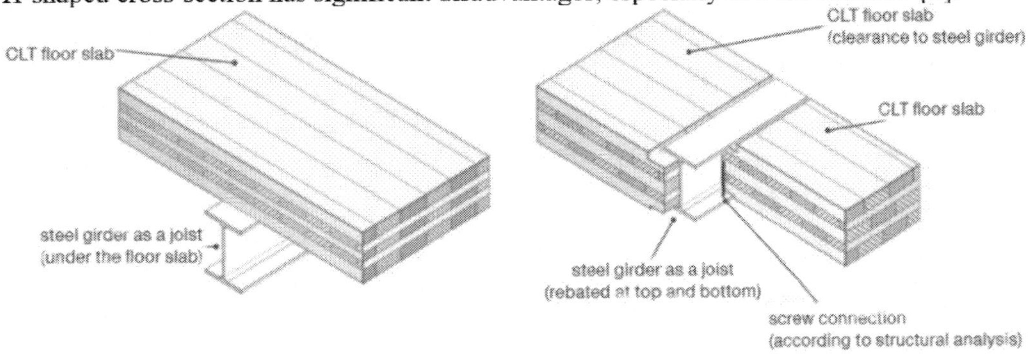


**Figure 1:** Comparison of regular configuration of hybrid ceiling system and slim-floor system. [5]

## 2.1 Classification of hybrid structure

Steel cross-section replaced by composite cross-section significantly increases the durability of the structure. Namely DELTABEAM is the steel girder infilled with concrete, that creates durable composite beam. The combination of CLT ceiling slabs and DELTABEAM beam from Peikko has been used since 2011 [6]. Within the system, panels can be applied as simply laid decks or as continuous decks. However, the question arises as to how to interconnect individual decks on the opposite sides of the beam, so that the system can be considered as continuous deck.

## 2.2 Previously examined research

Research program focused on DELTABEAM with timber floor solutions was initiated by Peikko [7]. Investigation was focused on the load transfer from the slab to the beam through a compression arch against inclined web it is laboratory research. Full-scale load transfer tests were carried out [8]. Three main groups of specimens were examined. In the first group of samples (A1-A4), a floor joint consisting of DELTABEAM and a timber slab was applied. DELTABEAM composite timber slabs were used as samples in the second (B1-B4) and third assembly of the group (C1-C2). As shown in the Figure 2 CLT panels are laid on the downstand within the C series. During this set of experiments was inclination of timber edge examined to. Verification has occurred by using specimens with vertical and also inclined edge of panel. In latter case was fire situation examined. The supporting lower ledge with the inclined timber edge in order to simulate the fire conditions was removed. In fact, the steel ledge loses a major part of stiffness and resistance when exposed to fire. Used cross-sections for all specimens are described in the Figure 2.

| | Specimen | Timber edge | Ledge |
|---|---|---|---|
| DELTABEAM® Timber slab | A1 | Vertical | yes |
| | A2 | Inclined | yes |
| | A3 | Inclined | no |
| | A4 | Inclined with pockets | no |
| DELTABEAM® Composite Timber slab | B1 | Vertical | yes |
| | B2 | Inclined | yes |
| | B3 | Inclined | no |
| | B4 | Inclined with chamfers | yes |
| | C1 | Vertical | yes |
| | C2 | Inclined | yes |

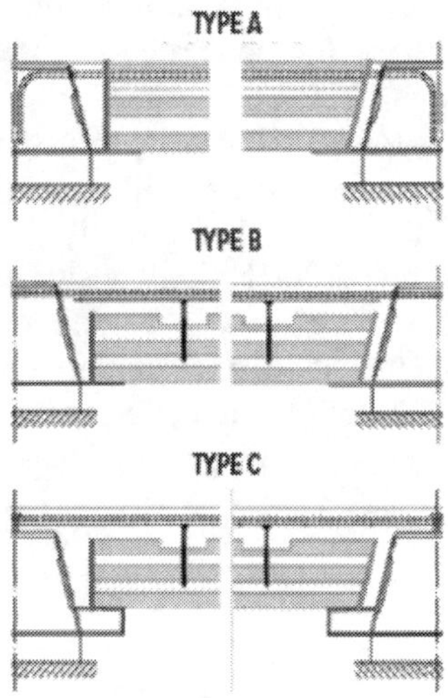


**Figure 2:** Overview of the specimens [8].

Group A has an applied the design load level corresponding to a 6 x 6 m grid with 3 kN/m$^2$ live load. The design load level corresponding to a 7 x 7 m grid with 3 kN/m$^2$ live load was applied during examination of B and C group of specimens [8]. The test set-up is shown in Figure 3.

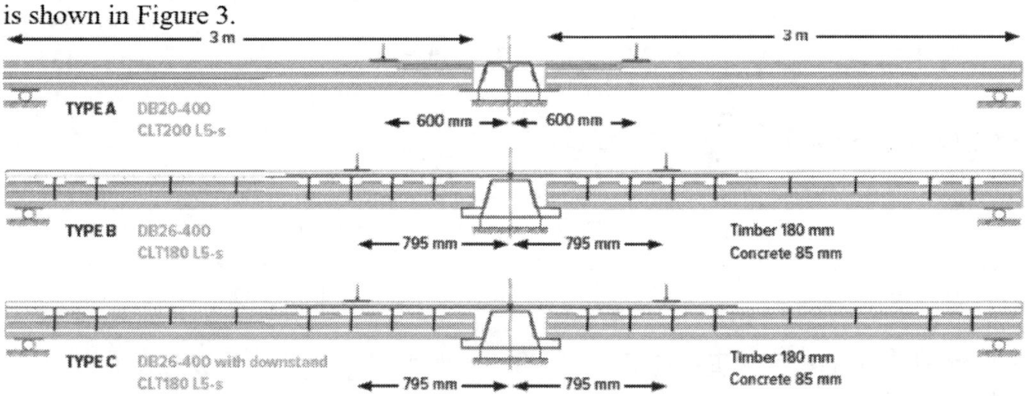


**Figure 3:** Test set-up [8].

The results of the experiments proved that each case are able to obtain to design load level and moreover get the peak load approximately three times greater. Indeed, the results confirmed the hypothesis about application of inclined edge of the panel. This confirmation is clearly visible in graphs (Figure 4).

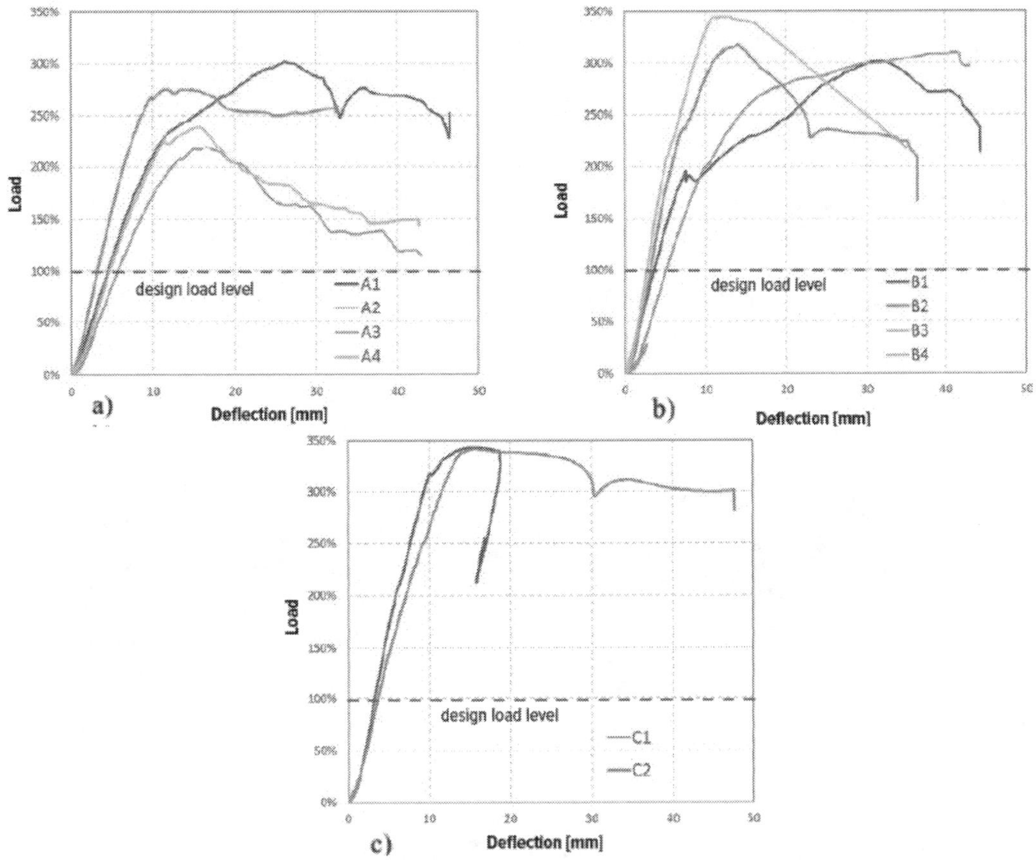


**Figure 4:** Load- deflection curves of specimens. a) Timber slab (A series) b) Composite timber slab (B series) c) Composite timber slab with downstand (C series) [8].

Based on this hypothesis, it can be said, that choosing an inclined edge is an effective solution, also due to more effective load transfer mechanism through a compression arch (Figure 5).

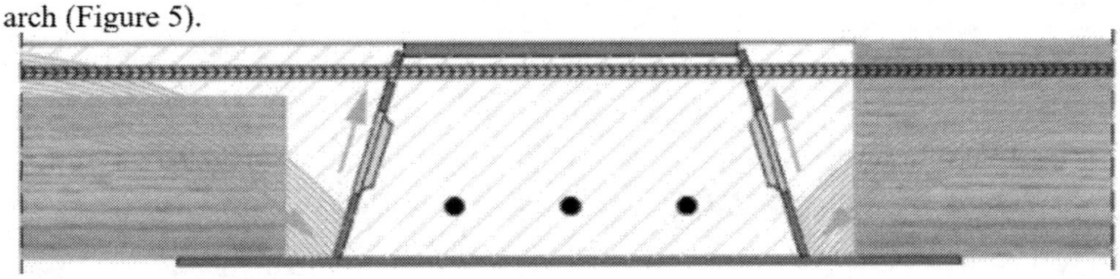

Figure 5: Load transfer mechanism in final situation [7].

## 3 Forthcoming research

Inclined edge will be used in this case, based on the assumption of the appropriateness from previous research mentioned in previous chapter. This part of the experiment will be focused on verifying the behaviour of the connection between slab and beam. Based on this, it is possible to reduce the overall dimensions of the specimens. Reinforcement would be centre of focus. Effectivity of ability of carrying tension forces will by observed. Grooves were cut into the CLT panels for placing the rebars. Effectivity of interaction between rebar, concrete and the shape of groove will be examined.

### 3.1 Test setup

Within of the prepared experiment, it will follow up on the knowledge obtained from the previously mentioned research. In the first phase of case will be used two types of specimens. The first sample is assembled as in the previous research, exactly sample A2. In this case it will have label A (Figure 6 a)). Based on the similarity of the specimens, comparable results are expected. This sample will continue to be considered as a reference sample. CLT panels have similar thickness (200 mm), each of them with a 5-layer assembly. This panels have grooved longitudinal notches (in the direction of span) for rebars, reinforcement is going through the DELTABEAM D20-400 to the panel on the other side of the beam. At the end, the hollow parts are filled with concrete C 20/25. By growing notch in shape U is created wedge part of CLT panel and is activated too. Rebars are bend in corresponding shape with notches too. Based on this hypothesis, sample B was created (Figure 6 b). Each specimen will be examined with corresponding test set-up.

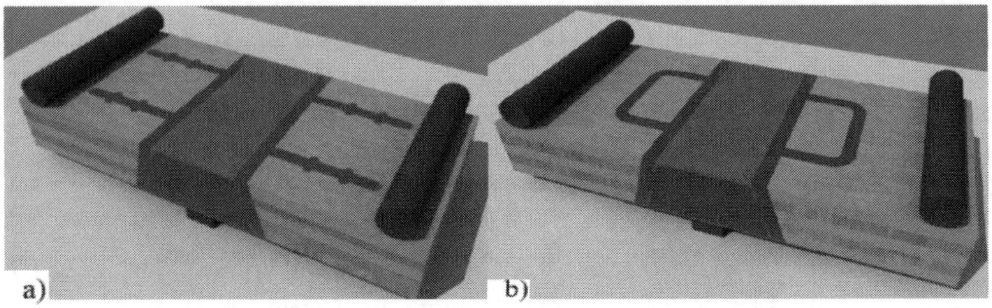


**Figure 6:** Specimens set-up. a) sample A b) sample B.

Support is applied centric under the beam. The samples will be loaded with line load applied at a distance from the centre line of DELTABEAM that equals three times the thickness of the slab, so to maximize the shear in the area close to the beam (Figure 7). Each specimen has an applied the design load level corresponding to a 6 x 6 m grid with a load of 3 kN/m². Load was two times increased to final load of 9 kN/m² during numerical analysis.

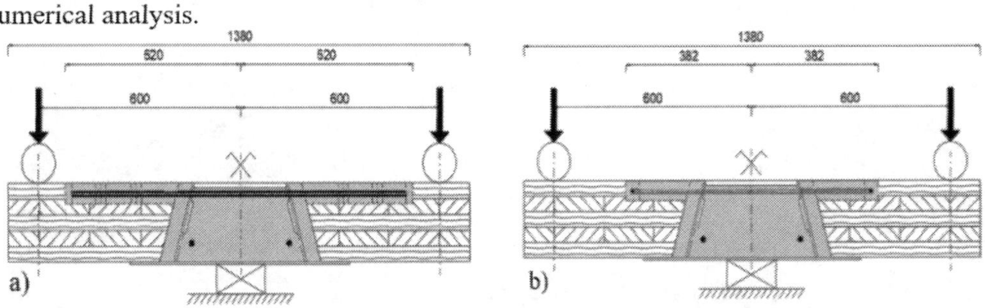


**Figure 7:** Test set- set up and specimens dimensions. a) Sample A b) Sample B.

### 3.2 Numerical analysis

Numerical models have been created for each variant. Detailed analyses were made in program ANSYS Workbench [9]. Equal methodology of test set-up was used. By monitoring their behaviour through deformation is the most suitable way of comparing the specimens. Specimen A was used as reference (Figure 8), and it is primary used to compare samples. During analysis was examined that application of wedge (B) was not so effective as it was predicted (Figure 9).

**Figure 8:** Global deformation of specimen A.

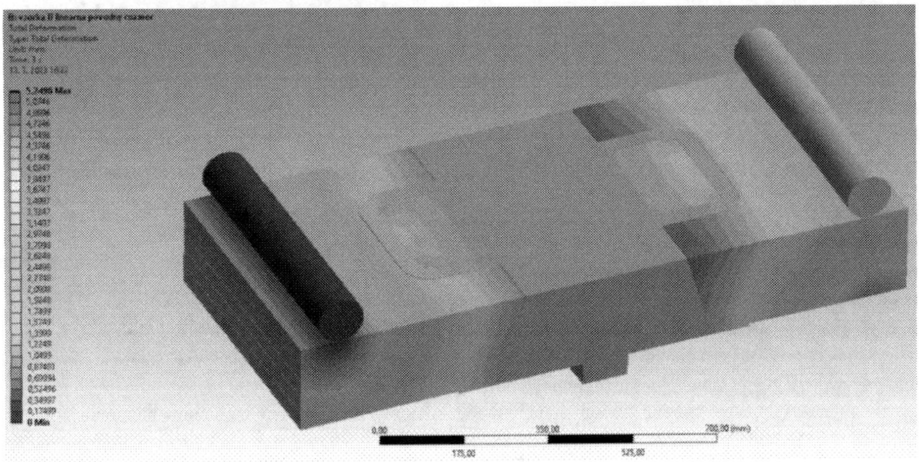

**Figure 9:** Global deformation of specimen B.

Due to this result the length of wedge was modified to maximal its length. This editing had as consequence to increase of effectivity of connection. It is possible to observe in Figure 10.

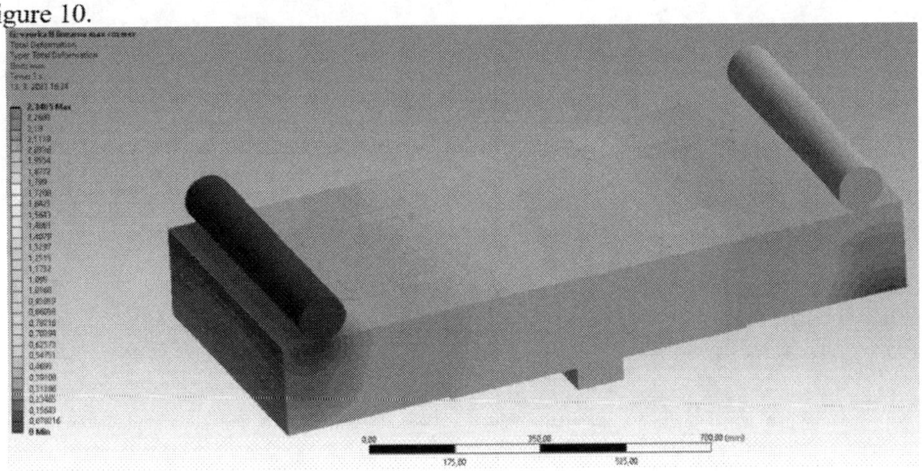

**Figure 10:** Global deformation of modified specimen B (B2).

# 4 Results and discussion

The effectiveness of the inter-connection within the sample A was confirmed. Longitudinal rebars are acceptable, provided that the role to take horizontal tension forces is provided. Especially it is necessary to take to account the requirement of sufficient anchoring length (Figure 11).

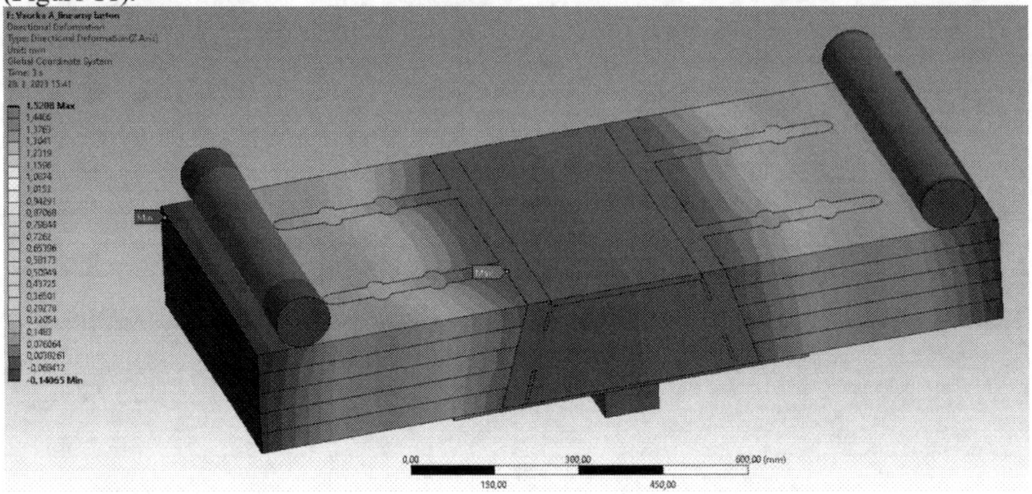


**Figure 11:** Vertical deformation of specimen A.

During the research process it is possible to state that the initial considered shape of sample B is not very effective (Figure 12). However, by adjusting the position of the groove, it is possible to increase the efficiency of the connection (Figure 13). Ensuring that the groove is occupied even beyond the shear area achieved better connectivity in the composite joint. As it is visible in Figure 13, deflection decreased beside to primary B sample. After modification of sample B, i.e. B2, it is possible to state similar resistance of the connection as in the case of sample A.

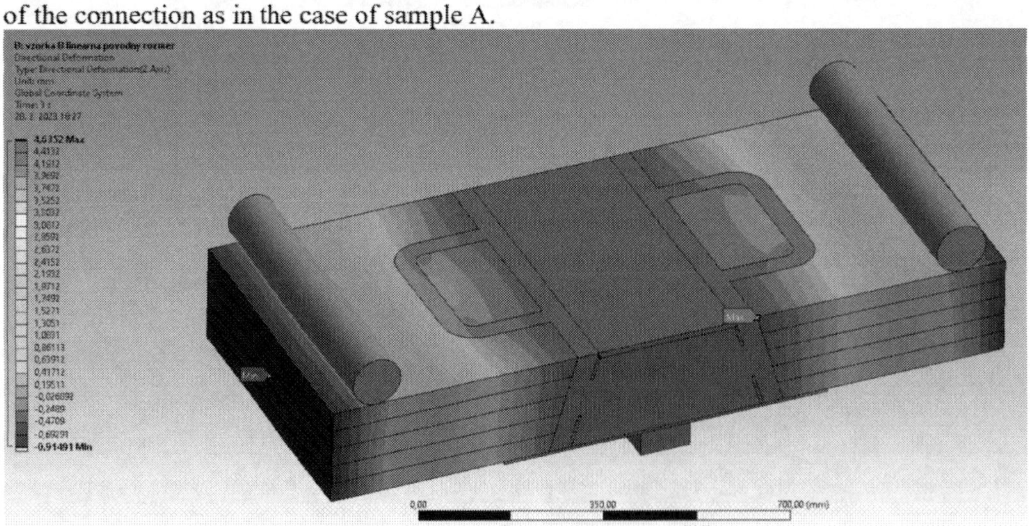


**Figure 12:** Vertical deformation of specimen B.

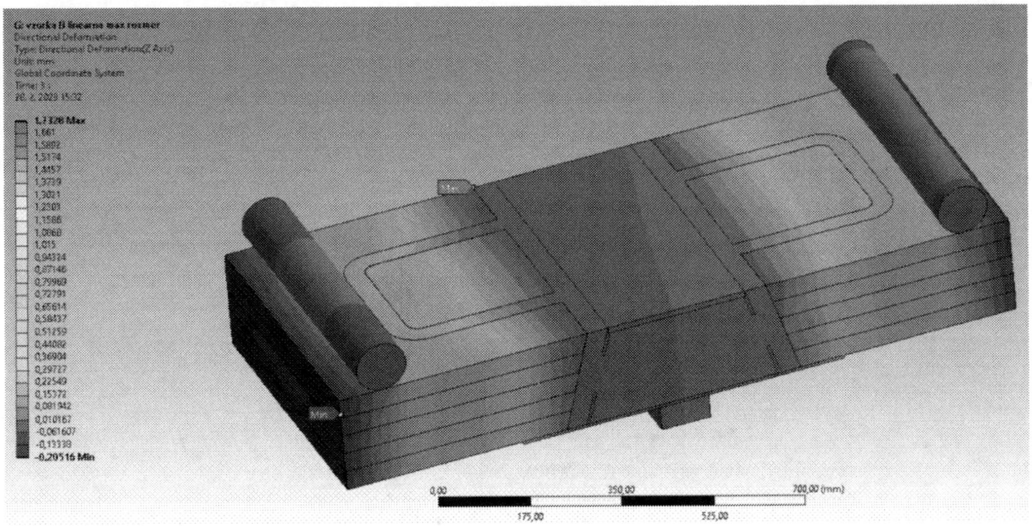


**Figure 13:** Vertical deformation of modified specimen B (B2).

As it is obvious from previous Figures, specimen A has maximal deflection 1.52 mm and when it is compared to sample B with vertical deformation 4.64 mm, it is obvious that sample A is more tough. After the adjustment of sample B can be stated that the vertical deformation decreased to 1.73 mm. Approximately similar values can be stated as predictor of functional joint connection for updated sample B. Specimen B2 can be stated as legit alternative for slim floor solution. This type of connection sufficiently demonstrates the capability to create inter connection of panels of each side of the DELTABEAM, and effect of continuous deck can by stated.

Based on these results, it can be concluded that the resistance of each joint variation is able to withstand the applied loads. However, it will be necessary to verify these hypotheses based on laboratory experiments which are in process at this time. However, the modification of the connection in the joint demonstrated the possibility of creating an adequate alternative join connection able to withstand during load transfer process which they occur in section around middle support for continuous deck. Of course, the statement has to be verified by series of laboratory experiments.

## 5 Conclusion

This paper is dealing with problematic of creating an alternative of joint in hybrid ceiling structure. The intention was to achieve a sufficient inter-connection and obtain considerable stiffness so that the effect of the continuous deck could be considered as relevant.

The result of the analysis can be evaluated from several points of view. First of all, creating alternative of the original joint connection with higher shear resistance in longitudinal direction with beam was main reason of creating sample B. Also, was necessary that alternative was able to withstand other loads applied during life of structure. It has been proven, that this is an option to pursue. However, it will be necessary to complete these analyses and continue with the laboratory experiments so that it is possible to say with certainty, that this joint connection is suitable for further use in practice.

This paper has been supported by the grant for motivation and support increasing the quality and efficiency of the scientific research activities of young researcher workers 2022.

## References

1. A. Ataei, A. A. Chiniforush, M. Bradford, H. Valipour, Cyclic behaviour of bolt and screw shear connectors in steel-timber, Journal of Constructional Steel Research, pp. 328-340, (2019)
2. S. Pargáč et al., Efektívny návrh stropných konštrukcií nemocnice v Českých Budejoviciach (Effective design of ceiling constructions of the hospital in České Budějovice), STATIKA STAVIEB 2020: collection of contributions 26th conference, pp. 168-173, (2022)
3. ArcelorMittal - LongProducts Sections and Merchant Bars, Slim-Floor - An innovative conncept for floors, (2017)
4. F. Y. Moritani, C. E. J. Martins, A. M. P. G. Dias, A Literature Review on Cold-Formed Steel-Timber Composite Structures, BioResources, (2021)
5. StoraEnso, StoraEnso Wood Products - Building Solutions, rethink, (2015)
6. S. Pargáč, Použitie DELTABEAM nosníkov v stropoch drevostavieb (Usage of DELTABEAM beams in ceilings of timber buildings). STATIKA STAVIEB 2020: collection of contributions 25th conference, pp. 292-296, (2020)
7. Peikko, DELTABEAM Slim Floor Structure - With Timber Construction, (2022)
8. Peikko, PEIKKO WHITE PAPER - DELTABEAM WITH HYBRID TIMBER FLOORS - LOAD TRANSFER TESTS, (2022)
9. ANSYS, Ansys Workbench 2021 R2, www.ansys.com, (2021)

MATEC Web of Conferences **385**, 01029 (2023) https://doi.org/10.1051/matecconf/202338501029
*Young Scientist 2023*

# The calibration of NLFE model for the assessment of the maximum punching capacity of flat slabs

*Natalia* Gregusova[1*], and *Jaroslav* Halvonik[2]

[1]Faculty of Civil Engineering, Slovak University of Technology in Bratislava, Radlinskeho 11, 810 05 Bratislava, Slovak Republic
[2]Faculty of Civil Engineering, Slovak University of Technology in Bratislava, Radlinskeho 11, 810 05 Bratislava, Slovak Republic

**Abstract.** The requirements on structural engineers to design more subtle flat slabs are still increasing and therefore they must often design a large amount of reinforcement against punching. The maximum amount of transverse reinforcement is limited by the maximum punching capacity of slab, so the precise determination of the limit is essential. In the second generation of Eurocode 2, the model for the assessment of the maximum possible increase of the punching shear resistance of slab is less empirical and it considers the performance of the punching shear reinforcement. The main objective of the paper is to develop nonlinear finite element model, which would determine as closely as possible the maximum punching capacity of flat slab, strongly reinforced with punching shear reinforcement. So that, an effect of some transverse reinforcement's parameters on the maximum punching shear strength could be investigated quite easily. For this purpose, the software GID was used. The calibrated NLFE model showed quite good agreement with experiment. The difference is less than 8 %, even for slabs reinforced with double-headed studs it did not exceed 1%.

## 1 Introduction

Brittle failure mode of flat slabs is very dangerous type of collapse because it could occur without any warnings. Therefore, the completely understanding of flat slab's behaviour is significant and plenty of research regarding the flat slab's behaviour have occurred. The requirements for structural engineers to design subtler structural members slowly become a commonplace. Consequently, regarding the flat slabs, the requirement for the higher amount of punching shear reinforcement occurs. One of the most used alternatives how to increase punching shear resistance of flat slab is to add reinforcement against punching. However, at some point further increase of an amount of shear reinforcement does not lead also to the further increase in punching resistance. This limit is expressed as the maximum punching capacity of flat slabs. In the current version of Eurocode 2, it is determined as the $k_{max}$ multiple

* Corresponding author: natalia.gregusova@stuba.sk

of the punching strength without any shear reinforcement and the resistance of concrete diagonal strut. Factor $k_{max}$ is an empirical coefficient influenced only by the type of punching shear reinforcement. It could be a little bit disturbing that the determination of such a significant value as the maximum punching shear capacity of slab is still clearly empirical and it considers only the type of punching shear reinforcement. So, the performance of shear reinforcement and thus parameters as an anchorage efficiency or reinforcement detailing are not considered in the calculation. In addition, some countries modify its value in their national annexes and the differences are significant. In the second generation of Eurocode 2, which has been prepared, the determination of it is not clearly empirical anymore. The factor $k_{max}$ is replaced by coefficient $\eta_{sys}$, which considers the distance between the first row of punching shear reinforcement and the face of column and its anchorage efficiency.

The main purpose of this paper is to calibrate non-linear model, which would help to investigate the influence of punching shear reinforcement's performance on the maximum punching capacity. It would be also possible to modify previous mentioned parameters to investigate what effect it has quite easily.

## 2 Calibration of non-linear model according to the experiment

The calibration of NLFEA model was carried out according to the experiment, performed by Stefan Lips at École Polytechnique Fédérale de Lausanne (2012) [1]. In his dissertation thesis, he dealt with flat slabs, supported by square column, and reinforced with a large amount of punching shear reinforcement. He tested up to 16 specimens in total, but for this paper, the results only of six slabs were used. All slabs are in square shape with dimensions 3m x 3m. Top layer of bending reinforcement is ø 20/100 and bottom layer ø 10 / 200 mm. Each slab has a thickness of 250 mm. Two specimens, PV1 and PL3 are without any punching shear reinforcement, PL7 and PL8 are reinforced with double-headed studs, PF2 and PF3 are reinforced with stirrups shaped into cages. One of the variables is the column size $a_c$. Other characteristics of the compared specimens are shown in the Table 1.

**Table 1.** The specimens used for comparison [1].

| S | $a_c$ | $d_{eff}$ | $f_c$ | $f_y$ | $f_{yw}$ | $\rho_{eff}$ | $\rho_{w,eff}$ | PSR | PSR ø |
|---|---|---|---|---|---|---|---|---|---|
| | [mm] | [mm] | [MPa] | [MPa] | [MPa] | [%] | [%] | | [mm] |
| PV1 | 260 | 210 | 34.0 | 583 | - | 1.50 | - | - | - |
| PL7 | 260 | 197 | 35.9 | 583 | 519 | 1.59 | 0.93 | STUDS | 14 |
| PF2 | 260 | 208 | 30.4 | 583 | 536 | 1.51 | 0.79 | STIRRUPS | 10 |
| PL3 | 520 | 197 | 36.5 | 583 | - | 1.59 | - | - | - |
| PL8 | 520 | 200 | 36.0 | 583 | 519 | 1.57 | 0.85 | STUDS | 14 |
| PF3 | 520 | 209 | 37.1 | 583 | 536 | 1.50 | 0.79 | STIRRUPS | 10 |

**PSR – Punching shear reinforcement*

The specimens were supported by steel plate with dimensions of column. The load was applied by four hydraulic jacks, and it was distributed to the slab by eight steel rods. The used test set up is shown at Figure 1.

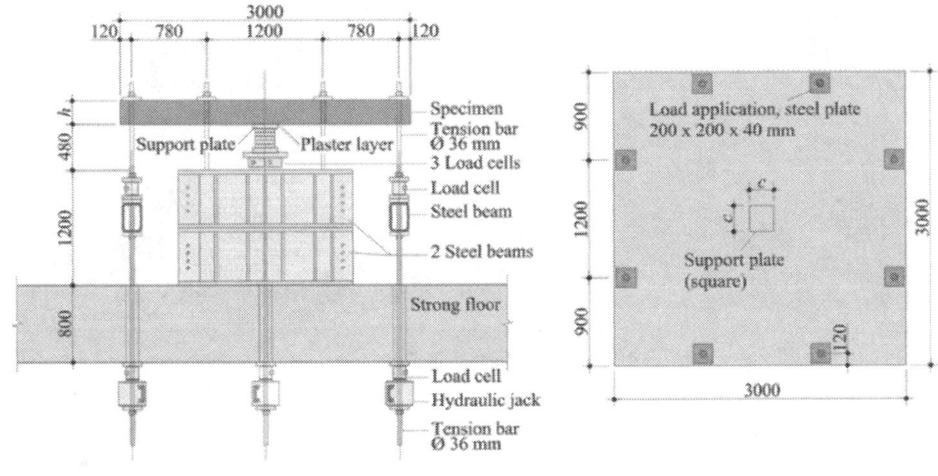


**Fig. 1.** Test – set up [1].

## 3 Nonlinear finite element analysis (NLFEA)

For nonlinear finite element analysis, the software GID was used. This software is used for prediction of behaviour of reinforced concrete structures, considering for instance formation of cracks, crushing of concrete, or yielding reinforcement. GID uses the volumes for creating 3D object, surfaces for 2D objects and lines for modelling 1D elements [2]. To every previous mentioned element can be assigned material with certain properties (for the calibration exact properties from Table 1 were used). The slab and plates were modelled as volumes, with assigned material SOLID Concrete. The most important variable input was cylinder concrete strength $f_c$, which influences fracture energy and tensile strength significantly. For steel plate, which presents the column support, SOLID Elastic material was used. The same material was also assigned to the next two steel plates on the top surface of slab. There were modelled only due to right distribution of load to the slab, so their self-weight was neglected.

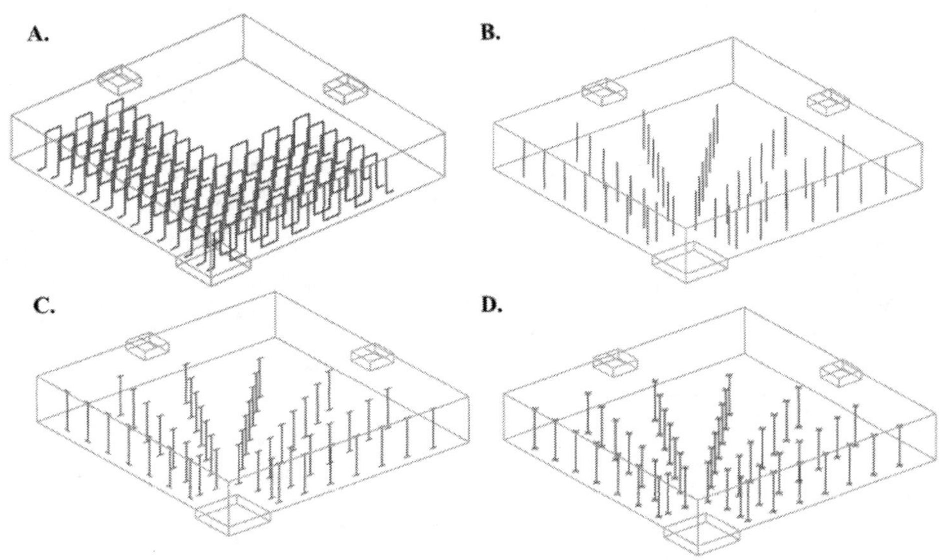


**Fig. 2.** Models of punching shea reinforcement, **A**-Stirrups, **B**-Studs (0), **C**- Studs (-), **D**-Studs (+).

Flexural and punching shear reinforcement was modelled as 1D reinforcement, see Figure 2. In total, three models were created for each specimen reinforced by double headed studs. The efficiency of modelling methods of double-headed stud is described in Section 4.

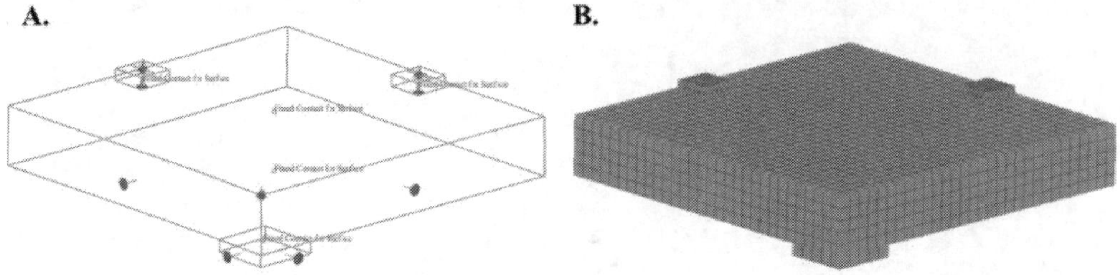


**Fig. 3. A** – support conditions and monitor points of slab, **B** – hexahedra finite element mesh.

Due to the double symmetry of specimens, only a quarter of it was modelled. In the two top steel plates, the prevention of movement in "z" direction is applied. The other support conditions of analysed slabs are based on the symmetry of slab, see Figure 3. For the definition of load history GID uses "Intervals"[2]. The application of load was divided into two intervals, with only self-weight of slab acting in the first one. In the second interval the load is represented by displacement, acting at the column in the bottom-up direction. To simulate the experiment as much as possible, the load was applied to the slab directly through the plate and thus column was not modelled. The last and very important step is the determination of finite element mesh. For the slab meshing the hexahedra finite element type was chosen. According to the reference [3], to obtain sufficiently reliable results, it is enough to divide the slab thickness into five finite elements layers and further mesh refinement does not have so significant influence on a quality of the analysis results. For the calculation the standard Newton-Raphson iterative technique is used.

## 4 Analysis of results

Compared specimens can be dived into three groups, the specimens without any punching shear reinforcement – Table 2, the specimens strongly reinforced by stirrups –Table 3 and the specimens strongly reinforced by double-headed studs – Table 4.

**Table 2.** The specimens without any punching shear reinforcement.

| SPECIMEN | $a_c$ | $V_{Exp}$ | $V_{GID}$ | $V_{Exp}$ / $V_{GID}$ |
|---|---|---|---|---|
| **PV1** | 260 mm | 974 kN | 899.2 kN | 1.08 |
| **PL3** | 520 mm | 1324 kN | 1279.7 kN | 1.03 |

**Table 3.** The specimens reinforced with stirrups.

| SPECIMEN | $a_c$ | $V_{Exp}$ | $V_{GID}$ | $V_{Exp}$ / $V_{GID}$ |
|---|---|---|---|---|
| **PF2** | 260 mm | 1567 kN | 1673.4 kN | 0.94 |
| **PF3** | 520 mm | 2242 kN | 2178.5 kN | 1.03 |

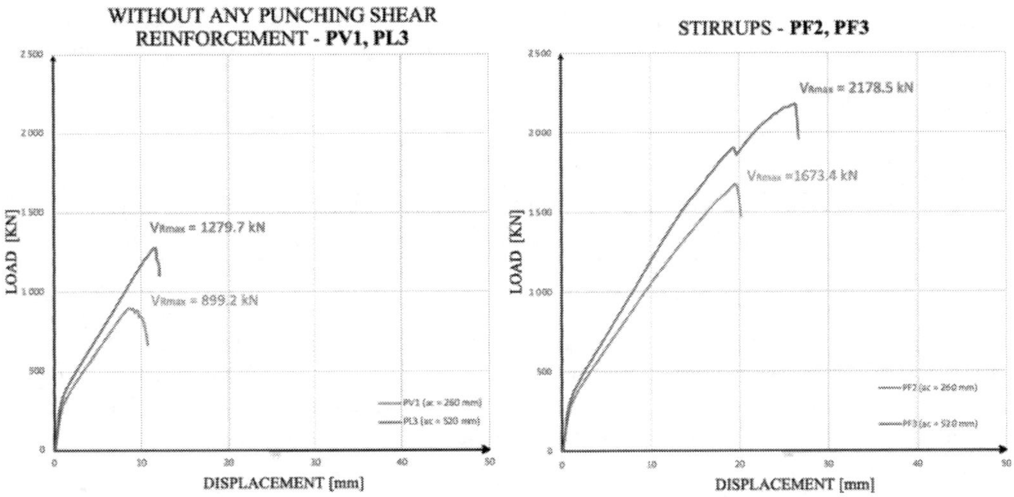


**Fig. 4.** Load – displacement diagrams of specimens without punching shear reinforcement and reinforced with stirrups.

During the calibration of model, the most challenging was modelling of double-headed studs, because of the consideration of head's anchorage. At first, only vertical part of stud was modelled – the specimens PL7 (0) and PL8 (0). This model neglected the contribution of the heads to the anchoring efficiency of the double-headed studs, which understandably led to underestimation of the slab's punching shear capacity. For the consideration of the anchorage of the heads of studs, the short rod was modelled perpendicular to the main part of stud. At first, the short bars were only in direction "x" – PL7 (-) and PL8 (-). Final model of double headed-stud has short bars in two orthogonal directions "x" and "y" – PL7 (+) and PL8 (+). Efficiency of individual methods of modelling double-headed studs can be seen at the Figure 5.

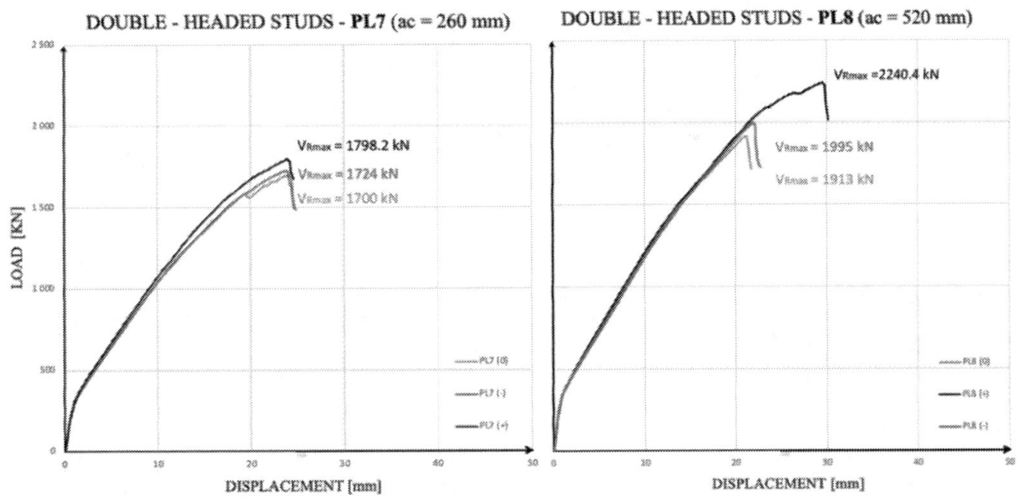


**Fig. 5.** Load -displacement diagram of specimens reinforced with double-headed studs.

Finally, the quite accurate model of slabs reinforced by double-headed studs was calibrated. As it can be seen in Table 4, the difference between experiment's results and NLFEA model is not more than 1%.

**Table 4.** The specimens reinforced with double – headed studs.

| SPECIMEN | $a_c$ | $V_{Exp}$ | $V_{GID}$ | $V_{Exp} / V_{GID}$ |
|---|---|---|---|---|
| **PL7** | 260 mm | 1773 kN | 1798.2 kN | 0.99 |
| **PL8** | 520 mm | 2256 kN | 2240.3 kN | 1.01 |

According to the reference [1], all specimens, except PL8 failed in punching shear. The test of specimen PL8 was stopped before flexural failure occurred. In the Figure 6, principal total strain and crack formation of analysed specimens is displayed. According to NLFEA, the first group of analysed slabs, PV1 and PL3 failed in punching. It was concluded because of Figure 6.C and the fact that bending reinforcement was not yielding. For stirrup-reinforced specimens PF2 and PF3 (Figure 6.B), it was not possible to determine an exact failure mode, but according to experiment the slabs failed in punching shear. For PL7 conical and very steep shear crack formed near the support (Figure 6.A.). This is typical for punching shear failure. As mentioned at the beginning of the paragraph, the slab PL8 failed in flexure it is confirmed by Figure 6.A2 and the fact that unlike the shear reinforcement, the bending reinforcement was yielding.

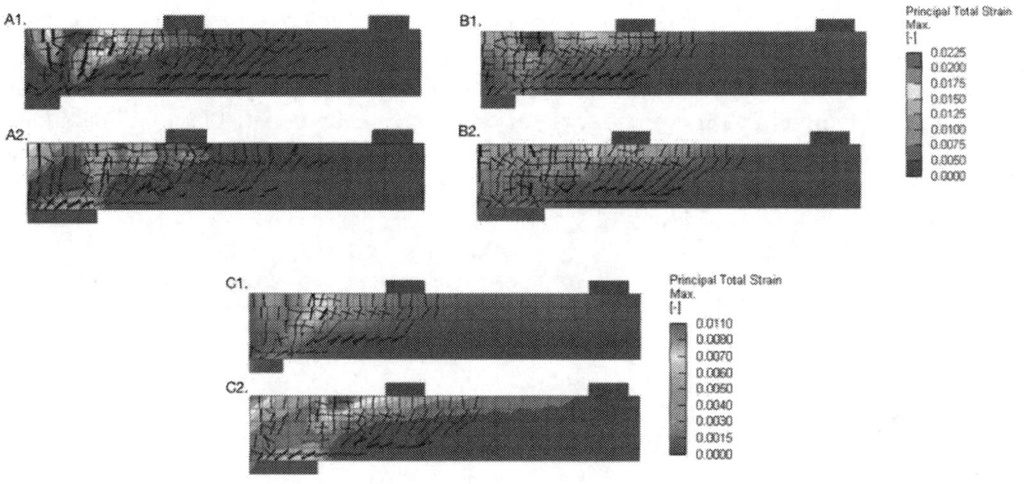


**Fig. 6.** Crack formation and principal total strain of failed specimens. 6.A – specimens reinforced with double-headed studs, 6.B – specimens reinforced with stirrups, 6.C – specimens without any punching shear reinforcement.

## 5 Conclusions

Based on comparison of NLFEA model's results and results obtained from the experiment, the following conclusions can be drawn:

- Regarding the flat slabs without any shear reinforcement, a deviation of up to 8% was found.
- There is a possibility to create non-linear model, in software GID, which can determine the maximum punching capacity of flat slab with punching shear

reinforcement quite accurately. For slabs reinforced with stirrups, the difference was maximum of 6%. Model for slabs reinforced with double-headed studs has shown to be the most accurate – the difference was maximum of 1%.

- Although, we can achieve pretty good agreement between the real maximum punching capacity of specimens and NLFE model specimens' maximum punching capacity, the rotation was in real experiment much larger than NLFE model showed.
- Comparison of deflections was not performed, because information about displacement tested specimens were not available. Therefore, it would be appropriate to develop another calibration, where deformations would be investigated, too.

This work was supported by the Scientific Grant Agency of the Ministry of Education, Science, Research and Sport of the Slovak Republic and the Slovak Academy of Sciences Contract No. VEGA 1/0310/22.

## References

1. S. Lips, Punching of Flat Slabs with Large Amounts of Shear Reinforcement Stefan LIPS (2012)
2. V. Červenka, J. Červenka, Z Janda, ATENA Program Documentation Part 8 User's Manual for ATENA-GiD Interface (2021)
3. Ľ. Kormošová, J. Halvonik, T. Augustin, Uncertainties in nonlinear modelling of punching, Engineering Mechanics, Institute of Thermomechanics of the Czech Academy of Sciences, 294–297 (2020)

# The assessment of the hydrological regime of the Torysa River by the flashlines index

*Patrik* Nagy[1*], *Katarzyna* Kubiak Wojcicka[2], *Miroslav* Garaj[1], and *Milan* Gocic[3]

[1]Department of Environmental Engineering, Faculty of Civil Engineering, Technical University of Kosice, Kosice, Slovakia
[2]Department of Hydrology and Water Management, Faculty of Earth Sciences and Spatial Management, Nicolaus Copernicus University, Torun, Poland
[3]University of Nis, Faculty of Civil Engineering and Architecture, Aleksandra Medvedeva 14, 18000 Nis, Serbia

**Abstract.** Climate change has a significant impact on the local and regional hydrological regime. In this study, the hydrological regime and variability of flows in the Torysa river basin were investigated for the period 1973-2020. The flows were evaluated by the RBI index. The results showed that the Torysa river has a large flow variability.

## 1 Introduction

Climate change has a significant impact on the local and regional hydrological regime, which affects social and environmental systems [1–3]. Researching the development of the frequency of hydrometeorological variables on the longest possible timescales provides information on hydrometeorological characteristics and, consequently, on the probability of their occurrence in the future [4–8] e.g. for water management, the period of low flows limits the use of surface waters, the abstraction of water for various purposes, the use of flows for energy purposes, navigation, etc.

This study evaluates the spatial and temporal variability of flow and identifies flood periods using daily flow values from the upper reaches of Torysa (Slovakia) for the period 1973-2020. The flashlines index, also known as the Richard-Baker index (RBI), was used to assess the hydrological regime and flow variability of the Torysa River.

## 2 Material and Methods

### 2.1 Study area

The investigated area covers the Torysy mountain basin in eastern Slovakia (Fig. 1). Torysa is a river in eastern Slovakia, an important left-hand tributary of the Hornád. It drains a basin of 1,349 km$^2$, has a length of 129 km and is an upland-lowland type of river. It springs in the Levočské vrchy below Škapová, about 5 km northwest of the village of

* Corresponding author: patrik.nagy@tuke.sk

Torysky, at an altitude of about 1,080 m a.s.l. m. At first it flows to the southwest, later it turns to the north and soon to the east. Near Lipany, it leaves the Levočské vrchy, gradually turns to the southeast and continues through the valley between Čergov (in the northeast) and Bachurňa (in the southwest). In Prešov, it takes on its most important tributary, the Sekčov, and flows through the Košice basin to the south, already in the form of a lowland river with many meanders. Southeast of Košice, near the village of Nižná Hutka, it flows to the left into Hornád. There are 3 hydrological stations on the river: Prešov, Sabinov and Brezovica – Torysa [9].

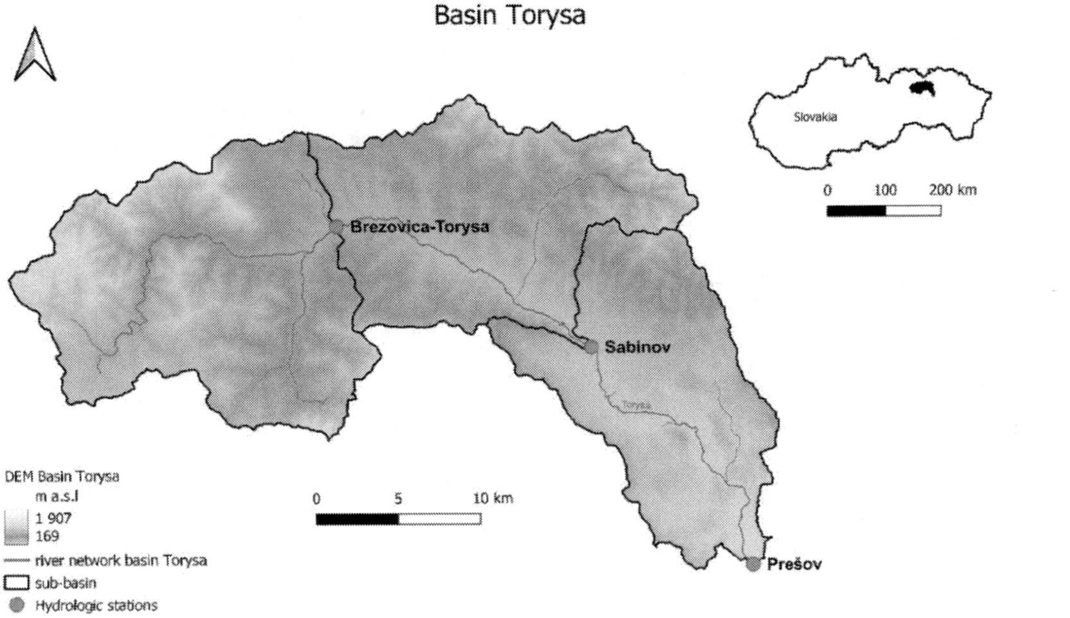


**Fig. 1** map basin Torysa.

**Table 1.** Basin Torysa elevation.

| Hydrological station | Catchment area (km²) | Elevation of the catchment area (m a.s.l.) | | |
|---|---|---|---|---|
| | | maximum | median | minimum |
| Prešov | 667.34 | 1261.08 | 599.59 | 236.28 |
| Sabinov | 476.19 | 1261.08 | 698.59 | 319.99 |
| Brezovica-Torysa | 263.50 | 1261.08 | 796.85 | 421.73 |

## 2.2 Methods

The variability of the flows of the upper section of the Torysa was investigated on the basis of data obtained from the Slovak Hydrometeorological Institute. The data included daily flow values obtained for 3 hydrological stations: Prešov, Sabinov and Brezovica-Torysa for the period 1973-2020. Based on daily values, quarterly values were calculated according to seasons. The degree of volatility of flows in seasonal terms was determined based on the volatility of flows *Cv* and the flashliness index, Richard-Bakes Index (RBI), which characterizes the short duration of the phenomenon. The flow variation index *Cv* was calculated as the standard deviation of all daily flow values for each season divided by the mean annual flow. The flamboyance index proposed by Baker et al. [10] was used to determine the rate of change in short-term flows, which according to the authors is called the RB Index (RBI).

$$RB\ index = \frac{\sum_{i=n}^{n}|q_i - q_{i-1}|}{\sum_{i=1}^{n} q_i} \quad (1)$$

Where: $q$ - average daily flow, $i$ - day, $n = 365$ (366)

RBI is a dimensionless index that ranges from 0 to 2 [10]. Zero represents constant flow, while a higher index value indicates increased flow frequency. Statistical calculations were performed using the 'R' program. The aim of this study was to analyze the long-term changes in the flow of the Torysy River for the period 1973-2020. The degree of seasonal flow variability was determined based on the flow variability index *C*v and the flare index (RBI), which characterizes the short duration of the phenomenon.

## 3 Results

The analysis of average periods of RBI values showed that in the period from 1973 to 2020, the RBI values at the Prešov, Sabinov and Brezovic-Torysa hydrological stations ranged from 0 to 0.53, while at the Prešov and Brezovica stations the RBI indicator ranged from 0 to 0.53, which indicated. high changes in short-term flows. In the analyzed multi-year period 1973-2020, the highest RBI values were recorded in the summer period. The lowest RBI values were recorded in the winter period.

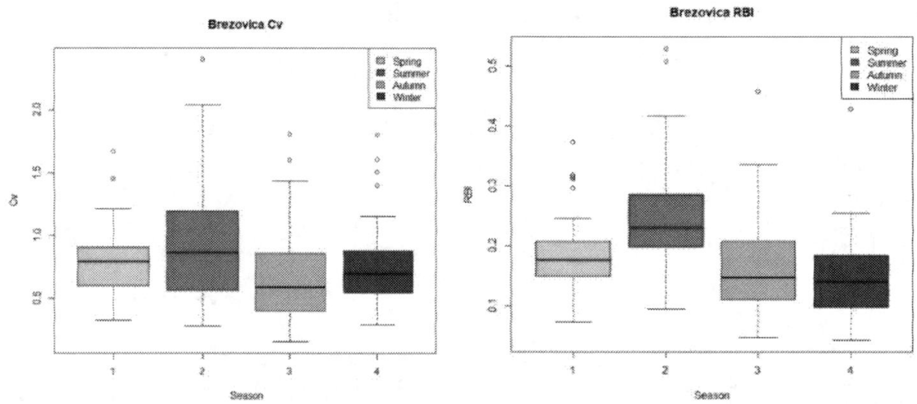


**Fig. 2.** Broezovica CV and RBI.

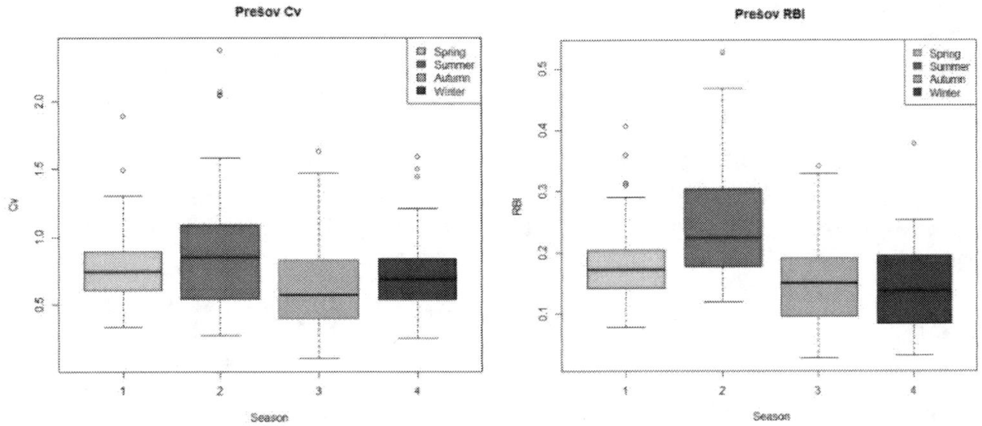


**Fig. 3.** Prešov CV and RBI.

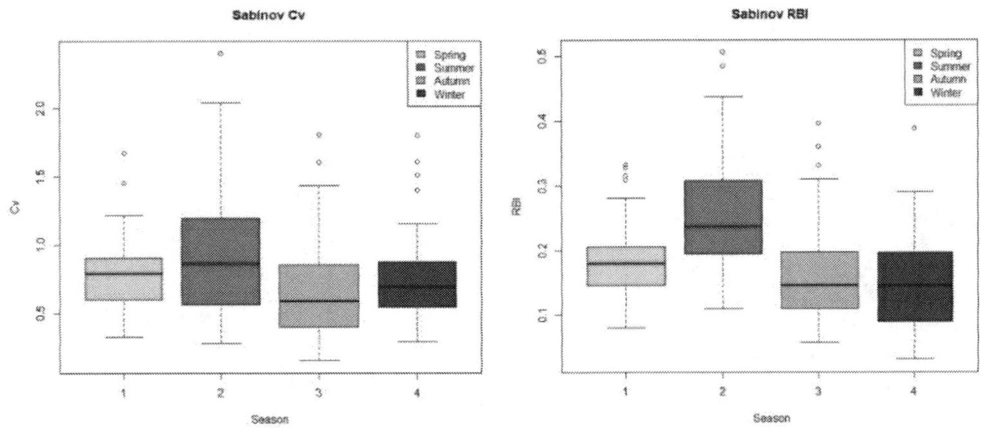


**Fig. 4.** Sabinov CV and RBI.

Figures 2-4 show the distribution of *C*v and RBI index for seasonal periods. y (RBI> 0.4) recorded in summer at all observed stations.

# 4 Conclusion

The study examined the variability of flows in the Torysy river basin in three hydrological stations where daily flows were studied, according to the seasons. The data covered the years 1973-2020, on a daily basis. The RBI index was used for evaluation. Higher annual amplitudes of RBI values were recorded in the summer period. The result shows that the Torysa river has the character of a hot river and has a great variability of flows.

This work has been supported by Slovak Research and Development Agency under the Contract no. APVV-20-0281 and SK-SRB-21-0052 Innovative approaches to the assessment and management of drought risk due to climate change.

## References

1. IPCC (International Panel on Climate Change). Managing the Risks of Extreme Events and Disasters to Advance Climate Change Adaptation. A Special Report of Working Groups I and II of the Intergovernmental Panel on Climate Change. Cambridge University Press, Cambridge, UK, and New York, NY, USA, **582** (2012)
2. C. Papadaki, E. Dimitriou, *River Flow Alterations Caused by Intense Anthropogenic Uses and Future Climate Variability Implications in the Balkans,* Hydrology, **8**, 7, (2021)
3. P. Nagy, I. Minea, M. Zelenáková, *Baseflow in Hornád basin on the eastern Slovakia*, (2021)
4. K. Kubiak-Wójcicka, P. Nagy, M. Zeleňáková, H. Hlavatá, H. F. Abd-Elhamid, *Identification of Extreme Weather Events Using Meteorological and Hydrological Indicators in the Laborec River Catchment, Slovakia, Water, 13*(10), 1413, (2021)
5. X. Huang, D. L. Swain, *Climate change is increasing the risk of a California megaflood,* Science Advances, **8**, 32, (2022)
6. B. G. Johnson, C. S. Morris, H. L. Mase, P.S. Whitehouse, C. J. Paradise, *Seasonal flashiness and high frequency discharge events in headwater streams in the North Carolina Piedmont (United States),* Hydrological Processes, **36**, e14550, (2022)
7. L. Holko, J. Parajka, Z. Kostka, P. Skoda, G. Bloschl, *Flashiness of mountain streams in Slovakia and Austria,* Journal of Hydrology, **405**, 392-401 (2011)
8. M. Hanel,A. Vizina, P. Máca, J. Pavlásek, *A multi-model assessment of climate change impact on hydrological regime in the Czech Republic*, J. Hydrol. Hydromech. 2012, **60**, 3, 152–161 (2012)
9. E. Mazúr, M. Lukniš, *Geomorphological units (Geomorfologické jednotky). M 1: 1 000 000*, Country Atlas of the Slovak Republic. Ministry of the Environment of the Slovak Republic, Bratislava & Slovak Environment Agency, **88** (2002)
10. D.B. Baker,R. P. Richards, T.T. Loftus, J.W. Kramer, *A new flashiness index: characteristics and applications midwesterns rivers and streams,* Journal of the American Water Resources Association, 503-522 (2004)

# Experimental testing of 3DCP reinforcement Wall segments

*Vojtech* Vanek[1*], *David* Beckovsky[1], *Ondrej* Nespesný[1], *Jan* Vystrcil[1], and *Jan* Pencík[1]

[1]Faculty of Civil Engineering, Institute of Building Structures, Brno University of Technology, Veveri 331/95, 602 00 Brno, Czech Republic

**Abstract.** The wall segments are formed using additive layer-on-layer technology. The concrete mix is therefore laid in layers and the final desired element is gradually formed. The manufactured columns are all of different shapes and reinforcements. The columns are adapted for lateral load testing, whereby they are subjected to three-point bending when loaded in service. A stress of 1 MPa is transmitted to these columns and the columns are progressively loaded with a horizontal force. The test is primarily to determine how much stress these columns can carry in the wall plane in tension bending. The results of the measurements are compared with brick masonry.

## 1 Introduction

Additive manufacturing (AM), or the incremental method, was developed in the early 1980s by Charles W. Hull [1]. The main invention, called stereography, is believed to be the first incremental method using UV light. Subsequently, the knowledge of this technology was deepened, resulting in the development of new AM methods. These may be the fabrication methods called Fused Deposition Modelling (FDM) [2], Selective Laser Melting (SLM) [3]. These methods are based on the 3D printing method. AM is an incremental method, and it is not only from 3D printing as it is promoted. An example of an incremental method is Laminated Object Manufacturing (LOM) [4].

These fabrication methods have recently been joined by a method of 3D printing concrete, the Paste Extrusion Modelling (PEM) method [5]. The fabrication method is based on robotic arms, frame printers and gantry printers. The first such printers were made in 2000 [6], [7]. Print heads with nozzles are then placed on the endpoints, either active [8] or passive [9]. Most of the methods are very similar, but the fabrication of different structural elements depends on the method used. Each method contains new possibilities and research in the field.

However, this is why creating structures using additive technologies is so popular in the world. In the literature Huang S. et al. [10], forty-two structures that were produced using 3D printing are listed. Mostly, these are single-storey civil engineering buildings that are manufactured on site at the future location of the building, or these elements are manufactured by companies inside the halls as prefabricated parts, which are then transported

* Corresponding author: Vojtech.Vanek2@vutbr.cz

to the construction site where they are subsequently connected to form a perfect uniform structure. They are also used in road structures such as bridge structures, where they can be printed to support bicycle roads and bridge irregularities in the terrain [11].

In civil engineering, additive technology is most used for wall structures. Vertical load bearing structures include all wall and column structures that are at a perpendicular or partially acute or obtuse angle to the vertical tangent of the structure. 3D printed concrete structures are mostly primarily used for these vertical structural elements and structures. When comparing these 3D printed concrete vertical structures with conventional methods of construction such as small- and large-scale masonry systems using mini-cranes, timber buildings, or solid concrete structures, emphasis must be placed on speed of construction, assessment for load-bearing capacity and reliability, thermal engineering, and more.

This paper focuses on determining the maximum forces on different geometries when compared with brick masonry. These comparisons will also include different reinforcement elements and other possible reinforcement.

# 2 Material and Methods

## 2.1 Material

The test specimens were made from precast mixtures that were mixed at a ratio of 1:2. The modified mix was applied to the outer perimeter of the structure, and the material properties are shown in Table 1-2. The mixing was done to add cellulose fibres in one of the mixes. Without this combination, the structure would have been weakened and larger shrinkage cracks or even destruction of the specimen could have occurred. The inner filler was made from a prefabricated material designed for 3D printing with the properties of see Table 3.

**Table 1.** Material mechanical characteristic of one peace prefabricate mixture.

| | |
|---|---|
| Grain size | <1.0 mm |
| Flexural tensile strength after 28 days | min. 8 MPa |
| Compressive strength after 28 days | min. 40 MPa |
| Density | 2150 kg.m$^{-3}$ |

**Table 2.** Material mechanical characteristic of two peace prefabricate mixture.

| | |
|---|---|
| Grain size | <1.0 mm |
| Flexural tensile strength after 28 days | min. 6 MPa |
| Compressive strength after 28 days | min. 30 MPa |
| Density | 2000 kg.m$^{-3}$ |

**Table 3.** Material mechanical characteristic of prefabricate mixture to the infill.

| | |
|---|---|
| Grain size | <0.5 mm |
| Flexural tensile strength after 28 days | min. 7 MPa |
| Compressive strength after 28 days | min. 50 MPa |
| Density | 2000 - 2200 kg.m$^{-3}$ |

The wall structures are reinforced with concrete ladder steel. The steel is marked B500B. The steel therefore has mechanical properties in tensile strength 550 MPa and yield strength 500 MPa. The steel members will form the reinforcement baskets located inside the columns.

## 2.2 Design and Manufacturing of Segments

The design geometry is based on test blocks of brickwork. The experimental wall segments were designed with basic dimensions of 340 mm thickness, 790 mm width and 1000 mm specimen height. The width of the printed footprint was set to 45 mm, with a contact area of 30 mm according to the output passive print head. The height of the printed trace based on the flow rate and the speed of the robotic arm was set to 15 mm. One printed layer consists of one trace, i.e. a single perimeter printing method was used. The figures given are the same for all test wall segments. The proposed variants are then subdivided according to the internal reinforcement elements. These are round reinforcement columns with an internal diameter of 120 mm or square columns with internal dimensions of 110 × 110 mm. The geometry of the variant segments can be observed in Fig. 1.

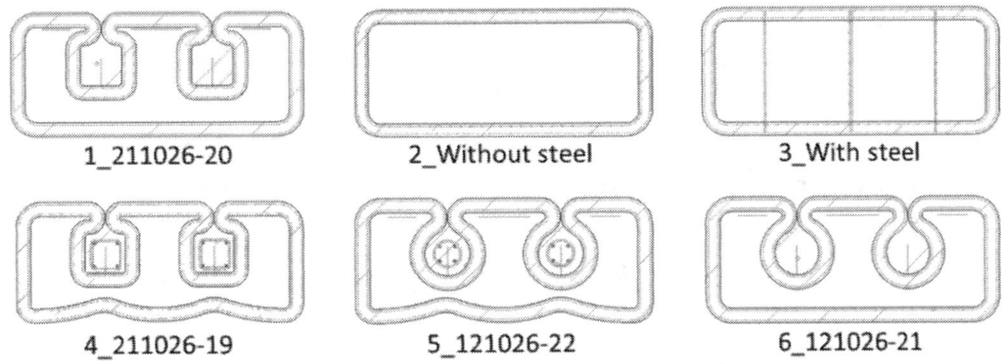


**Fig. 1.** Geometry of the 3DCP wall segments variants.

After creating the geometry in the Rhino modelling program, the modelled segments were inserted into a special program. Here the segments were divided and edited for printing. Printing was then carried out on a KUKA robotic arm. A so-called passive print head [9] was used for printing. The extrusion of the material was guided in a flexible pipe by a screw feeder, which was placed outside the robotic arm. The printing speed was set to an average of 150 mm.s$^{-1}$. The created and prepared segments were transported to the test laboratory.

Reinforcement baskets were fabricated for the internal columns using reinforcing steel with 8 mm diameter main members and 6 mm split reinforcement. The reinforcement baskets were made in two variants. The square baskets were made in the size of 85 × 85 × 1000 mm and the circular baskets were made with a diameter of 80 mm and a height of 1000 mm as can be observed in Fig. 2. Subsequently, they were inserted into the hollow reinforcement columns and were cast with precast compound after 3D printing.

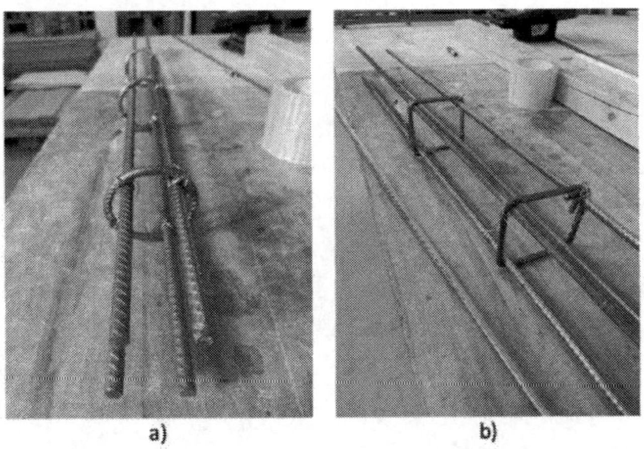


**Fig. 2.** Reinforcement cage made of concrete steel a) circle, b) square.

### 2.3 Testing Method

The test of the experimental segments was designed for three-point bending (3PB). The testing was carried out according to Czech norm ČSN 73 2030. The measurements were oriented vertically. The support points were made of I-shaped steel sections on which 50 mm wide steel washers were placed. The axial distance of the supports was set at 700 mm. The whole load assembly is shown in Fig. 3.

The loading was induced in the horizontal direction and the applied force was applied perpendicular to the wall plane. The applied force was applied by a hydraulic linear actuator SCA 65. The loading was carried out at a constant strain increment $s_l = 0.083$ mm.s$^{-1}$. The measuring range of the external pressure transducer at the end of the hydraulic actuator was 0-100 kN. Measurement of the deflection path was carried out using potentiometric cable path sensors. The transducers recorded the deflections of the experimental specimens at four points 15-18. As can be seen from Fig. 3, the sensors were located at half the distance of the supports in the pulling region and two were located above the supports in the pushing region. The potentiometers and the hydraulic press were connected to a central control panel which was connected to a PC. Eight readings were recorded for each sample during the measurement process. The sampling frequency of the control panel was set at 5 Hz.

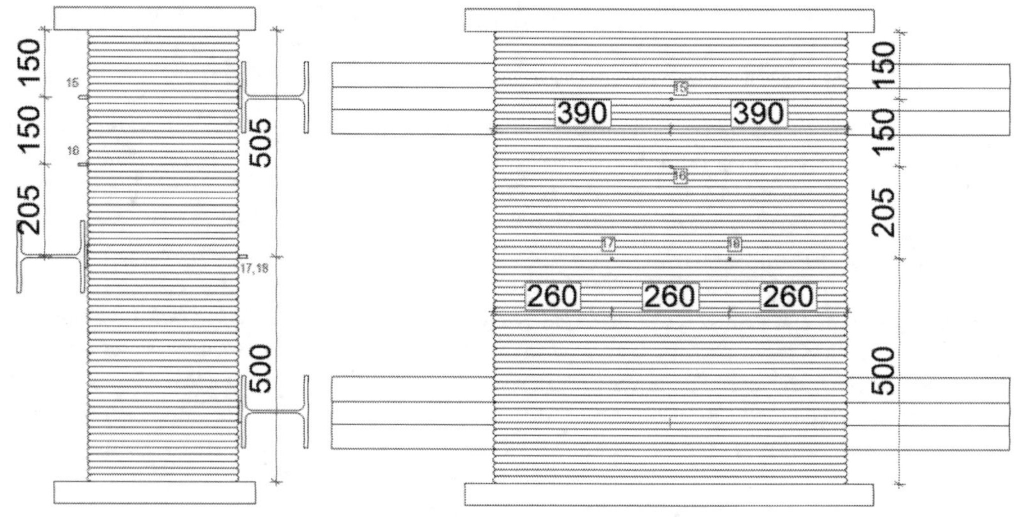


**Fig. 3.** Schematic of the 3DCP wall segment test load system.

## 3 Results and Discussion

### 3.1 Geometry of Specimens

Data evaluation was performed systematically. Input values such as height, width and length of segments were recorded before measurement. The recording of these values was done using a conventional tape measure with a preference to 1.0 mm. Since the segments are organic in shape for accuracy of measurement a 3D scanner was used. It was used to record the top printed area of the segments. The read area was then used in the calculation and determination of the force for the 1 MPa loading. Fig. 4. shows the loading scheme and the real prepared test just before the measurement.

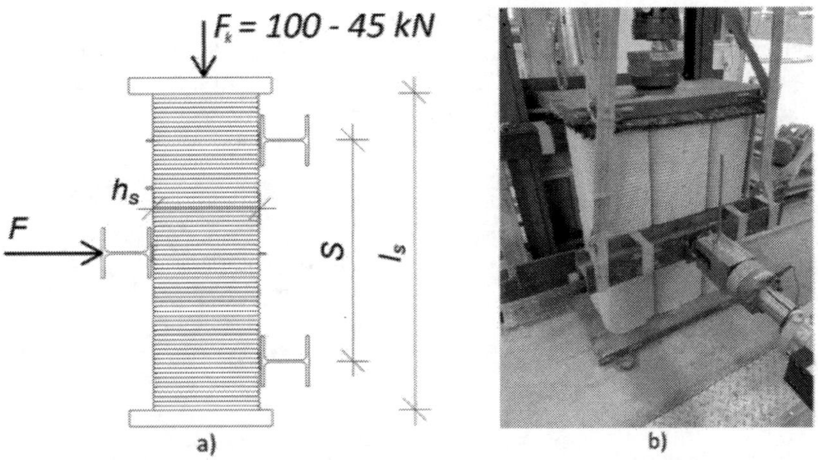


**Fig. 4.** Loading style, a) Geometry diagram; b) Real measurement; where $F$ [kN] is force, $h_s$ [mm] is hight of specimen, $S$ [mm] is distance of supports, $l_s$ [mm] is length of specimen and $F_k$ [kN] is force of attraction.

### 3.2 Segment Behaviour Characteristic

The results of the test were recorded in a chart. In Fig. 5, the measured data from all experimental segments are shown. The curves of each test sample are different at first glance. The diagram describes on the horizontal axis the deflection of the specimen in the middle of the span, with the measurement points 17/18. The values from the measured points 17/18 were averaged. Segment 02 and 03 have many times lower strength than segments 01,04-06. This behaviour is due to the absence of internal reinforcing columns. The graphical representation shows the different behaviour of segments 02 and 03, where it is confirmed that the internal reinforcement has an effect on the force transmitted by the stiffeners and thus the strength of the specimen itself is increased. The behaviour is confirmed by other samples. Segments 01,02,04-06 did not contain horizontal reinforcement and therefore did not have as much deflection of the segments on the drawn part as test specimen 03.

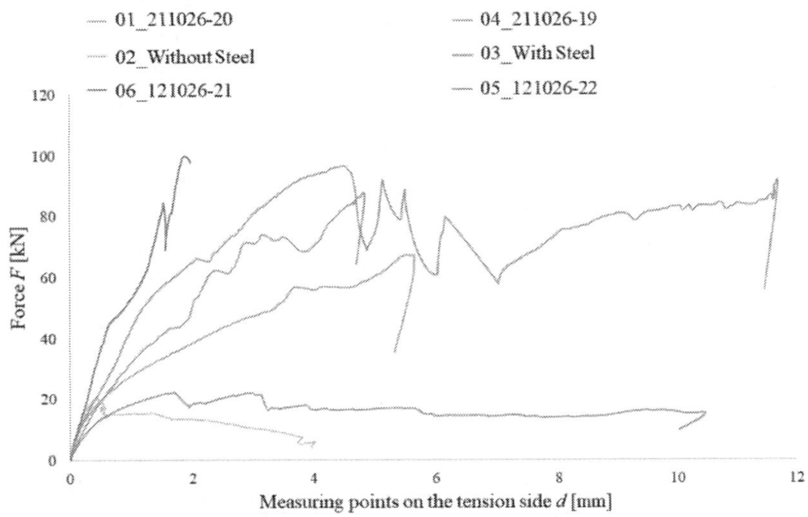


**Fig. 5.** Graphical representation of the dependence of the applied force on the deflection of the specimens. The dependence of horizontal reinforcement in the specimen is evident from test specimen 03 compared to 01,02,04-06.

The maximum achieved forces from the measurements were determined. In the initial loading condition, linear behaviour can be observed for all tested segments. The initial behaviour is linear in 80 % of the cases up to 16 kN. Subsequently, microcracks and certainly cracks are likely to occur which weaken the segments, as can be seen in Fig. 6. During further loading of the structures, stiffening elements start to appear.

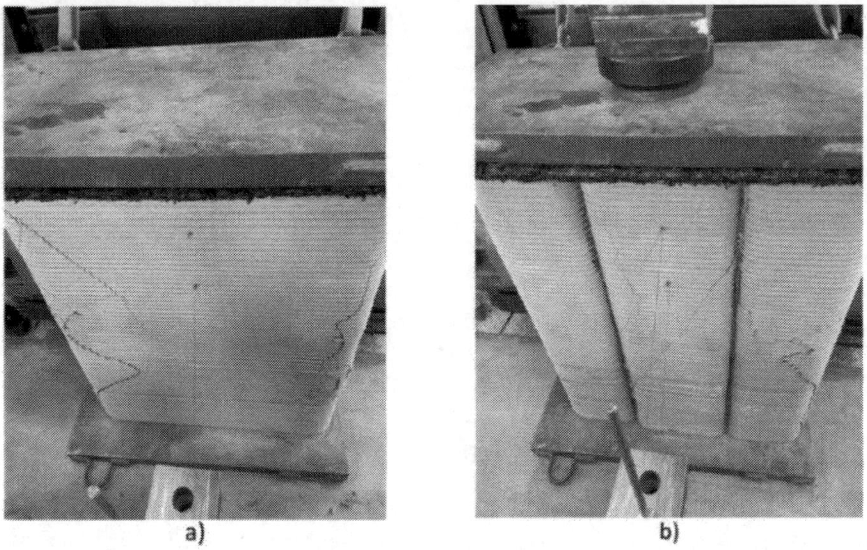

**Fig. 6.** Experimental specimens after completion of 3PB test a) without internal reinforcing columns 02/03; b) with reinforcing internal columns 01/04/05/06.

The experimental specimens labelled 2 and 3 are several times weaker in terms of strength relative to the segments with internal columns. The characteristic behaviour of these segments is in their internal reinforcement. A higher deflection can be observed for the segment with steel reinforcing bars. The maximum loads differ by a percentage difference of 10 % when comparing specimens 2 and 3.

The reinforcing bars present in segments 1,4,5,6 contribute to several times higher strength of the segments. Four times higher force absorption can be observed than in samples 2 and 3. Therefore, Table. 4 has been created, where the maximum loads achieved are presented in conjunction with the maximum deflection of the specimen on the tensile side and the force at initiation of the first cracks.

**Table 4.** Maximum and initiating forces applied to the test specimen segment.

| Lable | Max Force $F_{max}$ [kN] and Deformation $d_{max}$ [mm] | | Initiating force $F_{inc}$ [kN]: |
|---|---|---|---|
| 4_211026-19 | $F_{max}$ = | 87.88 kN | $F_{inc}$ = 73.04 kN |
| | $d_{max}$ = | 4.8367 mm | |
| 1_211026-20 | $F_{max}$ = | 67.27 kN | $F_{inc}$ = 56.64 kN |
| | $d_{max}$ = | 5.5123 mm | |
| 6_121026-21 | $F_{max}$ = | 99.75 kN | $F_{inc}$ = 45.13 kN |
| | $d_{max}$ = | 1.9143 mm | |
| 5_121026-22 | $F_{max}$ = | 96.60 kN | $F_{inc}$ = 65.71 kN |
| | $d_{max}$ = | 4.4988 mm | |
| 3_With steel | $F_{max}$ = | 22.26 kN | $F_{inc}$ = 22.26 kN |
| | $d_{max}$ = | 1.6989 mm | |
| 2_Without steel | $F_{max}$ = | 20.29 kN | $F_{inc}$ = 20.29 kN |
| | $d_{max}$ = | 0.4299 mm | |

Brick masonry was used to compare the results. The masonry achieved a characteristic flexural tensile strength at failure in the plane parallel to the bearing joints of $f_{xk} = 0.36$ N.mm$^{-2}$. Percentage comparisons between the values of the brick masonry and 3DCP wall segments are shown in Table 5.

**Table 5.** Calculation and comparation of flexural tensile strength from maximum force.

| Lable | $w_s$ [mm] | $S$ [mm] | $h_s$ [mm] | $F_{max}$ [kN] | $f_{mean}$ [N.mm$^{-2}$] | 3DCP $f_{xk}$ [N.mm$^{-2}$] | Brick $f_{xk}$ [N.mm$^{-2}$] | Procentage |
|---|---|---|---|---|---|---|---|---|
| 4_211026-19 | 790 | 700 | 340 | 87.88 | 1.01 | 0.67 | 0.36 | 185% |
| 1_211026-20 | 790 | 700 | 340 | 67.27 | 0.77 | 0.52 | | 140% |
| 6_121026-21 | 790 | 700 | 340 | 99.75 | 1.15 | 0.76 | | 210% |
| 5_121026-22 | 790 | 700 | 340 | 96.60 | 1.11 | 0.74 | | 200% |
| 3_With steel | 780 | 700 | 335 | 22.26 | 0.27 | 0.18 | | 45% |
| 2_Without steel | 780 | 700 | 335 | 20.29 | 0.24 | 0.16 | | 40% |

where $w_s$ [mm] is wight of specimen, $f_{mean}$ [N.mm$^{-2}$] is the average bending strength.

# 4 Conclusion

Measurements were made on 3DCP wall segments made by ICE Industrial Services a.s. The testing was performed in the AdMaS centre where the test stand was set up. The load was applied perpendicular to the wall plane. Constant loading was performed by a hydraulic press. The segment loading was set at 1MPa.

Experimental measurements showed that the internal reinforcing columns have a great influence on the maximum load of the test elements. When loaded, wall segments without internal reinforcing columns show up to four times lower strength than reinforced columns.

Wall segments without internal reinforcing columns were created with the same geometry. The internal reinforcement of the structure is formed by the internal reinforcement of one of the columns, which makes them different from each other. This fact is reflected in the deformation of the wall. The specimen with the reinforcement has a higher deflection in the tensile region. The maximum force applied to the segment was determined by percentage difference with a value of 10%.

The test specimens with internal reinforcement columns, behave very similarly in failure. The results were compared according to round and square internal reinforcing columns. The two variants were divided according to the reinforcement. The segments with round columns were judged to be within 5% difference in maximum load when comparing the reinforcement variation. The square reinforcing columns were judged to be within 25% percentage difference when the same assessment was made. All percentage values were based on one measurement. Multiple test bodies need to be tested for stronger static value and probability.

This paper was written with the financial support of specific university research project FAST VUT, FAST-S-23-8413 (Data support for the design of green facades and roofs with integrated photovoltaic panels) and FAST-J-23-8344 (Investigation of the characteristic behaviour of reinforced elements and structures made by additive technology using a 3D concrete frame printer).

## References

1. I. Hager, A. Golonka, R. Putanowicz. Proc. Engi. **151**, 292-299 (2016)
2. J. Lluch-Cerezo, R. Benavente, M.D. Meseguer, S.C. Gutiérrez. Proc. Manu. **41**, 890-897 (2019)
3. E. Yasa, J. Deckers, J. Kruth. Rap. Prot. Jour. **11**, 312-327 (2011)
4. B. Dermeik, N. Travitzky. Adva. Engi. Mate. **22**, 24 (2020)
5. F. Hu, T. Mikolajczyk, D.Y. Pimenov, M.K. Gupta. Mate. **14**, 22 (2021)
6. I. Aydin, F.R. Biglari, B.J. Briscoe, C.J. Lawrence, M.J. Adams. Comp. Mate. Scie. **18**, 141-155 (2000)
7. S. Lim, R. Buswell, T. Le, R. Wackrow, S. Austin, A. Gibb, T. Thorpe. Proc. o. t. 28 ISA. 665-670 (2011)
8. A. Kazemian, X. Yuan, R. Meier, E. Cochran, B. Khoshnevis. Addi. Manu. **2**, 5 (2017)
9. T.T. Le, S.A. Austin, S. Lim, R.A. Buswell, A.G.F. Gibb, T. Thorpe. Mate. A. Struc. **45**, 1221-1232 (2012)
10. S. Huang, W. Xu, Y. Li, Fron. o. Arch. Rese. **11**, 653-669 (2022)
11. T.A.M. Salet, Z.Y. Ahmed, F.P. Bos, H.L.M. Laagland. Virt. a. Phys. Prot. **13**, 222-236 (2018)

# Sensing technologies for construction productivity monitoring

*Daria* Kempecova[1*], and *Maria* Kozlovska[1]

[1]Department of Technology, Economics and Management in Civil Engineering, Faculty of Civil Engineering, Technical university of Kosice, Vysokoskolska 4, Kosice 042 00, Slovakia

**Abstract.** Construction is a complex and dynamic industry that requires constant monitoring and evaluation to ensure project success. In recent years, the need for accurate and real-time performance data has become more pressing, making traditional evaluation methods based on manual data collection and direct observation less effective. To address this challenge, sensor technologies have emerged as a promising solution for construction performance monitoring. The objective of this article is to present a comprehensive overview of the current state of sensor applications for construction performance monitoring, including recent advancements in this field. Additionally, the article will examine the advantages and disadvantages of sensor technologies in the context of construction performance monitoring.

## 1 Introduction

The issue of construction productivity has become increasingly pressing in recent years. Conventional methods of performance evaluation in construction, which are based on manual data collection and on-site observations, are both time-consuming and susceptible to human error. To address this problem, the utilization of sensor-based monitoring and documentation technologies has gained significant attention in the construction industry and academic circles [1].

Sensing technologies refer to a range of devices and systems that can be used to gather and transmit data from construction sites. These technologies include wireless sensors, GPS systems, laser scanning devices, and computer vision technologies. They work by capturing data on various aspects of construction, such as material movement, machine operation, and human activity. This data is then processed and analysed to provide valuable insights into construction performance [2].

Historically, activity performance information has been gathered manually through the use of stopwatches and manual observation, or by analysing recorded video footage of the operation. These manual methods are tedious, time-consuming, and susceptible to errors. Additionally, with the advent of new construction and assembly techniques, previous data becomes obsolete and additional manual effort is required to collect new cycle time data. Automated activity monitoring methods can greatly reduce manual effort in data collection

* Corresponding author: daria.kempecova@tuke.sk

and provide support for effective production planning and monitoring systems. The establishment of an automated activity monitoring and classification system requires the capability to recognize activities occurring in each work area [3]. The hardware for activity classification includes components such as accelerometers, gyroscopes, magnetometers, Bluetooth, battery, Wi-Fi, memory, GPS, and light and temperature sensors. While combining sensors can improve accuracy, it also increases the cost of the hardware [4]. Automated progress monitoring frequently entails incorporating information from various sensors and sources based on the final goal [5].

The physical condition of the workers can be monitored through various sensors, such as electromyogram, electrocardiogram and heart rate can be checked using electrodes, body temperature using adhesive sensors, breathing rate using piezoelectric or piezoresistive sensors, body movement using accelerometers, body orientation using accelerometers, sweat levels through galvanic response sensors, oxygen saturation using pulse oximeter [6]. In the one hand, construction worker fatigue leads to decreased productivity, reduced service quality, and potential accidents, therefore understanding workers' physiological aspects is crucial for improving both safety and productivity [7]. In the other, implementing electronic performance monitoring for workers comes with the challenge of meeting GDPR standards [8]. Before monitoring, workers should be informed and trained on the use of the devices to be used. They must also give their consent for the collection of data regarding their performance during daily tasks, as required by legal and ethical guidelines [4].

In order to increase the success rate of sensor applications, electronic data collection devices must fulfil numerous requirements, such as durability and resistance, return on investment (ROI), energy efficiency and independence, ergonomic size and weight, user-friendly design, scalability, reliability, high data transmission rate, minimal communication infrastructure and data storage, and secure data handling [9]. There are two main methods for recognizing construction equipment activity, which depend on how the primary sensors are utilized. The first approach is called the "active sensor-based approach," where sensors are attached to the equipment or worker being monitored. The second approach is the "passive sensor-based approach," where sensors are not mounted on the equipment being tracked but are external, such as using video and audio information to classify activities [3]. Furthermore, most of the case studies, proposed methods that includes the integration of these technologies with Building Information Modelling (BIM). This state of affairs provoked that the construction industry is presently investing and will continue to invest in three key technologies in the near future: analysis models for BIM dimensions, sensing technology, and business information models. This emphasizes the importance of using sensing techniques and model-based methodologies to achieve better business results [4].

Despite the existence of sensory approaches relating to productivity, there is still a significant gap between the available sensor technologies and the understanding of their application in terms of construction performance evaluation. This gap has not been thoroughly examined, and as a result, there is still a poor understanding of sensors application in construction. Therefore, the aim of this article is in-depth examination of the modern sensing technologies and techniques used for monitoring productivity in construction projects. It provides a classification of sensors, systems, their advantaged and disadvantages, and approaches used for real-time monitoring in both indoor and outdoor construction environments.

## 2 Methodology

A review of the scientific literature was conducted to create a summary of all sensor-based technologies suitable for real-time monitoring of construction. The Web of Science electronic database was used as the main source, with the main keyword for this article being "Sensing technologies." The search was refined in three steps: 1) setting journal selection criteria to choose the most recent articles from 2013 to 2023; 2) adding the keyword "Construction" to the search to focus on technologies used in the construction industry; 3) using the keywords "performance monitoring" and "productivity monitoring" because they are related concepts that are equally important for the success of a construction project. This review only uses open access articles without limitations on language or science category, and documents were identified and selected based on their relevance to this study. The initial search generated up to 135,398 results due to the broad concept of "Sensor technologies," but after the refining steps, 109 articles remained.

During the in-depth review phase, content analysis and interpretation were carried out to reflect the latest developments in the field of sensing technology and fulfill the purpose of the article. After that, only the technologies that were used to monitor construction process productivity, workers, and machines were selected for detailed analysis.

## 3 Sensing technology classification

Classifying sensing technologies can be difficult due to the diversity of techniques and methods involved, as well as the variety of applications and contexts in which they are used. Additionally, many sensing technologies overlap and can be used in combination with each other, further complicating the classification process. As a result, it is important to have a clear understanding of the specific sensing needs and requirements for a given application in order to select the appropriate technology. The study conducted by A.S. Rao et al. provides a comprehensive overview of the different sensing technologies and techniques utilized for real-time monitoring of construction projects [5]. The authors used this classification as a basis for future analysis and summarized it in a Figure 1.

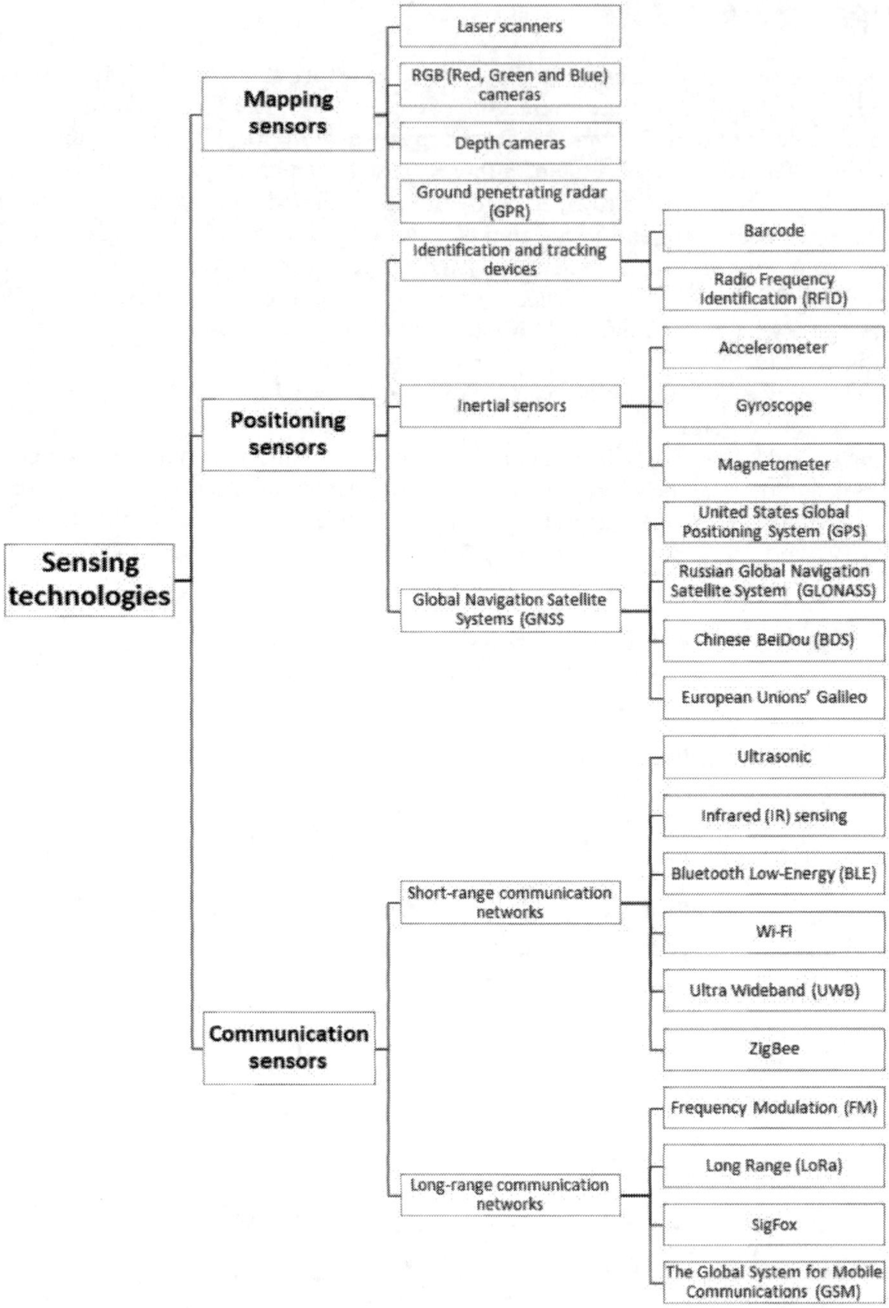


**Fig. 1.** A summary of sensor-based technologies for real-time monitoring of construction.

Since this article deals primarily with the issue of construction productivity, the authors selected and reviewed only those technologies that, according to the literature, is mainly used in the case studies for the purpose of monitoring the performance and productivity of construction. These sensing technologies are included in Table 1 with an analysis of their advantages and disadvantages in relation to construction productivity monitoring.

**Table 1.** The detailed overview of selected sensing technologies.

| **Sensing technology** | **Reference** | **Description** |
|---|---|---|
| Inertial Measurement Unit (IMU) | [5, 10, 11, 12, 13] | IMU is commonly composed of gyroscopes for measuring and transmitting angular rate data, along with accelerometers for measuring and transmitting specific force data.<br>Advantages:<br>• gyroscopes are utilized for measuring angular rate,<br>• accelerometers are utilized for measuring specific force or acceleration,<br>• small size of data<br>Disadvantages:<br>• requires attaching sensors to each resource involved in the activity.<br>• problem of error accumulation, that leads to considerable drift over time<br>• sensitivity to magnetic interference. |
| Bluetooth low energy (BLE) | [14, 15, 16, 17] | BLE sensor technology is a tag/beacon that can operate for many months on a coin cell battery. This technology, can be complemented by online mapping tools, for indoor and outdoor tracking options at reasonable prices. Furthermore, devices can be moved from one place to another at the end of the project or can be easily replaced.<br>Advantages:<br>• low power consumption,<br>• extended range,<br>• compact size,<br>• reduced latency,<br>• superior energy efficiency,<br>• infrastructure flexibility.<br>• cost-effectiveness and ease of installation.<br>Disadvantages:<br>• vulnerable to rapid fading and significant fluctuations in the received signal strength,<br>• need for relocating<br>• measured data size is too large. |
| Light Detection and Ranging (LiDAR) | [18, 19, 20, 21, 22, 23] | LiDAR employs laser light pulses to gauge the distance between the sensor and an object by measuring the time it takes for the light to reflect back. The laser pulses, which are emitted in nanoseconds, enable the rapid scanning of regions and the generation of detailed, precise 3D images that can be used to produce elevation maps.<br>Advantages:<br>• quick data collection,<br>• can be mounted to vehicles,<br>• can survey inaccessible areas,<br>• 3d models can be created from scan data,<br>• get accurate measurements within the general scenes of the site.<br>Disadvantages:<br>• too inflexible for indoor use,<br>• objects and weather can interfere with scanning,<br>• does not work well with difficult shape of objects,<br>• point cloud data registration process cannot be done automatically,<br>• high cost of purchasing and deploying. |
| RGB-Depth | [18, 20] | RGB-D is a technology that combines both color RGB (Red, Green, Blue) and depth information to provide a more comprehensive 3D view of the environment. It is commonly used in computer vision, robotics, and augmented reality applications.<br>Advantages:<br>• provides both depth (D) and color (RGB) data,<br>• better visibility in low-light conditions or areas with visual obstructions.<br>Disadvantages: |

| | | |
|---|---|---|
| | | • equipment cost,<br>• limited range,<br>• privacy concerns. |
| Radio frequency identification (RFID) | [18, 20, 24, 25] | RFID is an example of RSSI-based tracking technology, that utilizes radio waves to recognize and track individuals or items. Can be used for material and tool tracking in indoor construction sites to enhance productivity and control processes.<br>Advantages:<br>• able to detect the presence of objects through walls<br>• can be used indoors.<br>Disadvantages:<br>• chip sensitivity,<br>• metal shielding,<br>• restricted read range,<br>• not suitable for intricate environments,<br>• limited to presence detection or localization,<br>• not allow determining the construction progress status. |
| Ultra-Wideband (UWB) | [14, 26] | UWB is a technology that uses low power radio waves to transmit and receive data over short distances. It has a high level of accuracy in tracking the location of objects, people, and equipment within indoor and outdoor environments.<br>Advantages:<br>• high read range,<br>• low power consumption,<br>• the ability to function both indoors and outdoors.<br>Disadvantages:<br>• material-dependent propagation,<br>• Limited range (up to 100 meters),<br>• multipath interference,<br>• sensitivity to line-of-sight,<br>• difficulties in deploying systems in dynamic environments. |
| Global Navigation Satellite System (GNSS) | [27] | GNSS is a technology that uses a network of satellites to determine the precise location of objects, people, and equipment. It is commonly used in construction productivity tracking to provide accurate location data for equipment and personnel.<br>Advantages:<br>• large coverage area,<br>• provide high-accuracy positioning data,<br>• real-time monitoring,<br>• cost-effective,<br>• easy to install and maintain,<br>• non-intrusive.<br>Disadvantages:<br>• high equipment cost,<br>• limited indoor coverage,<br>• require a significant amount of power to operate. |

| | | |
|---|---|---|
| ZigBee | [5, 14, 28, 29] | Wireless technology based on the IEEE 802.15.4 standard that is used for low-power, low-data rate communication between devices. It is commonly used in the Internet of Things (IoT) and building automation. The introduction of ZigBee wireless networks made it possible to control on-site construction resources. This technology has various applications such as tracking and monitoring the activities of workers on construction sites; tracking resources in indoor construction environments; reducing the risk of accidents due to falling objects to improve safety and prevent accidents.<br>ZigBee consists of three main devices:<br>1. Coordinator (ZC) - the central device that manages the network and coordinates communication between devices. It selects a channel and network number to initiate network operation. If the Coordinator is not present, the network can still operate normally.<br>2. Router (ZR) - a device that relays data between other devices in the network.<br>3. End Device (ZED) - a device that is connected to the network but does not forward data to other devices, such as a sensor or light switch. It does not need to maintain the network structure.<br>Advantages:<br>• low-power,<br>• low-data rate.<br>Disadvantages:<br>• limited by battery life,<br>• line-of-sight issues,<br>• primary appropriate for outdoor use. |

## 4 Results and Discussion

One of the key advantages of sensor technologies is that they can provide real-time data that is much more accurate and reliable than manual data collection methods. This is particularly important in the construction industry, where labour costs are a major factor. Also helps to reduce the risk of human error and enables construction managers to make more informed decisions. Furthermore, the use of sensors can help to improve safety on construction sites, by monitoring hazardous conditions and alerting workers to potential dangers. Additionally, the use of sensors allows for remote monitoring of construction sites, making it possible to obtain real-time information from anywhere, at any time [3].

Despite the many benefits of sensor technologies, there are also some disadvantages to consider. One challenge is the cost of installing and maintaining these systems. The intricacy and dynamism of construction sites, resulting from high levels of worker and material movement, and variations in building types and locations, present significant challenges for tracking technology, both in terms of hardware and algorithmic requirements. There are various limitations of sensors from a hardware perspective, for instance signal strength fluctuation, range, interference, measurement accuracy, and cost. While the cost of sensors has come down in recent years, they are still a significant investment for many construction companies. Additionally, there may be some resistance from workers and construction managers who are used to traditional data collection methods and are hesitant to adopt new technologies. These limitations hinder the deployment of tracking systems in real-world construction environments. The algorithmic perspective of tracking technology may entail the fusion of data from several sensors, the application of noise filtering on the data, and the incorporation of environmental changes. Furthermore, some tracking techniques demand a time-consuming offline configuration phase for the initial system setup and depend on stable environments to ensure consistent tracking [14, 30].

Regardless of this disadvantages, researchers continue to develop frameworks for the application of sensors in construction. For example, automated equipment tracking framework have been proposed by Ergen et al. [31], which utilize radio frequency identification (RFID) and global positioning system (GPS). Vahdatikhaki and Hammad [32] further enhanced this framework by incorporating motion data with equipment location to identify activities. Furthermore, both shallow and deep machine learning methods have been applied to identify equipment activities using IMUs [3].

A hybrid-tracking system, which incorporates Bluetooth Low Energy (BLE) technology, motion sensors, and Building Information Model (BIM), has been developed by Park et al. The study leverages the unique characteristics of each tracking information source, where BLE beacons offer absolute position information but can be impacted by high levels of noise, motion sensors provide relative position information but are susceptible to drift, and building geometry information represents a novel type of knowledge that can enhance tracking accuracy. The tracking system's motion sensor component is not exempt from operational limitations. During position tracking, the motion sensors in mobile devices that are freely reoriented by the user produced noisy data, making it challenging to obtain reliable heading direction and step counts. As a result, in the experiments, the test subjects were instructed to maintain a fixed position and orientation of the mobile device with respect to the body by using a belt clip or an armband throughout the study [14].

Bao et al. [33] utilized long-sequence videos to automatically detect, identify, and track activities performed by an excavator and a dump truck. Additionally, unsupervised classifiers were used to expand the bag-of-video-feature-words model into the construction domain, allowing for the learning and classification of labour and equipment activities [3].

Worker activity recognition has been approached using both active and passive methods, much like equipment activity recognition. Worker's velocity, gravitational forces, and orientation can be determined by electronic devices equipped with a battery or power source and a combination of sensors such as accelerometers, magnetometers, and gyroscopes [4]. However, the leading deep learning method for this task uses large complex neural networks to process data from wearable devices [12]. For instance, in an experimental setup, Akhavian et al. [35] employed machine learning methods and mobile sensors to recognize the activities carried out by a human crew. Nath et al. [34] also utilized mobile technology and machine learning techniques for the automated assessment of ergonomic risk posed to workers. Additionally, a framework based on motion tensor decomposition using IMU was suggested to identify the awkward postures of workers [3].

Rashid et al. conducted a study that presented a system for identifying activities through audio signals. The findings of the research highlight the potential of the proposed system to be utilized for automated monitoring and data gathering in modular construction factories, along with other recognition frameworks that are based on inertial measurement units (IMU) or computer vision (CV) [3].

Cheung et al. developed a system that merges Building Information Modeling (BIM) and Wireless Sensor Networks (WSN) to visually monitor safety conditions on construction sites through a spatial, color-coded interface. The system is capable of automatically eliminating hazardous gases. Numerous wireless sensor nodes were deployed on an underground construction site to collect data on hazardous gas levels and environmental conditions (temperature and humidity). The sensor nodes consist of a gas sensor, temperature and humidity sensor modules, a system main board, a Zigbee wireless transmission module, and a power supply module. In the event of an abnormal status detection, the BIM model will trigger an alarm and activate a ventilator on-site to alert and eliminate the hazard [36].

As tools advance and become more costly, tracking the status of mobile and compact devices is becoming increasingly crucial. However, existing sensors alone cannot intelligently extract meaningful information from their collected data. Thus, the data must be transmitted to the cloud for further processing, resulting in increased communication power requirements, detection latency, and heightened security and privacy concerns [12]. Another conclusion is that much of the research is focused on identifying hazardous conditions at construction sites. Safety is often given more priority over performance and quality in the construction industry as unsafe conditions pose immediate risks to workers, and it's easier to detect and address them. This observation is supported by other sources, including [5]. However, balancing safety and performance is crucial for the success of a construction project, as high-quality work is important to meet client expectations and avoid costly mistakes and delays.

## 5 Conclusion

Sensor technologies have the potential to revolutionize the way construction performance is monitored and evaluated. By providing accurate, real-time data, they can help to improve decision-making, reduce labour costs, and improve safety on construction sites. However, there are also challenges that must be overcome, such as the cost of implementation and data privacy and security concerns. To fully realize their benefits, they must be used in combination with other tools and methods, such as process optimization and project management software. Furthermore, it is important to carefully consider the specific needs of each construction project and select the sensor technologies that are best suited to those needs. Nevertheless, the future of construction performance monitoring is likely to be heavily influenced by sensor technologies. The use of sensor technologies is a promising step forward for the construction industry and is likely to become increasingly important in the years to come.

Acquiring data from a construction site in real-time is just the initial challenge. The next challenge is to interpret and analyse this data. Manually reviewing this data is unrealistic and there is limited domain knowledge. Hence, there is a pressing need to create automated, sturdy, and scalable data analysis methods to extract crucial information from big data for Internet of Things based construction industry applications. Future research should focus on developing advanced data analysis methods specifically tailored to the construction industry. This includes techniques such as machine learning, artificial intelligence, and data mining algorithms that can automatically process and interpret the sensor data collected from construction sites. These methods should be scalable and able to handle big data generated by Internet of Things (IoT) applications. Furthermore, as the use of sensor technologies increases, it is imperative to address data privacy and security concerns. Future research should focus on developing robust security protocols and encryption methods to ensure the integrity and confidentiality of the collected data.

The activity presented in the article is part of the VEGA research grant No. 1/0336/22 "Research on the effects of Lean Production/Lean Construction methods on increasing the efficiency of on-site and off-site construction technologies".

## References

1. A.A. Khairadeen, O.J. Lee, D. Lee, C. Park, *Remote Indoor Construction Progress Monitoring Using Extended Reality*, Sustain. **13**, 2290 (2021).
2. C. Kroppa, C. Koch, M. König, *Interior construction state recognition with 4D BIM registered image sequences*, Autom. in Constr. **86**, 11-32 (2018).
3. M. R. Khandakar, J. Louis, *Activity identification in modular construction using audio signals and machine learning*, Autom. in Constr. **119**, 103361 (2020).
4. D. Calvetti, P. Mêda, M.C. Gonçalves, H. Sousa, *Worker 4.0: The Future of Sensored Construction Sites*, Buildings, 10, 169 (2020).

5. A.S. Rao, M. Radanovic, Y. Liu, S. Hu, Y. Fang, K. Khoshelham, M. Palaniswami, T. Ngo, *Real-time monitoring of construction sites: Sensors, methods, and applications*, Autom. in Constr. **136**, 104099 (2022).
6. U.C. Gatti, S. Schneider, G.C. Migliaccio, *Physiological Condition Monitoring of Construction Workers*. Autom. Constr. **44**, 227-233 (2014).
7. A. Aryal, A. Ghahramani, B. Becerik-Gerber, *Monitoring Fatigue in Construction Workers Using Physiological Measurements*. Autom. Constr. **82**, 154-165 (2017).
8. D. Calvetti, P.N.M. Magalhães, S.F. Sujan, M.C. Gonçalves, H.J. Campos de Sousa, *Challenges of Upgrading Craft Workforce into Construction 4.0—Framework and Agreements*, Proc. Inst. Civ. Eng. Manag. Procure. Law, 1-7 (2020).
9. T. Cheng, U. Mantripragada, J. Teizer, P.A. Vela, *Automated Trajectory and Path Planning Analysis Based on Ultra Wideband Data*, J. Comput. Civ. Eng. **26**, 151-160 (2012).
10. L. Joshua, K. Varghese, *Accelerometer-based activity recognition in construction*. J. Comput. Civil. Eng. **25(5),** 370-379 (2011).
11. G. S. Maruthi, V. Hegde, Application of MEMS accelerometer for detection and diagnosis of multiple faults in the roller element bearings of three phase induction motor, IEEE Sensors J. **16(1)**, 145–152 (2016).
12. M. Giordano, N. Baumann, M. Crabolu, R. Fischer, G. Bellusci, M. Magno, *Design and Performance Evaluation of an Ultralow-Power Smart IoT Device With Embedded TinyML for Asset Activity Monitoring*, IEEE Trans. Instrum. Meas. **71**, 2510711 (2022).
13. F. Li, C. Zhao, G. Ding, J. Gong, C. Liu, F. Zhao, *A reliable and accurate indoor localization method using phone inertial sensors,* in Proc. ACM Conf. Ubiquitous Comput. – (UbiComp '12), 421. (2012).
14. J. Park, J. Chen, Y. K. Cho, *Self-corrective knowledge-based hybrid tracking system using BIM and multimodal sensors*, Adv. Eng. Inf. **32**, 126-138 (2017).
15. J. Zhao, O. Sepp¨anen, A. Peltokorpi, B. Badihi, H. Olivieri, *Real-time resource tracking for analyzing value-adding time in construction*, Auto. Construct, **104** 52–65 (2019).
16. L. Mohanty, S. Chae, Y. Yang, *Identifying productive working patterns at construction sites using BLE sensor networks*, Dev. in the Built Env. **4**, 100025 (2020).
17. R. Faragher, R. Harle, *An analysis of the accuracy of bluetooth low energy for indoor positioning applications*, in: Proc. 27th Int. Tech. Meet. Satell. Div. Inst. Navig. (ION GNSS+ 2014), 201-210 (2014).
18. S. El-Omari, O. Moselhi, *Integrating automated data acquisition technologies for progress reporting of construction projects*, Auto. Construct. **20 (6)**, 699–705 (2011).
19. S. Dixit, *Impact of management practices on construction productivity in Indian building construction projects: an empirical study*, Org., Tech. and Manag. in Constr. **13**, 2383–2390 (2021).
20. Ch. Kropp, Ch. Koch, Markus König, *Interior construction state recognition with 4D BIM registered image sequences,* Autom. in Const. **86,** 11–32 (2018).
21. Ch. Zhang, D. Arditi, *Automated progress control using laser scanning technology*, Autom. n Constr. **36**, 108–116 (2013).
22. T. Omar, M.L. Nehdi, *Data acquisition technologies for construction progress tracking*, Autom. Constr. **70**, 143–155 (2016).
23. Z. Pučko, N. Šuman, D. Rebol, *Automated continuous construction progress monitoring using multiple workplace real time 3D scans*, Adv. Eng. Inf. **38**, 27–40 (2018).

24. P.M. Goodrum, M.A. McLaren, A. Durfee, *The application of active radio frequency identification technology for tool tracking on construction job sites*, Autom. Constr. **15**, 292–302 (2006).
25. E. Ergen, B. Akinci, R. Sacks, *Tracking and locating components in a precast storage yard utilizing radio frequency identification technology and GPS*, Autom. Constr. **16**, 354–367 (2007).
26. Q. Wang, M.-K. Kim, *Applications of 3d point cloud data in the construction industry: a fifteen-year review from 2004 to 2018*, Adv. Eng. Info. **39**, 306–319 (2019).
27. J.S. Alvares, D.B. Costa, R.R.S. de Melo, *Exploratory study of using unmanned aerial system imagery for construction site 3d mapping*, Construct. Innov. **18 (3)**, 301–320 (2018).
28. P. Lin, Q. Li, Q. Fan, X. Gao, *Real-time monitoring system for workers' behaviour analysis on a large-dam construction site*, Int. J. Dist. Sensor Net. **9 (10),** 509423 (2013).
29. Lei Hu, *Design of Remote Monitoring Iot System Based on 5G and New Optical Fiber Intelligent Structure*, Adv in Mat. Sc. and Eng. 12 (2022).
30. R. Faragher, R. Harle, *An analysis of the accuracy of bluetooth low energy for indoor positioning applications*, in Proc. 27th Int. Tech. Meet. Satell. Div. Inst. Navig. (ION GNSS+ 2014), 201–210 (2014).
31. E. Ergen, B. Akinci, B. East, J. Kirby, *Tracking components and maintenance history within a facility utilizing radio frequency identification technology*, J. Comput. Civ.Eng. **21 (1)**, 11–20 (2007).
32. F. Vahdatikhaki, A. Hammad, *Framework for near real-time simulation of earthmoving projects using location tracking technologies*, Autom. Constr. **42,** 50–67 (2016).
33. R. Bao, M.A. Sadeghi, M. Golparvar-Fard, *Characterizing construction equipment activities in long video sequences of earthmoving operations via kinematic features*, Constr. Res. Congr, 849–858, (2016).
34. N.D. Nath, T. Chaspari, A.H. Behzadan, *Automated ergonomic risk monitoring using body-mounted sensors and machine learning*, Adv. Eng. Inf. 514-526 (2018).
35. R. Akhavian, L. Brito, A. Behzadan, *Integrated mobile sensor-based activity recognition of construction equipment and human crews*, Conf. on Aut. and Robotic Constr. of Inf., 1–20 (2015).
36. W.-F. Cheung, T.-H. Lin and Yu-Ch. Lin, *A Real-Time Construction Safety Monitoring System for Hazardous Gas Integrating Wireless Sensor Network and Building Information Modeling Technologies*, Sensors **18**, 436 (2018).

# Assessment of the influence of the hydroaccumulation layer on the initial moment of rainwater runoff from the green roof structure

*Ondrej* Porwisz[1*], and *Jan* Jilek[1]

[1]Brno University of Technology, Faculty of Civil Engineering, 601 90 Brno, Czech Republic

**Abstract.** The article deals with the assessment of the influence of the hydro-accumulation layer on the initial moment of rainwater runoff from green roof. The authors deal with a comparison of two basic types of green roof compositions, where they assess the initial moment of rainwater runoff following the use of a hydroaccumulation layer.

## 1 Introduction

Together with colleagues, we verified and compared the efficiency of the hydroaccumulation layer in the composition of the green roof. The goal was to find out how much influence the hydroaccumulation layer has on the runoff conditions of the roof. We were led to the idea by the reconstruction of the green roof on Bratislavská Street (fig. 1), which had to be carried out precisely because of the absence of this layer and the negative effects caused by this absence. These were mainly visual shortcomings, when no greenery took hold. This led us to create two models on which we performed measurements using simulated rain.

**Fig. 1.** Roof with non-functioning green layers.

* Corresponding author: 136054@vutbr.cz

The original goal of the measurement was to determine whether the hydroaccumulation layer will significantly change the rate of rainwater runoff from the roof. However, this was not confirmed. We obtained but two additional values from the experiment. We will deal with these values in the following lines.

# 2 Methodology

## 2.1 Preparation and implementation of the test roof

For the purposes of the experiment, we assembled two identical experimental roofs in the AdMaS research area. The existing roof structure made of wooden beams, OSB boards and waterproofing made of modified asphalt strips was used as the basic building element of the experiment. OSB boards were used for the formwork of the model green roofs, which were stabilized with lumber struts. The material was also used for the composition of the green roof. Geotextiles, retention plates Aquadesk Retex and substrate, as we can see in Fig. 2. Lumber, garden hoses, spray valves and an adjustable water flow meter were used on the structure to simulate rain. The measurement itself took place on an electric scale and plastic tubs were used to catch the water.

**Fig. 2.** Constructing of experimental roof.

On the already prepared roof, we assembled formwork from OSB boards and lumber, which we arranged in two identical squares with an area of 1.44 m$^2$. We placed a geotextile on the waterproofing asphalt layer in the formwork. Next, we placed a substrate on model roof No. 1. (Tab. 1) After the geotextile, a hydro-accumulation layer of retention plates was placed on sample roof No. 2 and a substrate was placed on top of them. (Tab. 2) The resulting state is shown in Fig. 3.

**Table 1.** Layers of roof No.1.

| | thickness |
|---|---|
| Substrate | 40 mm |
| Geotextile | - |
| Asphalt HI | - |

**Table 2.** Layers of roof No.2.

| | thickness |
|---|---|
| Substrate | 40 mm |
| Aquadesk Retex | 40 mm |
| Geotextile | - |
| Asphalt HI | - |

**Fig. 3.** Completed experimental roof with final layer of substrate.

The construction for rain simulation consisted of a simple grid made of lumber with a square plan of 1.2 m in length, as we can see in Fig. 4. A garden hose was placed on the grid, in which spray valves were placed, so that the surfaces of the test roofs were contained by a uniform intensity of the test rain. An adjustable flow meter was placed at the inlet end of the hose.

**Fig. 4.** Rain simulator raster.

In the last row, an electric scale was prepared. This was placed under a gutter, from which rainwater was led during the experiment into a tub, which was placed on the measuring device. The display of the scale, which we see in Fig. 5, was placed next to a stopwatch and a watch, in order to monitor the temporal continuity of the experiment.

**Fig. 5.** Measuring equipment.

**Fig. 6.** Water flow meter.

## 2.2 The course of the experiment

The experiment was started by placing the rain simulation structures on the formwork and starting the water flow. The rate of simulated rain was based on 27 l/m²/ for 15 minutes. For an area of 1.44 m², the flow rate was therefore 2.59 l/min, as we can see in Fig. 6. The total amount of rainwater was 38.88 L for each roof tested. In Fig. 6 shows the moment of the beginning of the experiment, when both roofs were subsequently exposed to fifteen minutes of rain. The moment we started the simulated rain, the measurement of the volume of water in relation to the elapsed time was started. The value of runoff rainwater was recorded continuously.

**Fig. 7.** Start of rain simulation.

# 3 Results

## 3.1 Original data

As we mentioned at the beginning, the primary measurement was to confirm or disprove the thesis that the hydro-accumulation layer will significantly reduce the rate of rainwater runoff. However, the experiment did not confirm this thesis. The flow rate did not differ for most of the time, a significant difference occurred only between 11 and 13 minutes.

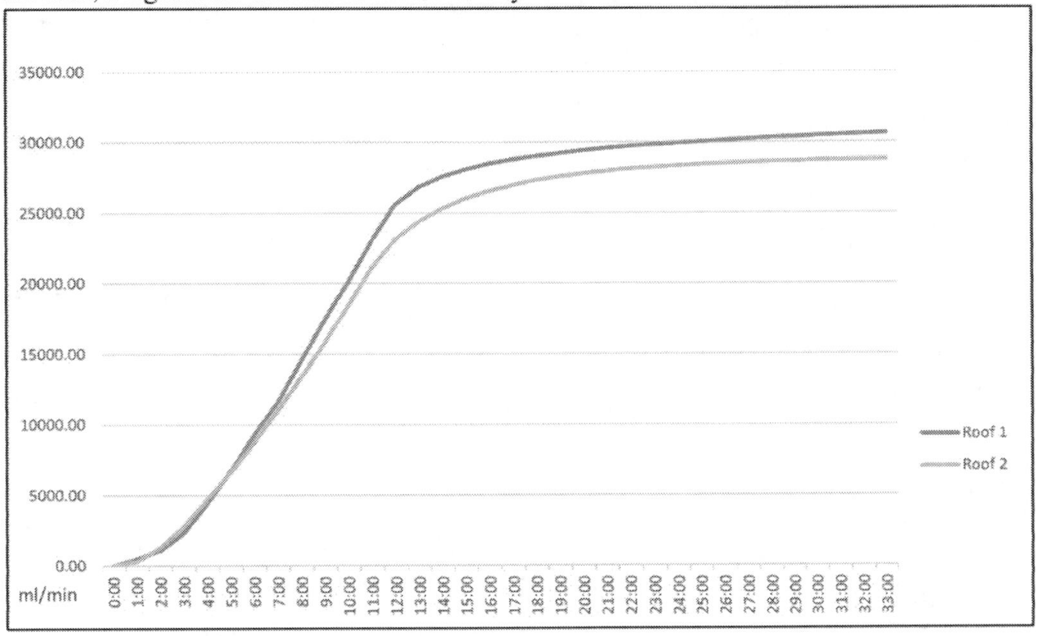


**Fig. 8.** Graph presenting the original results.

## 3.2 Data including the initial flow moment

However, while working with the measured data, we observed and recorded another data. From the graph below, it can be read that the initial moment of roof runoff without a hydro-accumulation layer is at 2:25. The initial moment of runoff of the roof with hydroaccumulation layer is 4:50 after the start of the simulated rain. For clarity of the experiment, we converted the measured weight of rainwater to volume in the ratio 1kg = 1 l.

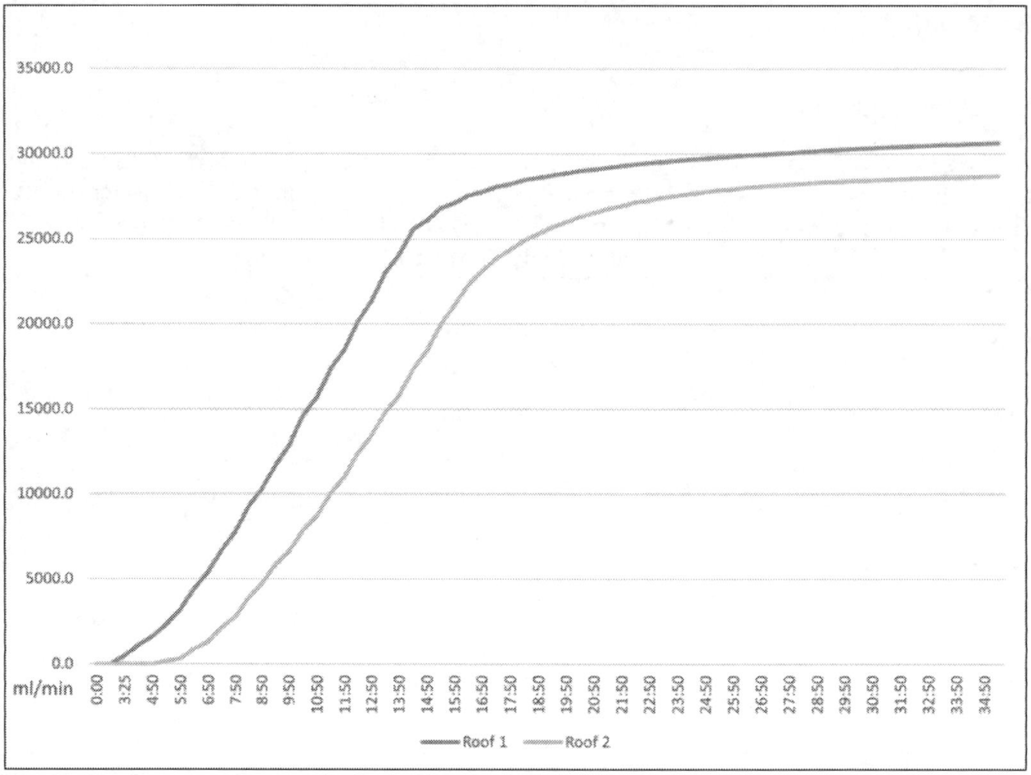

**Fig. 9.** Graph presenting results including the initial flow moment.

## 4 Results

We can see a significant difference compared to the first interpretation of the data right at the beginning of the graph. In this case, the measuring time was started at the same moment when the 15 minutes of artificial rain started. In contrast to graph 1 (fig. 8), when the measurement was started only at the moment when the rainwater started to flow away. In graph number 2 (fig. 9), we can therefore observe the values of the initial moment of runoff, when the rainwater began to flow from roof no. 1 at 2:25 and at roof no. 2 at 4:50. The difference in the amount of rainwater runoff increases for almost the entire 15 minutes of simulated rain. At minute 13, the increase slows down slightly, but still reaches a difference of 8.2 l. After 15 minutes, the difference starts to decrease, and after less than 35 minutes, it stops at 1.9 l. At this time, the experiment is finished because no more water flows out and retroactively, no detention is observed either.

## 5 Conclusion

Compared to the original intention of the measurement, the real differences between the roof without the hydroaccumulation layer and the roof with the hydroaccumulation layer were highlighted due to the initial moments of the outflows. In the original measurement, where we examined only the speed of water outflow, the biggest difference was 2.44 l in 13 minutes. After taking into account the delay caused by the initial flow moment, the largest difference on 1.44 m2 of experimental roof is 8.2 l between 13 and 15 minutes.

This interpretation of the data offers a more realistic model of possible behavior than the originally intended interpretation. A more pronounced difference in the speed of rainwater runoff confirms the ability of the hydroaccumulation material to slow down the initial runoff of rainwater and thereby more evenly load the use of rainwater drainage. From the measurements and working with the data, it was also found that none of the tested roofs had any detention, i.e. a condition where even the smallest amount of water would run off after a long time (hours) after the end of the rain.

Working with data and measured values, we came to the conclusion that even if the hydroaccumulation layer does not significantly change the course of rainwater runoff, it does significantly change the initial moment of runoff. This can, for example, result in a more even distribution to the storm sewer in a development where there is partial use of green roofs.

## References

1. P. Selník, D. Bečkovský, T. Rebrova, The use of recycled textiles in the construction of green roofs (Novopress s.r.o., Brno, 2018)
2. K. Lucket, Green Roof Construction and Maintance (The McGraw-Hull Companies, Inc., 2009)
3. J. Yan, S. Zhang, J. Zhang, S. Zhang,C. Zhang, H. Yang, Stormwater retention performance of green roofs with various configurations in different climatic zones, **319** (Journal of Environmental Management, 2022)
4. A. Raimondi, G. Becciu, Performance of Green Roofs for Rainwater Control, **189** (Water Resources Management, 2021)
5. D. Pumo, A. Francipane, F. Alongi, LV. Noto, The potential of multilayer green roofs for stormwater management in urban area under semi-arid Mediterranean climate conditions, **326** (Journal of Environmental Management, 2023)
6. Fig. 1-9 Author of the article

MATEC Web of Conferences **385**, 01034 (2023) https://doi.org/10.1051/matecconf/202338501034
*Young Scientist 2023*

# Biodiverse green roof: Phase 5 of the greenIZOLA project

*Alena* Vargova[1*], and *Marian* Vertal[1]

[1]Department of Architecture Engineering, Faculty of Civil Engineering, Technical University of Kosice, Vysokoskolska 4, 04001 Kosice, Slovakia

**Abstract.** Cities represent a massive area of dense materials that accumulate heat, do not retain rainwater, and therefore create an unpleasant condition for life. Many countries apply green roofs as an effective and practical tool to mitigate the adverse effects of urbanization. They are often designed as intensive and extensive. In recent years, a lot of research has been done on green roofs for a better understanding of their behavior in different conditions. In recent years, experimental green roofs have also started emerging in Slovakia. At the same time, the idea of restoring vegetation to industrial zones with help of such construction was made. The paper focuses on the latest planned phase of the project for the greening of an office building with an adjacent area in an industrial zone. Phase five consists of the design of a biodiverse green roof that will provide an alternative habitat for fauna and flora. The main elements of the design are the compositions of green and wetland roofs. Moreover, the paper summarizes the already implemented phases of the project.

## 1 Introduction

Vegetation in cities is an essential element that improves living conditions in metropolises. However, in some cases, it is not easy to preserve it and support urban development at the same time, which is why green roofs have been used frequently in recent years. Placing vegetation on the roof partially helps to restore the vegetation previously built over. Green roofs are often designed as extensive, for a smaller substrate thickness of 80 – 150 mm and minimal maintenance. Semi-intensive roofs are designed with a substrate thickness of 150 – 300 mm and must be irrigated, especially in hot and dry months. The selection of vegetation is more diverse with an intensive green roof, thanks to the greater substrate thickness of 300 mm and more, although such a roof requires more maintenance [1]. Special green roofs have started to be implemented nowadays, such as a biosolar roof (the combination of solar panels and a green roof) or wetland roofs (the application of wetland vegetation). The green layer is designed so as to create suitable conditions for vegetation growth and, at the same time, to impose the least possible load on the structure [2].

Many countries are taking a solid initiative to apply green roofs to new and existing buildings through reward or penalty. They use them as an effective and practical tool for

* Corresponding author: alena.vargova@tuke.sk

urbanisation. The composition of a green roof can retain a large amount of water and, as a result, delay stormwater runoff. Several researchers have studied the issue of rainwater accumulation in their region and found out that this ability of the roof varies mostly from 55% to 88% [3]. Besides, using the retained water in evapotranspiration is possible, which produces a cooling effect on the building and the surroundings. The substrate and vegetation have a higher albedo than, for example, asphalt strips; thus, part of the incident radiation can also be used for photosynthesis [4]. Hence, the waterproofing layer is protected from UV radiation and climatic influences. In terms of greening, a green roof can increase the abundance and diversity of birds, arthropods, and gastropods more than a conventional roof in the city [5]. It is this added value of the roof biotope that is the reason for the implementation of green roofs in many European cities [6] with declining vegetation. Last but not least, such a roof also brings aesthetic benefits to the building and the surroundings. Its spaces can serve for relaxation and psycho-hygiene in a built-up city.

Green roofs are often applied in city centres, on houses, apartment buildings, offices, and commercial buildings. However, they are only found in small numbers in industrial zones. These zones are growing along with the developing city. They represent a relatively large part of the city built-up primarily for the production, storage, and handling of materials. Therefore, it is necessary to design and introduce green infrastructure elements even in such parts of cities.

The paper presented is dedicated to the next planned phase of the greenIZOLA project, where vegetated structures are used for the green conversion of a site in an industrial (grey) zone. Once all the planned phases are completed, an increase in vegetation area from 14% to 41% should be achieved. The subject of the solution is the most extensive terrace in the building, which by its dimensions, offers space for the creation of a combined green roof, providing an alternative habitat for flora and fauna.

## 2 greenIZOLA project

The greenIZOLA project aims to return vegetation to the industrial zone using green structures. The project is supported by the APVV-18-0360 ACHIEve Active Hybrid Infrastructure for a Sponge City project. The subject of the solution is an office building with adjacent area in an industrial zone in Košice. The building is four-storey, made as a skeleton structure with infill masonry and finished with a flat roof. There are four terraces on the building (two, western and northern, on the second floor and another two, southern and northern, on the fourth floor). The composition of the roof and terraces is single-skin with layers in reverse order (from bottom to top: a reinforced concrete slab, a foam concrete gradient layer, a vapour barrier of asphalt strips, thermal insulation based on XPS and geotextile). In addition to the main building, there are two storage halls, three storage sheds, a fire tank, and a car park with 36 parking spaces.

The project is divided into nine phases, as shown in Figure 1. Various compositions of green roofs are used, including their special types, such as biosolar and wetland roofs. Moreover, the creation of a green wall on the eastern and southern parts of the building façade is considered. The resulting spaces can be used to research the interaction of vegetation layers with the building and the surroundings.

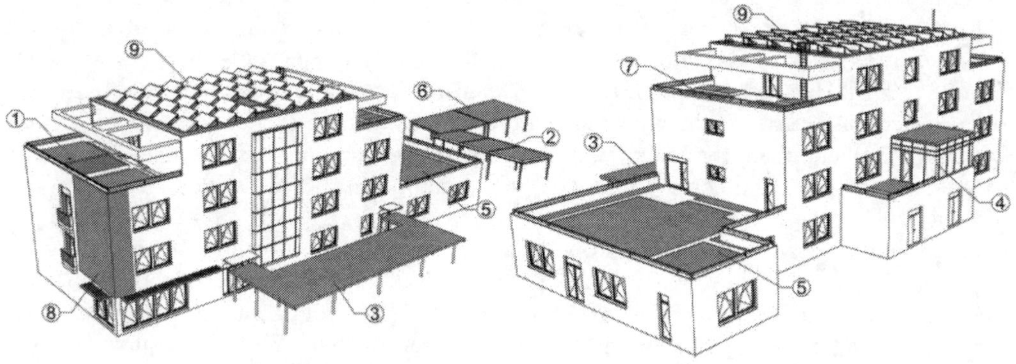

**Fig. 1.** Division of the project into phases for the green conversion of the building with the adjacent areas: 1 – Experimental green roof (green roofs and wetland roof); 2 – Existing shelter with three different compositions of a green roof; 3 – Carpor with a green roof; 4 – Extension of the western terrace on the 2nd floor; 5 – Biodiverse green roof – park on the terrace; 6 – Existing shelter with a green roof; 7 – Green roof of the northern terrace on the 4th floor; 8 - Green wall on the east and south side of the building; 9 – Biosolar roof.

The first phase, consisting of the construction of an experimental green roof on the southern terrace on the fourth floor, was implemented in May 2020. The terrace was divided into three test segments (TS I – a green roof with 120 mm thick substrate, TS II – a reference roof finished with a gravel layer, and TS III – a green roof with 240 mm thick substrate) [7]. In May 2022 (after two years), TS II, representing the reference roof, was replaced with a wetland roof composition. The next phase (Phase 2) was the construction of a green roof on the existing shelter in June 2021. The roof was divided into three parts: a composition with a mineral wool-based growth medium, a composition with mineral wool and a layer of substrate, and a composition with mineral wool and a substrate supplemented with a biochar admixture. In July 2021, a carport with a green roof was added to the area (Phase 3). The carport protects six parking spaces and the main and side entrances to the building from rain, hail, and snow. The last completed phase (Phase 4) was the extension of the western terrace on the second floor in April 2022. The roof of the extension was finished with an extensive green roof. The remainder of the western terrace will be supplemented with a portion of the wetland and green roof.

## 3 Biodiverse green roof (Phase 5)

### 3.1 Current condition of the northern terrace on the 2nd floor

The northern terrace has an area of approximately 171.82 m$^2$, is directly accessible from the office and the meeting room and has the best view of the terrace on the second floor. It is currently covered by a green roof made in 2010 (Figure 2). The implementation and design of the roof were conducted without in-depth knowledge of the subject or consultation with a botanist. The original roof design corresponded to the requirements for an extensive green roof at that time. It was made of layers of a drainage system, an 80 mm-thick substrate, and stonecrop plants. In addition to the vegetation part, the terrace has a walkway made of concrete tiles and concrete planters with shrubs, and the remaining surface is finished with gravel. Roof maintenance was minimal; the roof prospered relatively well in the first years; however, after a few years, the defects of the improper roof design and maintenance began to show, namely through the mortality of the vegetation.

**Fig. 2.** View of the existing condition of the northern terrace on the second floor.

## 3.2 Proposed condition of the northern terrace on the 2nd floor

The northern terrace is the largest on the property and offers space to create a "park" with a variety of vegetation. This space should mainly serve a relaxing purpose, as intended from the beginning. Thanks to the acquired experience and knowledge, a new composition of the green roof on the terrace was proposed in collaboration with a horticultural company (layout and landscape integration), a botanist (correct selection of vegetation species, possible experimentation with non-typical species of vegetation for our climate) and a construction company (for a correct structural design with no damage to the building). Each member is necessary for the correct and efficient design of a green roof that is intended to represent a space with added value in terms of greening. The idea was to design an intensive green roof to possibly plant taller vegetation. Since it is an existing building with static limitations, the designed thickness of the substrates must be smaller than for an intensive roof. Still, the design included several elements that will be incorporated on the terrace. Figure 3 shows the terrace layout in the current and proposed condition.

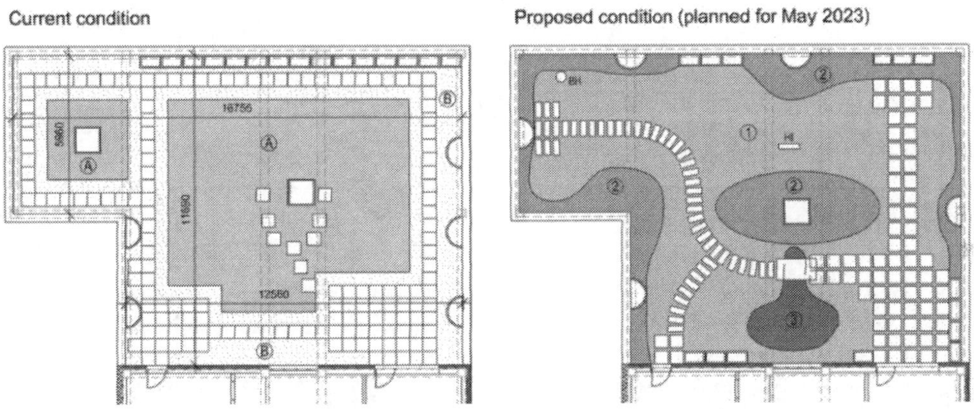


**Fig. 3.** The floor plan of the northern terrace on the second floor – the division of the current and proposed states - Phase 5 of the project; (A-a green roof with an 80 mm thick substrate layer, B-part of the terrace finished with gravel; 1-a green roof with a 100-120 mm thick substrate layer, 2-green roof with a 150-180 thick substrate layer, 3-wetland roof, HI-insect hotel, BI-birdhouse).

The terrace area will consist of three vegetation sections: a green roof with a 100 – 120 mm thick substrate layer, a green roof with a 150 – 180 mm thick substrate layer, and a wetland roof (Figure 4). The substrate is to be distributed unevenly to elevate the terrain gradually. The terrace will be finished with a root-proof waterproofing layer. A protective layer, drainage and water accumulation layer, filter layer, substrate, and vegetation will follow. A wetland roof shall comprise a pond liner, substrate mixture, wetland vegetation, and water. Walkway will lead through each section so visitors can walk along the terrace and use the space for rest and recreation. The terrace also includes benches, an insect hotel, a birdhouse, and concrete planters for planting shrubs.

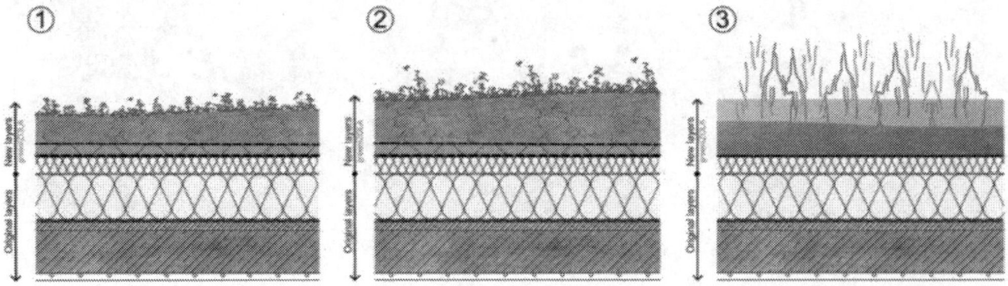

**Fig. 4.** Compositions of green roofs on the terrace: 1 – Green roof with substrate layer th. 100-120 mm; 2 – Green roof with substrate layer th. 150-180 mm; 3 – Wetland roof th. 150-200 mm.

The vegetation layout was adapted to the load-bearing skeleton system of the building, with beams in the longitudinal direction and around the perimeter of the building. The part of the green roof with a substrate thickness of 180 – 150 mm would generate a greater load (Table 1); therefore, the substrate distribution is proposed along the edge of the terrace near the columns and beams. This area represents 28% of the entire terrace. The second part of the green roof with a substrate thickness of 120 – 100 mm would generate less load (Table 1) and cover 64% of the terrace area. The two compositions overlap and build each other to create a gradual segmentation (land forming) of the terrace. For the smallest module, where the columns and beams are axially spaced 3.2 m apart, a wetland roof is proposed, which will be formed in the shape of a 'pond'. The area of this water element represents 4% of the terrace area. A bridge will cross the wetland, connecting the wooden and concrete walkways.

**Table 1.** The area and load of the proposed compositions on the terrace.

| | Roof type | Area ($m^2$) | Weight ($kg/m^2$) |
|---|---|---|---|
| 1 | Green roof with a 100 – 120 mm thick substrate layers | 110 | 125 – 150 |
| 2 | Green roof with a 150 – 180 mm thick substrate layer | 48 | 190 – 225 |
| 3 | Wetland roof | 7.7 | 270 – 350 |

Unwanted terrace layers such as those of the old green roof (the vegetation, substrate, and drainage system), paving stones, planters and gravel must be removed before the implementation of the new roof. The terrace needs to be insulated to meet all thermal-technical requirements. Once the layers have been removed and the surface cleaned, the composition will be supplemented with thermal insulation based on a PIR board (due to its good thermal insulation properties with a smaller thickness) and a waterproofing layer resistant to root overgrowth. Subsequently, the frame will be installed into which the individual layers of the wetland roof are to be placed. At the same time, the green roof layers (protective, drainage, storage, and filtration layers) will be constructed, and the planters installed. The next step will be the filling of the substrate, shaping of the terrain, installation of the paths, and the actual planting of the vegetation. The last step will be the installation of the birdhouse and the insect hotel.

## 4 Conclusion

The greenIZOLA project is a vision for restoring vegetation in an industrial zone using green structures. It focuses on an office building with adjacent areas into which elements of green infrastructure will gradually be incorporated in such a quantity that the statics of the building is not compromised. The project is divided into individual phases, which have been gradually implemented since 2020. The fifth phase envisages the transformation of the building's terrace using elements that should ecologically enhance the environment, contribute to improving the climate, protect the layers of the roof structure, increase the biodiversity of the environment, and aesthetically beautify the building and its environment. The original green roof with a substrate thickness of 80 mm made up only 40% (70 $m^2$) of the entire terrace area. After the modifications, the vegetation area will make up 96% of the terrace area. As the green area increases, so does the accumulation of rainwater, slowing down the water runoff, increasing the cooling effect and the diversity of flora and fauna (insects and birds). Moreover, the terrace can serve as an example of incorporating a wetland roof into the design structure of biodiverse green roofs. At the same time, it can be used for experimental purposes to link research in engineering and natural science disciplines.

This work was supported by a project run by Slovak Research and Development Agency APVV-18-0360 "Active hybrid infrastructure towards to a sponge city".

## References

1. Burian S. et al., *Vegetation composition of green roofs (standards for design, implementation, and maintenance)* (*Vegetačné súvrstvie zelených striech (štandardy pre navrhovanie, realizáciu a údržbu))*, in Asociácia pre zelené strechy a zelenú infraštruktúru o.z., Bratislava, (2022)
2. Vijayaraghava K., *Green Roofs: A Critical Review on the Role of Components, Benefits, Limitations and Trends*, in Renewable and Sustainable Energy Reviews, vol. **57**, pp. 740-752
3. Shafique M. et al., *Green Roof Benefits, Opportunities and Challenges – A Review*, in Renewable and Sustainable Energy Reviews, **90**, 757-773, (2018)
4. Gaffin S. et al., *Development of Green Roof Environmental Monitoring and Meteorological Network in New York City*, in Sensors (Basel), **9**, pp. 2647-2660, (2009)
5. Wooster E.I.F. et al., U*rban green roofs promote metropolitan biodiversity: A Comparative Case Study*, in Building and Environment, **207**, p. 207 (2022)
6. Dunnett N., *Green roofs for biodiversity: Reconciling aesthetics with ecology*, in Greening rooftops for sustainable communities, (2006)
7. Vertal' M. et al., *Experimental biodiverse green roof of the greenIZOLA project (Experimentálna biodiverzná vegetačná strecha projektu greenIZOLA)*, in Strechy-Fasády-Izolace, **11-12/2020**, pp. 25-29, (2020)
8. Čakyová K. et al., *The Concept of Green Industrial Zones*, in Proceeding of EcoComfort 2022, **290**, (2023)

# How to prepare for a greener future?

*Zuzana* Vranayova[1], and *Daniela* Kaposztasova[1*]

[1]Technical university of Kosice, Faculty of Civil Engineering, Institute of Architectural engineering, Department of Building services, Vysokoskolská 4, Kosice 042 00

**Abstract.** The global buildings industry has an enormous responsibility to transform towards a regenerative and decarbonized future. The new project SECOVE 101056201 aims to establish a network of cooperation among Centre of Vocational Excellence (CoVEs) in the sector of clean and sustainable energy. The CoVEs will be established in 5 European countries (Greece, Spain, Italy, Slovakia, and Portugal) and will all cooperate transnationally with each other through common structured forms of dialogue and communication. The paper deals with the main activities and goals.

## 1 Introduction

Sustainable energy centers of vocational excellence (SECOVE) aims to establish a cooperation platform for Centres of Vocational Excellence (CoVEs) across Europe in the sector of renewable and sustainable energy. It will be locally based, with a CoVE in each partner country. The platforms' main goal is to promote the uptake of innovative and qualitative lifelong learning opportunities, oriented towards the development of skills, competencies and the achievement of qualifications, including the promotion of mobility opportunities. The project will adopt and support a holistic approach to promote creativity, inclusion and entrepreneurial mindset together Figure 1[1].

SECOVE is a project designed over 4 years of implementation. Its ambition to achieve a positive impact in VET and the sustainable energy sector can be reasonably understood thanks to its long duration, involvement of a large number of partners from five different countries, representing different types of stakeholders and substantial financing. SECOVE team is indeed committed to monitoring, measuring and most importantly acting to ensure that the activities of the SECOVE network bring about a positive long-term change. It is co-funded by the Erasmus + Programme of the European Union [1].

* Corresponding author: daniela.kaposztasova@tuke.sk

**Fig. 1.** SECOVE project in general.

## 1.1 Challenges and objectives of SECOVE project

The project will organise local and national workshops, capacity building events, mobility, summer academies for female students, exchange programmes for students and teachers, innovation competitions and national conferences. In Slovakia, the project will translate into better links between vocational education and science, research institutions and the private sector, strengthening dual and inclusive education and supporting students beyond their vocational studies. The project started in July 2022 (Fig.2).

SECOVE addresses the following challenges:

- sustainable energy and its different applications such as energy efficiency in buildings, renewable energy and smart energy systems are integrated in national and regional policies of the involved countries
- development of new and relevant learning programs that tackle major challenges of sustainable development
- promotion of work-based learning and engagement of the labor market in VET
- attraction of more women in technical qualifications is apparent in all countries
- promotion of innovation and entrepreneurship at all levels of education and training, in line with regional smart specialization strategies and using the existing capacities of partners and local networks.

• establishment of a quality culture in vocational training that goes beyond conformity, with standards and regulations that address common and shared values for excellence, inclusiveness and sustainability [2]

Project **objectives** are as follows:

- to ensure the sustainability of SECOVE platform, through engagement of stakeholders, alignment with Smart Specialization Strategies and securing governance and financing
- to promote learner-centred and inclusive COVEs, bringing holistic approaches to the individual, labor market-driven qualifications under the umbrella topic of sustainable energy
- to promote creativity, entrepreneurial mindset and innovative learning opportunities through the cooperation between education, industry and incubators.
- to establish innovation hubs and support work-based learning at national and transnational level.
- to develop work-based schemes following the principles of ECVET and arrange transnational certification and recognition.
- to attract more women in technical occupations.
- to promote quality culture and support VET providers to adopt European and international quality standards [1].

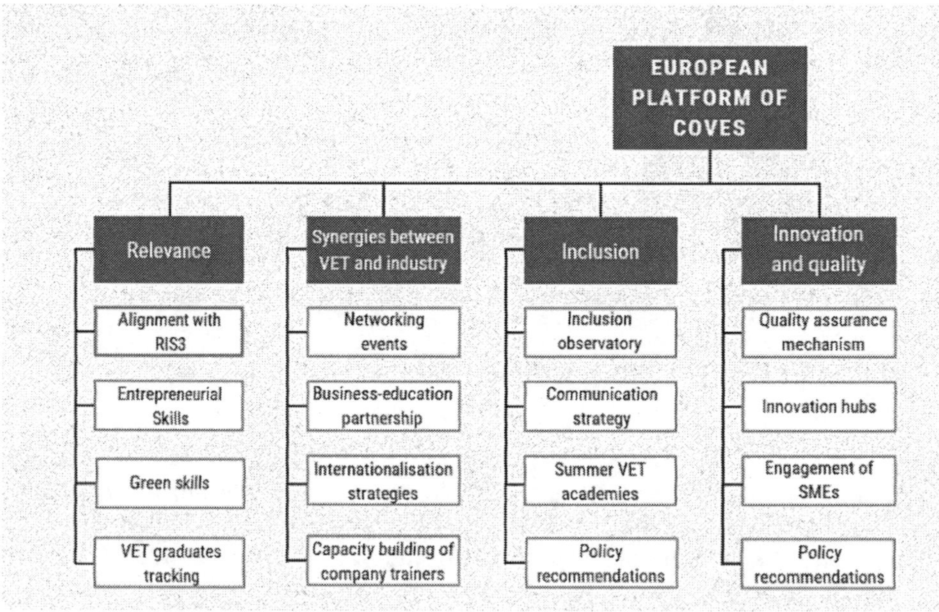


**Fig. 2.** SECOVE project flow diagram [1].

## 2 Green building industry

Construction is one of the most heavily regulated industries and is increasingly subject to more stringent rules about the environmental impact of constructing buildings. The World Green Building Council estimates that construction and the energy required to operate buildings account for around 40% of global carbon emissions [4].

Driven by the combined energy and climate crises, the EU is becoming more and more aware of the need to switch to a green economy. But even as Europe looks to ramp up the deployment of solar panels, heat pumps, and electric vehicles, it must face the fact that it simply does not have the people to get the job done.

"The best technology is only as good as the skilled workers who can install and operate it," said European Commission President Ursula von der Leyen and she added: "And with a huge growth in new technologies, we will need a huge growth in skills and skilled workers in this sector,"

2023 is the European Year of Skills and the EU's new Green Deal Industrial Plan listed skills as its third pillar.

The green building industry encompasses various practices aimed at reducing environmental impact while promoting sustainability (Fig.3). Green Industry means economies striving for a more sustainable pathway of growth, by undertaking green public investments and implementing public policy initiatives. The energy sector has responded to the shift towards green buildings by recognizing the critical role that buildings can play in meeting climate change ambitions. The implementation of energy efficiency improvements in the building sector requires developing innovative, reliable and high-quality technologies and infrastructure [3].

**Most Important Approaches to Improve Sustainability in the Design and Construction Industry in the Next Five Years, by Company Type**

Dodge Data & Analytics, 2021

| Architects | Engineers | Contractors | Owners | Investors |
|---|---|---|---|---|
| 1. Creation of Net-Zero/Net-Positive Buildings (52%) | 1. Creation of Net-Zero/Net-Positive Buildings (46%) | 1. Prefabrication and Modular Construction (45%) | 1. Creation of Net-Zero/Net-Positive Buildings (48%) | 1. Design for Disassembly and Recovery (57%) |
| 2. Controlling Embodied Carbon (42%) | 2. Controlling Embodied Carbon (36%) | 2. Creation of Net-Zero/Net-Positive Buildings (35%) | 2. Strategies to Increase Resiliency (31%) | 2. Controlling Embodied Carbon (35%) |
| 3. Passive Building Design (40%) | 3. Strategies to Increase Resiliency (36%) | 3. Strategies to Increase Resiliency (32%) | 3. (Tie) Controlling Embodied Carbon & Use of Design for Disassembly and Recovery (29%) | 3. Strategies to Increase Resiliency (33%) |

**Fig. 3.** Most important approaches to improve Sustainability in construction industry [4].

The EU is implementing various policies and targets, legislative measures and initiatives on green buildings, such as the Green Deal and the Circular Economy Action Plan. Targets are being set, the feasibility of which can be debated. European countries are nowadays embarking on a transition towards climate neutrality, digital leadership and net zero technologies. Driven by the combined energy and climate crises, the European Union (EU) is becoming more and more aware of the need to switch to a green economy.

The ILO emphasizes the importance of developing skills for green jobs to ensure a just transition to a low-carbon economy [5]. There will be a need for people whose jobs are not considered green to receive training in skills to enable changes that will improve sustainability. For example, a change to building regulations designed to improve energy efficiency creates a one-off change in skills requirements across a range of occupations in the construction sector, and a consequential requirement for training to meet these skills need. Researching these occupations, skills and training and education needs means looking beyond the core green industries [5].

Learning green building construction skills is a smart career investment because it plays an important role in the construction industry, energy sector, and economy. It is expected to grow further, leading to more jobs and better jobs for green building construction workers all over the word. Therefore, investing in training programs that teach these skills can help individuals prepare for a career in this growing field.

Several drivers of skills change in green buildings have been identified:

- emphasis on soft skills, such as communication and collaboration
- shortage of skilled workers in the green building sector
- investing in leadership development for career progression in the green skills sector
- creating new roles and supporting them with the right team, infrastructure, salary, and timescales

- people with the right skills can be a driver of change.

The focus group for skills and competencies discussed this issue on February 24th, 2023 and stands that skills and qualifications will be most important in creation a new job profiles in green building industry. Personal characteristics such as desire to work, motivation, flexibility, professional knowledge, manual skill, evaluation of green building project proposals, critical thinking. There is a need of combining knowledge in the field of construction with IT and architecture - also for lower qualification jobs (e.g. maintenance worker) a very important is combining knowledge of design and botany to design vegetated roofs and green structures. Digitization of the territory - how green the territory will look after construction can help developers or building managers to see the design in 3D. Specific knowledge in various phases of the cycle of design and implementation of green buildings as well as during the entire life cycle of GBs helps mangers to see their impact on people and outwardly on the climate and how vegetation technologies /systems contribute to sustainability in energy sector.

The OECD's policy review of VET, Learning for Jobs, recommends recruiting sufficient teachers and trainers for VET institutions who are well-acquainted with industry experience. Changes to the education system will be important to ensure students are adequately trained for labor and employment opportunities in the future [6].

## 3 Conclusion

The project will considerably reduce the gap between job market needs and vocational education offer through enhanced level of innovation and quality in terms of cooperation, course design and delivery of educational content. At the same time, the CoVEs will mainstream an equality-oriented approach to VET - such as the sustainable energy one- which sees a very low level of participation from women of all ages and education levels. The new paradigm promoted by SECOVE will comprehensively remodulate the work and cooperation protocols put in place so far in favor of a quality-based, innovative, and overall meaningful VET offer in the sector of renewable energies.

This work was supported by the Erasmus+ project SECOVE 101056201 "Sustainable energy centres for vocational excellence".

## References

1. SECOVE ERASMUS + project: https://secove-project.eu/news/
2. EFQEA implementation: a Cedefop analysis and main findings, 2021: https://www.cedefop.europa.eu/en/publications/4195
3. UNIDO, 2023: https://www.unido.org/our-focus-cross-cutting-services-green-industry/green-industry-initiative
4. DODGE, data and analytics: https://www.construction.com/toolkit/reports/World-Green-Building-trends-2021
5. ILO, 2011. Skills for green job. A global view. https://www.ilo.org/wcmsp5/groups/public/@dgreports/@dcomm/@publ/documents/publication/wcms_159585.pdf
6. OECD, 2011. Learning for job: https://www.oecd.org/education/skills-beyond-school/LearningForJobsPointersfor%20PolicyDevelopment.pdf

# Condition of the extensive green roof of the Brno-Komín Kindergarten in winter 2 years after implementation

*Jan* Jilek[1*], *Ondrej* Porwisz[1], and *Martin* Mohapl[1]

[1]BUT, Faculty of civil engineering, Veveri 95, 602 00 Brno, Czech Republic

**Abstract.** The article describes the state of the vegetation layer of the extensive green roof on the kindergarten building in Brno-Komín. Due to the low temperatures, some plants are in a vegetative rest. On the contrary, some resistant plants are in good vegetative state even during adverse climatic conditions. The green roof was implemented in 2021 and as vegetation a pre-grown vegetation carpet was used, which was grown and transported on a coconut mat.

## 1 Introduction

The article discusses the state of the extensive green roof during the period of vegetation dormancy. The roof was implemented in 2021 and since then the vegetation has already gone through the vegetation cycle and after 2 years we can see the visual result of the implementation company's work on the structure. In the pictures we can see the differences in the amount of vegetation that was green in its entire area after laying the carpet and in the vegetation boom phase (Fig.1).

**Fig. 1.** Extensive green roof.

* Corresponding author: jan.jilek93@seznam.cz

### 1.1 Extensive green roof

The essence of an extensive green roof is vegetation with a maximum degree of self-regulation, capable of maintaining an adequate quality without regular watering and with only minimal human care (usually 1 to 2 times a year inspection, removal of unwanted vegetation, fertilization depending on the type of substrate and development phase of the growth). The selection of plant species used must be adapted to the habitat conditions as much as possible.

The vegetation of extensive green roofs consists of plants with a high regenerative capacity and able to adapt to the extreme conditions of the habitat. The plants must be sufficiently competitive in the given conditions to suppress the development of unwanted plants. The vegetation of the extensive roof consists of vegetation with a predictable successional development, which may also include spontaneous settlement by other species not used during the implementation.

The thickness of the vegetation layer of extensive green roofs is usually in the range of 60-150 mm. For appropriately chosen types of succulents, a layer thickness of only 40 mm (or less) may be sufficient, on the other hand, for steppe herbaceous vegetation types, a layer thickness of up to 200 mm can be used. Extensive green roofs are usually non-walkable, i.e. access to areas with vegetation is only allowed to trained persons for inspection and technical maintenance [1].

### 1.2 Heat island of the city

In cities, often used surfaces are concrete, asphalt, metal materials and glass. The ambient temperature is influenced not only by the type of material, but also by its darkness and structure. The dark colour of the material has a greater ability to absorb heat energy and subsequently radiate heat. Metal materials, which have an increased ability to accumulate heat, which subsequently radiate, behave similarly.

Human activity also contributes to the warming of cities. Buildings radiate heat from air conditioning and heating facilities. Transport and industry also produce thermal energy. Cars not only emit heat, but also emit large amounts of exhaust gases.

All plants have the ability to transpire, thanks to which they bring moisture from the soil into the air. Even water that sticks to the surface of leaves from rainfall or water in dew drops is later evaporated into the atmosphere.

Vegetation also has a significant effect on vertical air flow. During the day, plants create cooler air in their surroundings, which sinks to the ground and pushes the warm air to the sides. At night, this phenomenon is reversed. This flow ensures air exchange [2].

## 2 Reconstruction of the green roof of the Brno-Komín kindergarten

The reconstruction of the green roof at the kindergarten in Brno took place in May 2021. It was an addition to the original roof structure, which was covered with PVC foil. A nop film with a laminated geotextile was placed on the structure, then a hydroaccumulation layer, aggregates and finally a substratum, on which a pre-grown vegetation carpet was placed in the last phase. Photos from the implementation are inserted into the texts on how the implementation went on the timeline (Fig.2, Fig.3, Fig.4, Fig. 5).

**Fig. 2.** Waterproofing control.

Before actually laying the roof structure, it was necessary to check the existing waterproofing (Fig.2). A number of control measurements were performed. One of the tests was a spark test, which locates capillary leaks in the waterproofing. A few minor bugs were found and fixed. To prevent rainwater from entering the structure.

**Fig. 3.** Placement of geotextile and nop film.

When laying the geotextile, there was a problem with strong wind, and therefore it was necessary to gradually load the geotextile with rolls of nop film and bales with hydroaccumulation plates (Fig.3).

**Fig. 4.** Hydroaccumulation and gravel placement.

As part of depositing the substrate and gravel, aluminum perimeter profiles were also placed, which separated the substrate from the aggregate around the perimeter of the roof and also in the areas where the ventilation shafts and ventilation ducts penetrate.

**Fig. 5.** Substrate and green vegetation carpet placement.

During the implementation, a mobile truck crane was used to transport materials to the roof of the building. A loose mixture pump was used to deposit the substrate, because manual spreading would be time-consuming on a roof area that exceeded 700m$^2$.

The implementation itself did not take more than a week. Workers worked systematically in work squads.

The completion of the roof was prevented only by an insufficiently blooming vegetation carpet, which was supposed to be supplied by a Czech company, but in the end was imported from southern Europe where the climate in May was warmer than in the Czech Republic.

The resulting appearance of the roof can be seen in Fig. 1 at the beginning of the article. The vegetation layer looks intact and the roof makes a beautiful impression.

### 2.1 Used pre-grown vegetation carpet type TR K 20

The pre-grown vegetation mat on an extruding coir carrier interwoven with polypropylene (PP) mesh is intended primarily for flat vegetation roofs. The mats contain a layer of substrate in which a mixture of several species of the Sedum genus are rooted [3].

The pre-grown mat contains 5-8 species of plants from: Sedum Album, Sedum Album Coral Carpet, Sedum Sexangulare, Sedum Hispanicum Minus, Sedum Lydium, Sedum Lydium Glauca, Sedum Acre, Sedum Refl exum, Sedum Refl exum Angelina, Sedum Spurium Fuldagut, Sedum Hybridum Immergrunchen [3].

## 3 Condition of the roof and vegetation after 2 years

Since its implementation, a number of work activities have been taking place on the green roof of the kindergarten. First, the roof ridge was repaired, the wiring was repaired and the control segment of the green roof was installed 1 year after reconstruction.

**Fig. 6.** Control segment installation.

During continuous inspections during the year, differences in the density of vegetation can be seen on the roof. The conditions for the vegetation on the green roof are different throughout the year, because the roof is not artificially irrigated, but only uses rainwater.

So we can see in Figure 7 that in the dry season in July 2022 the roof was less green and the density was lower than when the first image was taken.

**Fig. 7.** Green roof in 07/2022.

The next imaging took place 3 months later in October 2022 (Fig.8). In the photos we can again see the reduction of the green area of the roof and small areas without vegetation. It can therefore be said that the appearance of the extensive green roof exactly copies the seasons and the dependence of the state of the vegetation.

**Fig. 8.** Green roof in 10/2022.

# 4 Thermal imaging and dormant season

One of the last imaging took place at the beginning of 2023. It was imaged with a thermal imaging device. The condition of the vegetation is recorded in detailed photos and we can see in them that the vegetation is dampened by the frost and especially very resistant plants and weeds are doing well.

**Fig. 9.** Green roof in 02/2023.

Green areas are mainly made up of mosses, which are well resistant to very low temperatures. Higher temperatures on the roof can only be measured on the ventilation openings and air conditioning chimneys (Fig.10).

**Fig. 10.** Thermovision of green roof in 02/2023.

**Fig. 11.** Perennials and hardy stonecrops.

The plants, which are waiting for more favourable conditions to grow, still look wilted (Fig.12).

**Fig. 12.** Perennials and hardy stonecrops.

## 5 Conclusion

The condition of the roof corresponds to the nature and type of the extensive green roof. In the summer months, the greenery on the roof is rather subdued due to the lack of water, but it still fulfils an aesthetic function. In contrast to the winter season, where the vegetation is dormant and looks unsightly. Control measurements of vegetated roofs in Brno show higher saturation percentages in the spring, and this is also reflected in the visual appearance of the roof. After a minimum year of data collection, it will also be possible to determine more precisely when and at what intervals it is optimal to perform green roof maintenance depending on the state of moisture of the substrate layer.

This article was created with the support of the Faculty of Civil Engineering of the University of Technology in Brno, Institute of Technology, Mechanization and Construction Management.

## References

1. J. Šimečková. *Green roofs – Hope for the future II,* **44**, 6 (2006)
2. K. Čechová, D. Kozáková. *Heat Island of the city. Nika: Nature magazine about environmental protection* [online]. **44**, 30 (2018)
   Available: http://www.nika-casopis.cz/data/files/18-12.pdf
3. Dek, *Greendek trávníková koberec TR K 20* DEK. **2**, 1 (2022) Praha,
   Available: https://cdn1.idek.cz/dek/document/808884467
4. M. Sněhota, J. Hanzlíková, M. Sobotková, P. Moravcik. *Water and thermal regime of extensive green roof test beds planted with sedum cuttings and sedum carpets.* **21**, 5 (2021).
5. Skala V, Dohnal M, Votrubova J, Vogel T, Dusek J, Sacha J, et al. Hydrological and thermal regime of a thin green roof system evaluated by physically-based model. Urban For Urban Greening 2020;48.

6. S. Konasová, The benefits of vegetated roofs in reducing the excess heat in three urban areas with different climate conditions. Advances and Trends in Engineering Sciences and Technologies III- Proceedings of the 3rd International Conference on Engineering Sciences and Technologies, ESaT 2018; 2019.
7. K. Nečadová, P. Selník, H. Karafiátová, Differential substrate subsidence of the EnviHUT project pitched extensive green roof. MATEC Web of Conferences; 2016.

# The assessment and treatment of waste from glass fibre production for use as a filler in composites

*Karolina* Hudec Jakubikova[1*], *Radek* Hermann[1], and *Rostislav* Drochytka[1]

[1] Faculty of Civil Engineering, Institute of Technology of Building Materials and Components, Brno University of Technology, Veveri 95, 602 00 Brno, Czech Republic

**Abstract.** Waste glass from fibreglass production is waste that is generated at various stages of production. In nature, it is a combination of glass fibres, fine dust, but also larger glass grains, for example also from leakages during production. This material is or may be contaminated by impregnation, lubricants or some impurities that prevent its recovery during production. The main objective of this study is to assess the suitability of glass fibre waste and to optimise the pre-treatment of the raw material for use in the composite. This paper describes the determination of chemical composition, mineralogical composition by X-ray diffraction analysis, determination of bulk density, water absorption and specific surface area. The results obtained are compared with the primary raw material - glass fibre. The results showed that the waste from glass fibre production is primarily composed of silica, calcium, alumina and boron. Minor components include magnesium oxide, sodium oxide, fluorine, iron oxide, strontium oxide, potassium oxide, titanium dioxide, sulphur dioxide and chromium oxide. The specific surface area of treated glass fibre waste correlates with the specific surface area of cement, which is a prerequisite for the use of this waste as a partial substitute for cement in the production of building materials.

## 1 Introduction

### 1.1 Waste from glass fibre production

Waste glass from fibreglass production is waste that is generated at various stages of production. It is a combination of glass fibres, fine dust, but also larger glass grains, for example also from leakages during production. This material is, or may be, contaminated by impregnation, lubricants or certain impurities which prevent its recovery during production [1].

Primary glass is a solid with a high content of amorphous $SiO_2$. It is a substance formed by solidification of a melt without crystallisation processes. The hardness of glass on the Mohs hardness scale is 5-7. The term glass recyclate refers to container glass, especially bottle glass, made from 50 % silica sand, 10 % $Na_2CO_3$ (soda), 12 % $CaCO_3$ (limestone), 18 %

* Corresponding author: jakubikova.k@fce.vutbr.cz

crushed waste glass shards and 4 % other substances. The glass melt, in the form of a red-hot viscous mass, is produced at a temperature of 1500-1700 °C when it is fed into blowing or press-blowing machines for bottle production. The glass thus produced may be clear or coloured [2,3].

Due to its properties, composition and versatility, so-called E-glass occupies a major share (>90 %) in the production of glass fibres. The approximate composition of E-glass is 52-56 % silica, 16-25 % calcium oxide, 12-16 % alumina, 5-10 % boron oxide, 0-5 % magnesium oxide and <1 % alkali ($Na_2O$, $K_2O$) [4].

The production of glass fibres consists primarily in the preparation of the charge and its transport to the melting furnace. This is followed by melting, shaping by mechanical drawing or centrifugation and sufficiently rapid cooling to prevent crystallisation.

In the production of glass fibres, various wastes are generated at different stages of production. In nature, it is a combination of glass fibres, fine dust but also larger glass grains, for example, also arising from leakages during production [2,5].

## 1.1 Purpose and issues of the research

Recycling of construction waste is one of the steps of sustainable construction, which leads to a reduction of environmental pollution as well as a reduction in the cost of building materials. In this regard, the aim of this work is to assess the suitability of waste fiberglass and to optimize the pre-treatment of the raw material for use in the composite.

The following properties are to be determined:

- Investigate the composition and properties of construction waste from the production of glass fibre.
- Determine the technological criteria for the suitability of the pre-treatment of the waste.
- Identify the possibilities of using the waste in construction materials in relation to its designated properties.

The results obtained are compared with the primary raw material - glass fibre. The annual production of glass waste including waste from glass fibre production in the Czech Republic is 3.45 tonnes (for 2017) [6].

Due to its composition and high $SiO_2$ content, waste from glass fibre production shows high pozzolanic activity, which in combination with cement promotes the formation of CSH gels in the cement matrix. Finely ground soda-lime glass shows high pozzolanic activity, consuming portlandite to form CSH phases. These coat the reaction edge of the glass grains and reduce the monosulphate level. This ability of fine glass can in turn reduce the risk of ACP. A prerequisite for such glass properties is the fineness of the grinding.

Pozzolans, are natural siliceous substances, silica-clay substances, or a combination of both. When mixed with water, pozzolans do not harden or set themselves, but when ground finely and suitably, they react in the presence of water and calcium hydroxide ($Ca(OH)_2$) to form silicate or calcium aluminate compounds, which are the carriers of strength. These compounds are similar to those formed during the hardening of hydraulic substances (e.g. cements). Pozzolans must contain at least 25 % silica ($SiO_2$) and also contain alumina ($Al_2O_3$), smaller amounts of iron oxide ($Fe_2O_3$) and other oxides. According to EN 197-1, pozzolans include, for example, natural calcined pozzolans, which are substances of volcanic origin, clays, shales or sedimentary rocks activated by heat treatment and which comply with the above. Furthermore, pozzolans include fly ash and silica fume [7–9].

Research studies indicate that if the glass grain size is less than 75 μm, the material exhibits sufficient pozzolanic activity. However, with the use of ground glass and glass in concrete in general, the problem of alkali-silica reaction (ASR) arises, a reaction whereby the reaction of alkali (cement) and amorphous silicates (glass) in the cement matrix results in the formation of new and larger volume formations over time, which can lead to structural failure of the entire structure. It is known from research that when the glass is ground to a higher specific surface area and the mean particle size is below 300 μm, ASR no longer occurs [10].

Limiting the recyclability of glass fibre waste is the possible contamination by impregnation, lubricants or some impurities that prevent its reuse in production. Impregnation or lubrication of glass fibres is a surface treatment of the fibres which provides surface protection during further processing and also ensures that the individual elementary fibres are bonded together. The most used chemicals for fibreglass lubrication are synthetic resins, waxes, oils or silicones [11].

## 2 Metods and Experiment

### 2.1 Determination of the pre-treatment of glass fibre waste

Within this part of the research, waste pre-treatment methodologies were developed together with the assumption of optimization of the sorting, crushing and grinding process.

#### *2.1.1 Verification of pre-treatment capabilities with commonly available laboratory and production facilities*

Figure 1 shows the grain size curve of the untreated waste glass, which was determined by sieving the raw material through a set of standard sieves. Pre-treatment of the waste glass fibre production was carried out first in a jaw crusher, then in a vibrating and ball mill. The specific surface area of the glass waste thus pretreated was then determined by the permeability method with a value of 330 $m^2 \cdot kg^{-1}$ using a Blain apparatus.

Pre-treated waste glass was of 0-0.8 mm fraction. The maximum grain size of 0.8 mm waste glass ensures better cohesiveness of the mass and also prevents the formation of silica gel in such a form that it damages the structure.

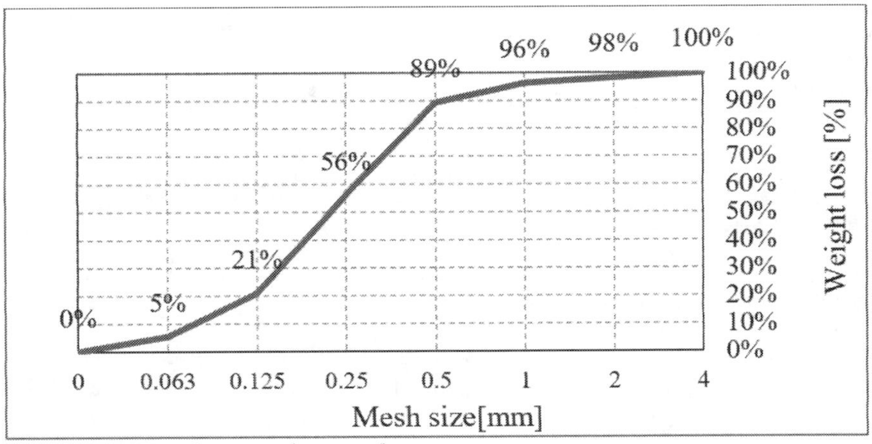


**Fig. 1.** Grain size curve of untreated waste glass.

### *2.1.2 Optimization of pretreatment processes to achieve suitable grain size curves*

When grinding glass, care must be taken to select an appropriate grinding time, as the time for additional grinding to a higher fineness increases exponentially with higher fineness. The optimum fineness of glass waste for partial cement replacement was determined by determining the specific surface area by the permeability method with a value of 330 $m^2 \cdot kg^{-1}$ using the Blain apparatus.

The use of waste glass as a substitute for the fine fractions of aggregate and sand in the production of composite materials is limited and may show negative results, mainly because of the risk of alkali-silica reaction between cement and waste glass aggregate and also because composite materials with higher amounts of glass replacing aggregate have lower flexural and compressive tensile strengths. The shape of the waste glass particles determines the degree of adhesion in the cement paste and therefore affects the compressive and flexural strengths.

In the optimisation of the pre-treatment processes, it was found that the use of pre-treated waste glass fibre production waste of 0-0.8 mm fraction as a filler to replace fine aggregate of the same fraction in the amount of 10-15 % of the total ballast, using a maximum grain size of 0.8 mm of waste glass, results in better cohesiveness of the mass and also does not lead to the formation of silica gel in such a form that it damages the structure. The resulting aggregate grain size curve used in the silicate composite is shown in Fig. 2 compared to the ideal grain size curve according to Fuller.

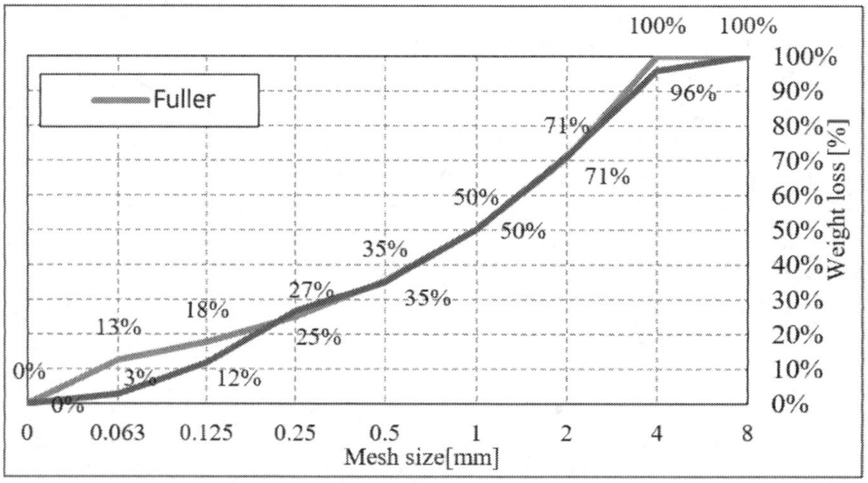


**Fig. 2.** Comparison of the ideal grain size curve according to Fuller and the used aggregate grain size curve in the composite rain size curve of untreated waste glass.

## 2.2 Research on glass fibre waste

The chemical composition was determined on a sample of the raw material - waste from glass fibre production. From the results of the chemical composition (Table 1) it can be seen that the waste from glass fibre production consists mainly of silica ($SiO_2$), calcium oxide (CaO), aluminium oxide ($Al_2O_3$) and boron oxide ($B_2O_3$). Minor constituents include magnesium oxide (MgO), sodium oxide ($Na_2O$), fluorine (F), iron oxide ($Fe_2O_3$), strontium oxide (SrO), potassium oxide ($K_2O$), titanium dioxide ($TiO_2$), sulphur oxide ($SO_3$) and chromium oxide ($Cr_2O_3$). From chemical analysis, it was possible to determine that this is so-called E-glass. E-glass occupies a major share (>90 %) in the production of glass fibres due to its properties, composition and versatility. The approximate composition of E-glass is 52-56 % silica, 16-

25 % calcium oxide, 12-16 % alumina, 5-10 % boron oxide, 0-5 % magnesium oxide and <1 % alkali ($Na_2O$, $K_2O$).

**Table 1.** Chemical composition of waste glass.

| $SiO_2$ | CaO | $Al_2O_3$ | $B_2O_3$ | MgO | $Na_2O$ | F | $Fe_2O_3$ | SrO | $K_2O$ | $TiO_2$ | $SO_3$ | $Cr_2O_3$ |
|---|---|---|---|---|---|---|---|---|---|---|---|---|
| 52,25 | 23,10 | 14,16 | 7,18 | 0,97 | 0,77 | 0,56 | 0,365 | 0,245 | 0,23 | 0,11 | 0,039 | 0,005 |

The results of the mineralogical composition show that the waste from the production of glass fibres is completely amorphous, which can also be confirmed by the presence of a typical curvature between the angles 2θ in the range 15-35°. The presence of iron oxides can also be confirmed from the results due to the typical interference between iron oxides when using a copper anode (CuKα), which appears as a discontinuous line in the resulting diffractogram.

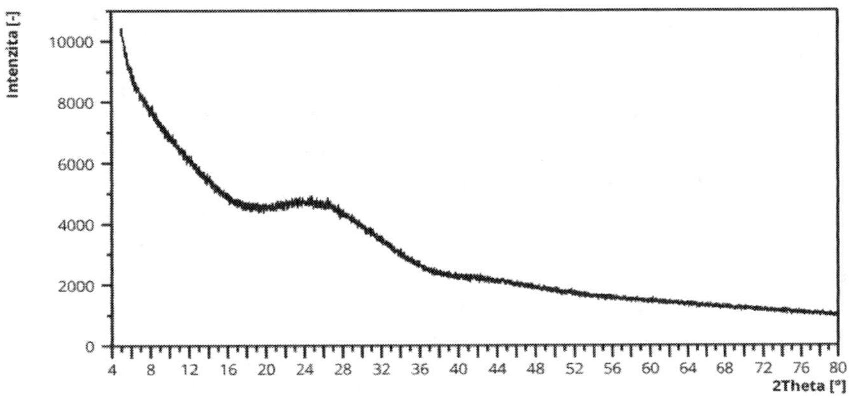


**Fig. 3.** X-ray diffraction analysis of raw material .

The results of the tests for the determination of bulk density, specific gravity, water absorption and specific surface area are given in Table 2 below. Since the raw material does not contain pores, the values for bulk density and specific gravity are the same. The raw material, waste from glass fibre production, is also non-absorbent. Pre-treatment of the raw material was carried out in a jaw crusher, vibrating and ball mill and then its specific surface area was determined by the Blaine method.

**Table 2.** Results of measured material properties of the raw material.

| Parameter | Value |
|---|---|
| Volume weight | 2.62 $g \cdot cm^{-3}$ |
| Specific gravity | 2.62 $g \cdot cm^{-3}$ |
| Absorbency | 0 % |
| Specific surface area | 330 $m^2 \cdot kg^{-1}$ |

The bulk density and specific gravity correlate with the density of ordinary glass. Because of the way the glass fibres are produced, which are completely amorphous, there are no pores in them and the absorbency of this waste from the production of glass fibres is therefore zero. The specific surface area of pre-treated glass fibre waste correlates with that of cement (300-350 $m^2 \cdot kg^{-1}$), which is a prerequisite for the use of this waste as a partial replacement for cement in silicate composites.

## 3 Results and Discussion

The research was carried out to determine the material properties of glass fibre production waste by means of a set of laboratory tests. The results of the chemical composition of the raw material are presented in Table 1, and the results of the mineralogical composition by X-ray diffraction analysis are presented in Figure 3.

The chemical composition was determined on the waste glass sample (Table 1), it was found that the waste glass fibre is primarily composed of silica, calcium, aluminium and boron. Minor constituents include magnesium oxide, sodium oxide, fluorine, iron oxide, strontium oxide, potassium oxide, titanium dioxide, sulphur dioxide and chromium oxide. Thus, it was possible to determine that the glass fibres are made of so-called E-glass.

The mineralogical composition was also determined by X-ray diffraction analysis (Fig. 3), from which it was found that the waste from glass fibre production is completely amorphous. As the raw material does not contain pores, the values of bulk and specific gravity are the same and also due to the absence of pores the raw material is also non-absorbent (Table 2). The specific surface area (Table 2) of the treated glass fibre waste correlates with that of cement, which is a prerequisite for the use of this waste as a partial substitute for cement in the production of building materials.

At the same time, a determination was made as to the method of pretreatment of waste from fibreglass production to maximise its use in the production of building materials. As part of the optimisation of the pretreatment process, it was found that the most advantageous is the pretreatment of waste from fibreglass production first in a jaw crusher, then in a vibrating or ball mill.

It was found that the pretreated waste glass can be used as a partial cement substitute due to its pozzolanic properties. The optimum value of specific surface area of pretreated waste glass fibre production is 330 $m^2 \cdot kg^{-1}$.

## 4 Conclusions

The study dealt with the testing and expert analysis of waste from glass fibre production. As part of the testing, the chemical composition of the raw material was determined. It was found that the waste from glass fibre production is primarily composed of silica, calcium, alumina and boron. From the determination of the mineralogical composition by X-ray diffraction analysis, it was found that the glass fibre waste is completely amorphous. Since the raw material does not contain pores, the values of bulk density and specific gravity are identical, and the absence of pores also makes the raw material non-absorbent. The specific surface area of the treated glass fibre waste correlates with that of cement, which is a prerequisite for the use of this waste as a partial substitute for cement in the production of building materials.

The second objective was to determine the optimum method of pretreatment of waste from fibreglass production to maximise its use in the production of building materials. It was found that the most advantageous method was pretreatment of waste from fibreglass production first in a jaw crusher, followed by grinding in a vibrating and ball mill. Waste glass can be used as a partial cement substitute due to its pozzolanic properties and the optimum specific surface area of the pre-treated fibreglass waste is 330 $m^2 \cdot kg^{-1}$. It was also found that the use of pre-treated glass fibre waste with a fraction of 0-0.8 mm as filler to replace fine aggregate of the same fraction in the amount of 10-15 % of the total ballast, better cohesiveness of the mass is achieved and the formation of silica gel in such a form as to damage the structure is avoided.

This paper was supported by the project of the Internal Grant Agency of Brno University of Technology, specific research project number FAST-J-23-8436 " Study of polymer composites with higher content of glass fibre waste."

## References

1. R. Gonçalves, A. Martinho, JP. Oliveira. *Recycling of Reinforced Glass Fibers Waste: Current Status.* Materials (Basel), **15**, p.159. (2022)
2. V. Hotař, V. Klebsa, B. Nikendey Holubová and V. Matoušek. *Automatic glass production technology,* Technical University of Liberec, p. 327. (2021)
3. K. L. Loewenstein, *The manufacturing technology of continuous glass fibres.* (1973)
4. A- Sissa, L. Delgado Sancho, S. Roudier, *Best available techniques (BAT) reference document for the manufacture of glass: industrial emissions Directive 2010/75/EU: integrated pollution prevention and control,* Joint Research Centre, Institute for Prospective Technological Studies. (2013)
5. D. Tucker, E. Ethridge a H. Toutanji, *Production of Glass Fibers for Reinforcement of Lunar Concrete.* Conference: 44th AIAA Aerospace Sciences Meeting and Exhibit. 7. (2006)
6. Department of Industrial Ecology, Section of Industry and Construction, Ministry of Industry and Trade; *analysis of the current state of selected commodities of secondary raw materials and their sources*, p. 219, Prague (2018)
7. A. Siddika, H. Hajimohammadi, A. Ferdous, Wahajwalla. *Roles of Waste Glass and the Effect of Process Parameters on the Properties of Sustainable Cement and Geopolymer Concrete*, Polymers, **13**, p.3935. (2021)
8. A. Aliabdo, A.E. Elmoaty, A. Y. Aboshama, *Utilization of waste glass powder in the production of cement and concrete.* Constr. Build. Mater., **124**, p. 866–877. (2016)
9. ČSN EN 197-1 Cement - Part 1: Composition, specifications and conformity criteria of cements for general use; The Office for Technical Standardization, Metrology and State Testing. (2012)
10. N. Kumarappan, *Partial replacement cement in concrete using waste glass.* International Journal of Engineering Research and Technology, **2.10**, p.1880-188. (2013)
11. L. Thomason, *Glass fibre sizing: A review, Composites Part A: Applied Science and Manufacturing*, **127**. (2019)

# The effect of cellulose fibres in an epoxy adhesive for bonding wooden load-bearing elements

*Petra* Lacikova[1*], *Jan* Vanerek[1], and *Adam* Betak[1]

[1]Brno University of Technology, Faculty of Civil Engineering, Veveri 331/95, Brno, Czech Republic

**Abstract.** The work deals with the modification of the properties of epoxy adhesive. The modification consists of reinforcing this adhesive with cellulose fibres. In the field of wood glued elements, the current research focuses on the possibility of increasing the ductile behaviour of the epoxy adhesive, which would significantly increase the ability of the adhesive to transfer the stresses induced by the volumetric changes of the wood adherend. For this purpose, the adhesive was modified with fibres having a reinforcing character, which may include cellulose fibres. Due to the epoxy resin's non-polar nature and the cellulose fibre's hydrophilic nature, which inhibits the fibre's homogeneous dispersion, the fibres have been modified with silanes. This paper describes the effect of adhesive modification by determination of tensile and shear strength at different amounts of filling and different types of silane pre-treatment of cellulose fibres. The optimum filling amount can be considered as 1.0% filling of nanofibrillated cellulose (NFC) with epoxy-functional silane pre-treatment.

## 1 Introduction

Epoxy resins possess versatility due to their ability to react with various substrates, making them widely used in industry for both protective coatings and structural applications, including laminates and composites. They are also commonly used for casting and bonding a variety of materials, such as metals, hard plastics, ceramics, and glass, with a particular emphasis on bonding wood [1-2].

These are two-component systems with extremely strong adhesive properties, allowing for the bonding of materials with high tensile and compressive strength. Epoxide resins are compounds that typically contain more than one epoxy (oxirane) group per molecule. This group is highly reactive, and its reactivity with a variety of substances (resulting in cross-linking of macromolecular products) is the basis for the applications of these resins as adhesives, laminates, and more. During the curing process, no by-products are released, and there is only slight shrinkage. These important thermosetting polymers exhibit high resistance to corrosion, a high modulus of elasticity, and are thermally stable. However, due to the highly cross-linked nature of epoxy resins, they tend to be brittle, limiting their use in many critical areas [1-5].

* Corresponding author: lacikova.p@fec.vutbr.cz

In today's world, where there is an increasing focus on producing high-quality products and constructions, epoxy adhesives are frequently modified in a variety of ways. One such method of modifying epoxy adhesives is filling them with fibre fillers to reinforce them [5-9].

The production of synthetic fibres requires large amounts of energy and threatens the quality of the environment, during production and recycling. Meanwhile, the processing of natural fibres is gaining popularity again due to their lower cost in comparison to synthetic fibres. Natural fibres have various benefits such as their lightweight, non-abrasive, and non-toxic nature, low acquisition costs, and biodegradability. Examples of natural fibres include flax, bamboo, sisal, hemp, jute, and especially cellulose fibres [7, 10-12].

Cellulose is the most abundant and important biopolymer on Earth. The potential uses of cellulose have developed and increased since the beginning of nanocellulose preparation. Nanocellulose has very good physical, mechanical, biological, and chemical properties. Nanocellulose can be divided into three groups, nanocrystalline cellulose (NCC), microfibrillated cellulose (MFC) and nanofibrillated cellulose (NFC) [10].

Cellulose is usually obtained by pulping from trees suitable for producing high-quality pulp. One of the disadvantages of cellulose fibres for their industrial applications is the strong hydrophilic character of their surface, which inhibits the homogeneous dispersion of these fibres in non-polar polymers. Because of the hydrophilic nature of the fibres, the adhesion is weakened, which in turn reduces the mechanical properties of such filled composites. Therefore, ways are being sought to chemically modify the natural fibres to make them compatible with the hydrophobic polymer matrix [7, 10].

The main objective of this research was to mitigate the brittle behaviour of the epoxy adhesive that was used by the addition of reinforcing fibres. The improvement of mechanical and other properties of composite materials depends mainly on the filler content, shape and size, surface properties and degree of dispersion [5]. Cellulose fibres were used to create a more flexible bonded joint. To ensure good dispersion of the fibres in the adhesive matrix, silane pre-treatment of the fibres was chosen. The silane pre-treatment was chosen to change the hydrophilic nature of the cellulose fibres to a hydrophobic nature, thereby providing better bonding to the epoxy adhesive matrix, which is non-polar in nature. The effects of the reinforcing fibres were verified by the resulting strength parameters and microscopic observation of filler dispersion. The bonded lap joints were exposed to hygrothermal stress. Epoxies, due to their brittle behaviour, are unable to accommodate the stresses induced by the volumetric changes in the wood adherend. The effect of modification with cellulose fibres was verified based on the tensile shear strength of the bonded samples after hygrothermal stress.

# 2 Materials and methods

## 2.1 Materials

A two-component epoxy system (Spolchemie, Czech Republic) was used for the experiment. The epoxy system consists of component A (epoxy resin: CHS-EPOXY 531) and component B (hardener: TELALIT 0492). The mass ratio of epoxy resin to hardener is 100:27. After curing, the manufacturer recommends post-curing at elevated temperature to achieve maximum final properties, specifically for 72 hours of exposure at 85°C. Post-curing will allow maximum mechanical and chemical properties to be achieved. The selected properties of the epoxy resin and hardener are shown in Table 1.

The surface silane pre-treatment of the fibres to improve the dispersion of cellulose in the non-polar polymer matrix and increase the interfacial bonding was used. For the silane treatment, the epoxy-functional silanes Silquest A-187 and Coatosil MP 200, amino-functional silanes Silquest A 1170, Silquest Y-1699 and Silquest A-1100 (Momentive, Grolman Group, Germany) were used.

**Table 1**: The selected properties of the components of the epoxy system.

| **Epoxy resin CHS-EPOXY 531** | |
|---|---|
| Epoxy index [eg/kg] | 5.5 – 5.7 |
| Epoxy mass equivalent [g/mol] | 175 - 182 |
| Viscosity at 25 °C [Pa.s] | 1.5 – 2.3 |
| Free epichlorohydrin [ppm] | max 10 |
| **Hardener TELALIT 0492** | |
| Amino number [mgKOH/g] | 550 - 600 |
| Viscosity at 23 °C [mPa.s] | 15 - 30 |

Fibrous cellulose powder TENCEL FCP 9/300 (Lenzing Fibers, Austria) was used as a reinforcing filler. The cellulose filler is white in colour, 100% pure and has a density of 0.35-0.60 g/cm$^3$. The sieve analysis (according to DIN EN ISO 4610) for fibrous cellulose powder is summarized in Table 2. To fill the epoxy adhesive for lap joints, microfibrillated cellulose (MFC) was transformed into nanofibrillated (NFC) cellulose. After microscopic analysis, the length of the NFC fibres was determined to be 180 μm with a diameter of 10 μm. For the shear-stressed glued lap joints, beech planks (*Fagus sylvatica* L.) with a density of (700 ± 50) kg/m$^3$ at a wood moisture content of (12 ± 1) % were selected.

**Table 2:** Sieve analysis by air jet sieve of fibrous cellulose powder (sieve analysis was obtained from TENCEL® FCP 9/300 material data sheet from Lenzing Fibers company, Austria).

| **Sieve [mm]** | **Remain [%]** |
|---|---|
| 0.200 | max. 3 |
| 0.100 | max. 40 |
| 0.032 | max. 85 |

## 2.2 Preparation of samples for tensile properties

Test samples of modified epoxy blends were prepared using a mechanical homogenizer with a constant filling amount of 1.0% MFC fibres with 0.6% and 2.4% silanes. A mechanical homogenizer was also used to disperse the epoxy adhesive with 1.0% MFC without silane and the reference mixture (epoxy resin + hardener). This procedure was chosen to determine the type and optimum amount of silane treatment of cellulose fibres for the next stages.

The amount of epoxy was heated to approximately 55°C. The silane was then poured into epoxy and MFC cellulose in the proper amount was added. The homogenizer (ULTRA-TURRAX T25, IKA, Germany) was set for 12,000 rpm for 20 minutes (Fig. 1). The hardener was added and the mixture was poured into the mould, where the mixtures were left to harden for 24 hours. After curing, the mixtures were consequently exposed to 80°C for 72 h for the post-curing process.

An ultrasonic homogenizer (SONOPULS, Bandelin, Germany) was used to defibre MFC to NFC (Fig. 2). The use of the ultrasonic homogenizer was intended to prevent the formation of air pores in the epoxy matrix and better dispersion of the fibres. A description of the pulping of the MFC to NFC and preparation of the modified adhesive by the ultrasonic homogenizer is given in the following subsection 2.3. For tensile strength, the epoxy resin with 1.0% NFC was prepared by the ultrasonic homogenizer. Test samples were subjected to a tensile strength test according to standard EN ISO 527-1 (type 1B) [13].

### 2.3 Preparation of test samples for tensile shear strength of bonded joints

After selecting the optimum amount and type of silane, the amount of cellulose fibre filling was optimized. The mixtures for lap joints are summarized in Table 3. MFC was diluted with ethanol at a certain fibre content. Selected silane (Coatosil MP 200) was added to the mixture (MFC in ethanol) in selected amounts of 2.4%. The mixture was stirred by the ultrasonic homogenizer (Fig. 2) at 50 Hz in 10 cycles for 2 min, with a 1 min delay between cycles to lower the probe temperature. After this pulping process, MFC becomes NFC.

Then the mixture is added to the heated epoxy resin at about 55 °C, which is mixed by the ultrasonic homogenizer at 50 Hz also in 10 cycles for 2 min, with a 1 min delay between cycles to lower the probe temperature. The mixture was stored for 2 hours in a vacuum oven at 50°C where the volatile ethanol was removed from the epoxy mixture. After two hours, a quantity of hardener was added to the mixture.

**Table 3:** Mixtures to optimise the amount of cellulose.

| Designation | Epoxy resin | Hardener | Silane/Amount | Cellulose/Amount of epoxy | Mixing |
|---|---|---|---|---|---|
| EP1 | CHS-EPOXY 531 | Telalit 0492 | - | - | Ultrasonic |
| EP2 | | | - | NFC 1.0 % | |
| EP3 | | | Coatosil MP 200/2.4 % | NFC/0.5 % | |
| EP4 | | | | NFC/1.0 % | |
| EP5 | | | | NFC/2.0 % | |
| EP6 | | | | NFC/5.0 % | |

The epoxy adhesive was applied to the rounded beech boards (150×130 ×5 mm) at a rate of 300 g/m$^2$. The adhesive was applied evenly to both the surfaces of the beech boards, which were glued together to form a 10 mm thick panel. The panel was then placed in a pressing assembly where a pressure of approximately 0.8 N/mm$^2$ was applied for 24 hours. Post-curing was not included in the preparation of these wood assemblies.

After three days, five 20 mm wide strips were cut from the bonded panels with close contact in the direction of the fibres. The test samples were prepared according to standard EN 302-1 [14]. Grooves were cut in the glued panel, transverse to the grain of the wood, to create a shear surface (20×20 mm) in the central part. The cuts separated the wood and glue layers but did not extend beyond the glue layer (Fig. 3).

**Fig. 1:** Homogenizer ULTRA-TURRAX T25, IKA during homogenization of the mixture with MFC.

**Fig. 2:** Ultrasonic homogenizer Bandelin SONOPULS during homogenization of the mixture with NFC.

The specimens (Fig. 3 (c)) were stressed in hygrothermal exposure environments (A4, A5) defined in EN 302-1. The test samples were placed horizontally into the water for soaking at all the sides. The standard climatic environment 20°C/65% (A1) was created by mixing $NH_4NO_3$ salt with water (63.9 g of salt per 100 ml of distilled water).

a) b) c)

**Fig. 3:** a) cutting grooves, b) notched samples, c) samples with 2.0% NFC (right A1, A4, A5) after the test.

### 2.4 Experimental procedures

#### *2.4.1 Determination of tensile properties of epoxy*

The determination of the tensile properties was performed according to the standard EN ISO 527-1 [13]. From each mixture with different types of silane pre-treatment, six specimens were prepared. The Testometric M350-20CT press device was used to determine the tensile strengths. A strain gauge (of 50 mm length) was placed on the working part of the samples, crosshead constant rate of 1 mm/min was set until the test specimen broke. The maximum applied force was recorded and then the tensile strength of the material was calculated.

#### *2.4.2 Determination of the tensile shear strength of lap joints*

The tensile shear strength of lap joints of beech wood was performed according to standard EN 302-1 [14]. The test specimen was symmetrically clamped in the jaws of a press device Testometric M350-20CT. The rate of force increase of the tensile testing machine was 1mm/min. The resulting value of shear strength was calculated as the average strength of ten valid measurements from each mixture of lap joints. The percentage of failure in the wood was visually estimated for each test body.

#### *2.4.3 Microscopic analysis*

Microscopic analysis of NFC dispersion in the adhesive matrix was carried out using a desktop scanning electron microscope MIRA3 (Tescan, Czech Republic). The instrument was set to an acceleration voltage of 20 kV, a magnification range of 200-5000×, and a secondary electron detector was used.

## 3 Results and discussion

### 3.1 Tensile strength

Different functional silanes and homogenization technologies were selected to verify which type of modified epoxy adhesive to use for application to wood lap joints. Based on the tensile strength results (Fig. 4), the homogenization technology, filler form and silane type were optimized.

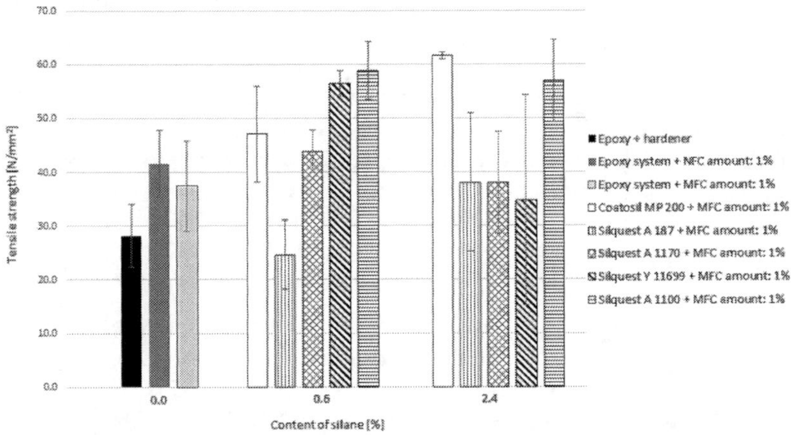


**Fig. 4:** Effect of type, amount of silane and filling in the epoxy adhesive on tensile strength.

For a silane content of 0.6%, the highest tensile strength was 58.9 $N/mm^2$ for Silquest A 1100 (amino-functional silane); in general, epoxy-functional groups achieved the lowest strengths. For a silane content of 2.4%, Coatosil MP 200 (epoxy-functional) had the highest tensile strength of 61.6 $N/mm^2$ and Silquest A 1100 (amino-functional silane) with a strength of 57.0 $N/mm^2$. It should be noted that high strengths were achieved even when the mixture (epoxy with MFC) was mixed with a mechanical homogenizer and the cured samples contained air voids (Fig. 6). The assumption that the strength increases with higher amounts of silane was confirmed identically for the epoxy-functional silanes (Coatosil MP 200 and Silquest A 187). Considering the assessment of the variance of the values, Coatosil MP 200 silane at 2.4% was chosen for further experimental purposes.

### 3.2 Tensile shear strength of glued joints

The wood glued specimens were exposed to hygrothermal stress (A4 and A5). All specimens showed a visible decrease in strength compared to reference environment A1. The results of the tensile shear strengths are shown graphically in Fig. 5. After determining the tensile shear strength, the degree of cohesive failure of the wood was visually assessed (the higher the value, the more stable the glued joint was).

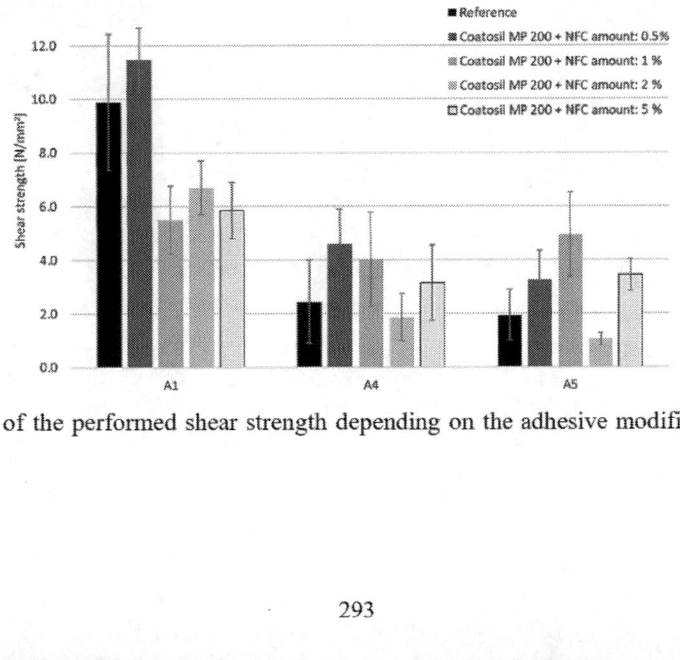


**Fig. 5:** Results of the performed shear strength depending on the adhesive modification and sample treatment.

The highest tensile shear strength was obtained in the acclimation environment (A1) at 0.5% NFC filling of 12.5 N/mm$^2$, which corresponded to 50% cohesive failure in wood. The reference specimens showed the highest cohesive failure rate in wood in A1. When the epoxy was filled with 1.0% NFC, the tensile shear strength in all three exposures (A1, A4 and A5) showed almost the same values, and the highest values of cohesive failure (70%) were obtained in exposures A4 and A5, indicating a stable bonded joint. In the reference case, there was almost a fourfold decrease in strengths after hygrothermal stress (A4 and A5) compared to the A1. In all cases, it was confirmed that the application of NFC fibres to the adhesive increased the stability of the bonded joint in exposures A4 and A5.

### 3.3 Microscopic analysis (SEM)

Fracture surface analysis (samples after shear strength determination and in the case of 1.0% MFC filling, samples after tensile strength determination) was carried out to assess the difference in mixing technology and to assess the dispersion of NFC filler in the epoxy adhesive matrix. Fig. 6 a) and Fig. 6 b) are micrographs of the mixtures that were processed by the mechanical homogenizer (MFC fibres). Air pores and cracks are observed in these images. In the fracture surface images of mixtures prepared by an ultrasonic homogenizer (NFC fibres) – see Fig. 6 c) to f), the presence of air pores was not evident. The use of different amounts of filling is visible in Fig. 6 c) and d). All the micrographs in Fig. 6 demonstrate that the dispersion of the fibres in the epoxy matrix was orderly.

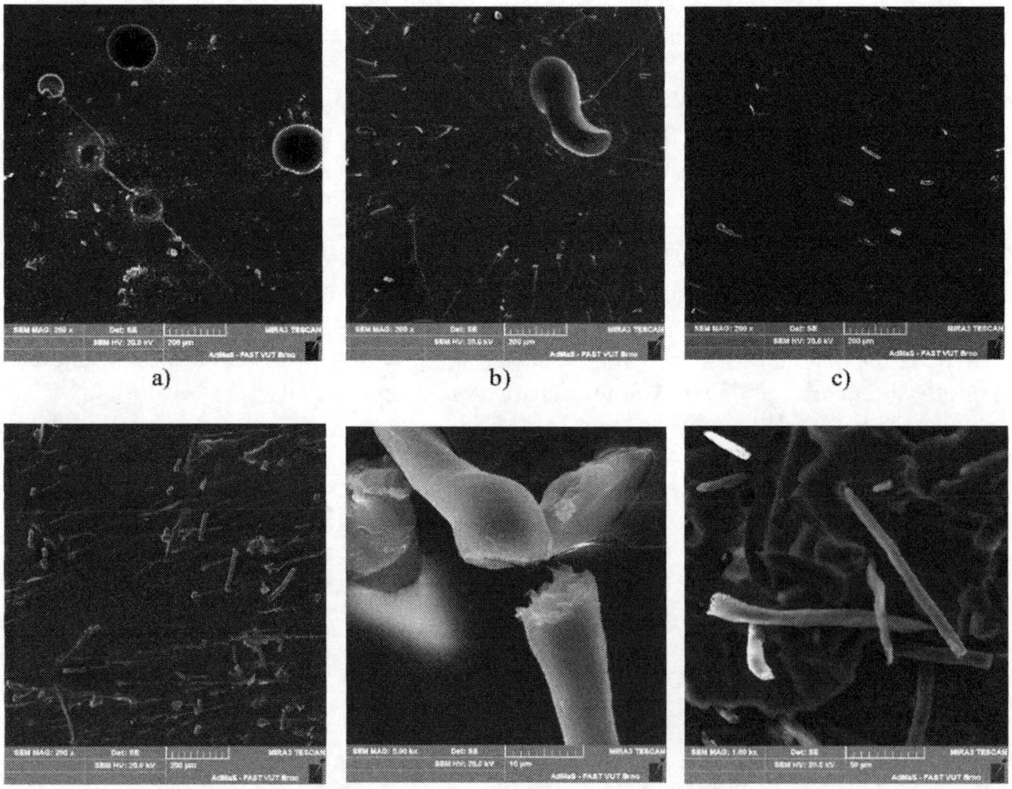


**Fig. 6:** Micrographs from SEM - a) reference epoxy (200x), b) epoxy with 1.0% MFC (200x), c) fibre dispersion at 0.5% NFC filling (200x), d) fibre dispersion at 1.0% NFC filling, e) detail of broken NFC fibre at 2.0% filling (5000x), f) fibre distribution at 5.0% filling (1000x).

# 4 Conclusion

The experiment evaluated the effect of pre-treatment of cellulose fibres with different types of silanes. Based on the tensile properties of the modified epoxy system (modification by 1.0% of MFC cellulose fibres), the amount of silane pre-treatment by 2.4 wt% appears to be the optimum amount for proper dispersion of the two-phase system. Coatosil MP 200 epoxy-functional silane had the most positive impact on this epoxy system. Based on the assessment of the influence of cellulose fibre size effect in epoxy adhesive, pulping cellulose into the NFC showed better strengths than MFC- filled epoxy specimens.

Due to the change in filler dimensions, an ultrasonic homogenizer was chosen for better dispersion. Based on SEM analysis, it fulfilled the assumption of avoiding the formation of air pores in the adhesive matrix. The micrographs from the SEM analysis also confirmed good dispersion of the NFC cellulose fibres in the adhesive, no clumps were observed, hence the silane pre-treatment was an appropriate step. Proper dispersion was also observed even at the highest amount of NFC at 5.0% using the same amount of silane coupling agent.

In all cases, it was confirmed that the application of NFC fibres to the epoxy adhesive increased the stability of the bonded joint under hygrothermal stress. Such modified adhesive can better adapt to the volume changes of the wood adherend. From the determined properties, 1.0% of NFC filling with silane pre-treatment of cellulosic fibres (Coatosil MP 200) appears to be the optimum filling rate for wood glued materials. The adhesive treated in this way can transfer the stresses in the glued joint induced by the volumetric changes of the wood adherend.

This paper was supported under the Internal Grant Agency of the Brno University of Technology, Faculty of Civil Engineering, specific junior research grant number FAST-J-23-8271with project name: *Improving the properties of the bonded joint using a fibre-filled isocyanate-based adhesive.*

# References

1. L. V. McAdams, J. A. Gannon, High Performance Polymers and Composites, 258-318, (1991)
2. C. A. May, Epoxy Resins Chemistry and Technology, (1988)
3. C. K. Riew, J. K. Gillham, *Rubber Modified Thermoset Resins*, Advances in Chemistry, **208**, (1984)
4. B. Ellis, Chemistry and Technology of Epoxy Resins. (1993)
5. A. Omrani, L. C. Simon, and A. A. Rostami, *Influences of cellulose nanofiber on the epoxy network formation*. Mater. Sci. Eng.: A. **490**, 131-137, (2008)
6. H.P.S. Abdul Khalil, A.H. Bhat, and A.F. Ireana Yusra *Green composites from sustainable cellulose nanofibrils: A review*. Carbohyd. Polym. **87**, 963-979, (2012)
7. J. Lu, P. Askeland, and L. T. Drzal, *Surface modification of microfibrillated cellulose for epoxy composite applications.* Polymer. **49**, 1285-1296, (2008)
8. M.H. Gabr, M.A. Elrahman, K. Okubo, and T. Fujii, *Effect of microfibrillated cellulose on mechanical properties of plain-woven CFRP reinforced epoxy*. Compos. Struct. **92**, 1999-2006, (2010)
9. H. A. Al-Turaif, *Relationship between tensile properties and film formation kinetics of epoxy resin reinforced with nanofibrillated cellulose*. Prog. Org. Coat. **76**, 477-481, (2013)

10. T. Abitol, A. Rivkin, Y. Cao, Y. Nevo, E. Abraham, T. Ben-Shalom, S. Lapidot, and O. Shoseyov, *Nanocellulose, a tiny fiber with huge applications.* Curr. Opin. Biotech. **39**, 76-88, (2016)
11. N. Lavoine, I. Desloges, A. Dufresne, and J. Bras, *Microfibrillated cellulose – Its barrier properties and applications in cellulosic materials: A review*. Carbohyd. Polym. **90**, 735-764, (2012)
12. J. Yuan, G. Du, H. Yang, et al. *Functionalization of cellulose with amine group and cross-linked with branched epoxy to construct high-performance wood adhesive*. Int. J. Biol. Macromol. **222**, 2719-2728, (2022)
13. EN ISO 527-1, Czech Office for Standards, Metrology, and Testing (2020)
14. EN 302-1, Czech Office for Standards, Metrology, and Testing (2013)

# Size effect of bonded area for bonded joints between ETICS with an acrylic thin-film plaster and sheet metal component

*Ales* Prucha[1*], *Barbora* Necasova[1], and *Jiri* Slanhof[1]

[1]Faculty of Civil Engineering, Brno University of Technology, Veveri 331/95, 602 00 Brno, Czech Republic

**Abstract.** The issue of anchoring sheet metal components has been part of construction practice for a very long time. Sheet metal components can be found on most contemporary buildings and due to the increasing thickness of insulation the possibility of anchoring to the supporting substrate is a problem. This paper discusses the issue of the bonded joint size dependence on the load capacity for fastening sheet metal elements to ETICS facades. A thixotropic silane modified polymer adhesive was used for this research. It was concluded that the size has a significant effect on the load capacity of the joint and their dependence was described.

## 1 Introduction

Plumbing elements are an important part of every modern building. These elements are mainly used to cover and prevent water from entering the structure. A correct and well-executed technological procedure is therefore very important. In these times of energy crisis, more and more attention is being paid to the energy self-sufficiency of buildings and one of the ways to achieve this goal is to insulate them. In practice, this insulation is most often done with ETICS systems with 200 mm thick thermal insulation [1,2] And the anchoring of plumbing elements into these insulation systems is highly inappropriate due to the perforation of the system and possible water seepage. Previous research has already shown that glued joints are a very suitable alternative for fastening through the thermal insulation system to the substrate and shows its high load-bearing capacity which exceeds the load-bearing capacity of the plaster guaranteed by the manufacturer [3].

The paper continues the initial series of tests and focuses on the effect of the bonded joint size and the possible negative effect of the thermal insulation on the tensile capacity and discusses the possibility of further research.

* Corresponding author: Ales.Prucha@vut.cz

# 2 Materials and used method

## 2.1 Materials

In a study of bonded joints for fastening sheet metal elements to modern facades, hot-dip galvanized steel was the most suitable material of the three tested with the highest adhesion 0,694 MPa. [3] For this reason, hot-dipped galvanized steel brand S235JR (1.0038) according to the EN 10025-2 standard [4] with a tensile strength of min. 360.0 MPa and a yield strength of 235.0 MPa were selected and manufactured for this size dependence test.

The thixotropic silane modified polymer adhesive ARDEX CA 20 P for external applications produced by Ardex Baustoff GmbH was chosen for bonding to the substrate.

ETICS was used as the base substrate with the following layer sequence: CETRIS BASIC - 16mm, Adhesive and screed matter UNI AM - 3 mm, EPS Polystyrene - 40 mm, Adhesive and screed matter UNI AM - 3 mm with Fabric R117 mesh 4x4.5 mm, 145 g/m$^2$, Universal primer UNI PUTZGRUND, Polished acrylic plaster KUNSTHARZ-PUTZ - 1,5 mm.

## 2.2 Methods

### *2.2.1 Preparation of Test Specimens*

For this test, samples were made from the following materials: base particleboard, cement screed, thermal insulation, cement screed with fiberglass fabric, polished acrylic plaster and a metal circular target. For the tensile test, the metal targets were made of hot-dip galvanized steel in the following sizes: 125 mm$^2$, 250 mm$^2$, 500mm$^2$, 1000 mm$^2$ and 2000mm$^2$. For the test was used a façade composition that corresponds with the actual composition commonly used on buildings with a contact insulation system. For the tests, thermal insulation which is soft and easily deformable was used on purpose. The intention was to measure the deformation of the adhesive as a dependence of its interaction with the thermal insulation, which has a significantly lower tensile strength than the adhesive under test. A 1250 mm x 200 mm test specimen was made, with 3 mm thick trowel cement and 40 mm thick thermal insulation. On top of the thermal insulator, a screed cement was applied in two layers, with fibreglass fabric fitted between the two layers. After the screed had cured, the plaster application process started. This process consists of two parts: the primer and the plaster itself, with a thickness of 1.5 mm. Then the metal circular target was bonded, with close attention to the exact size of the bonded area without internal bubbles and a constant adhesive thickness of 3 mm to avoid negative influence on the results [5]. The composition of the test specimen for the tensile test is shown in Figure 1.

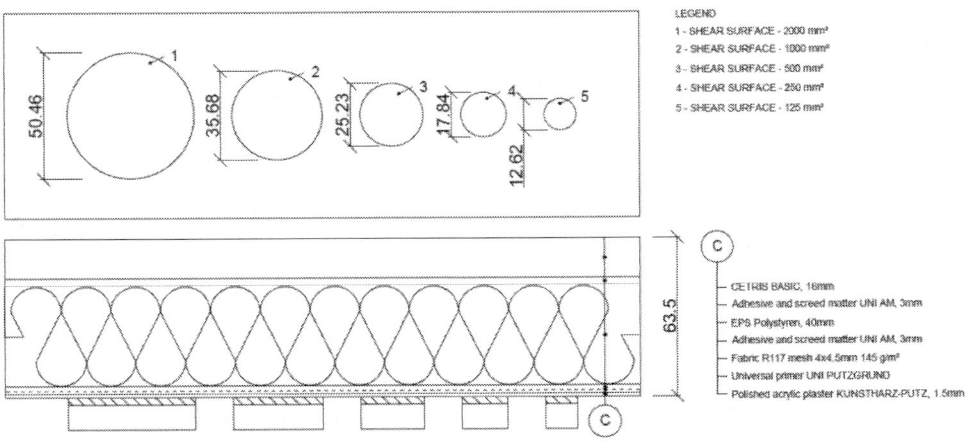


**Fig.1.** Test specimen for a tensile test.

### 2.2.2 Adhesion Tests

For this test, only tensile adhesion in the transversal direction was chosen, where the point of the test is to record the force course in relation to the elongation of the adhesive that is required to tear the adhered metal from the substrate. The tests were carried out on a LaborTech E.2 measuring device fitted with L06 vice jaws under normal laboratory conditions at a temperature of (+20±3) °C and relative humidity of (55±10) % with load bearing carried out at 2 mm per minut until failure, defined by a drop in force to 50% of the maximum achieved tensile force. The most important values for the evaluation were the maximum achieved force F (in N). The tensile test was based on the procedure defined by the Czech technical standard ČSN 732577 [6]. The tensile strength $\sigma_{adh}$ (in MPa) was subsequently calculated according to the following relation :

$$\sigma_{adh} \coloneqq \frac{F}{A} = \frac{F}{125;\,250;\,500;1000;\,2000} \tag{1}$$

## 3 Results

The results were used to generate two charts showing the dependence of the applied force F on the area A shown on Figure 2 and a tensile strength $\sigma_{adh}$ on the area A is in Figure 3.

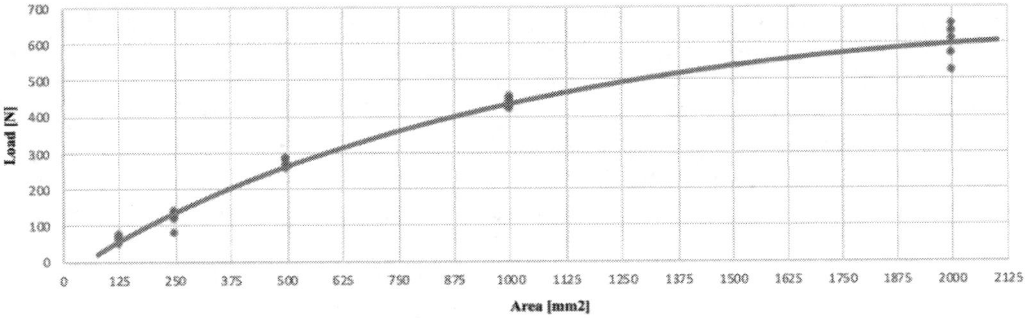


**Fig. 2.** Dependence of the applied force F on the area A.

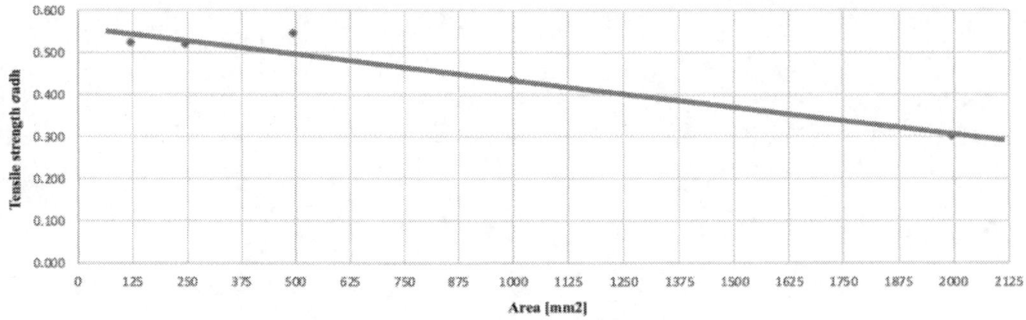


**Fig. 3.** Dependence of the applied tensile strength $\sigma_{adh}$ on the area A.

The data from the deformation diagrams were tabulated in a clear way, with the decisive point for the evaluation being the maximum tensile force. Based on this data, the tensile adhesion $\sigma_{adh}$ was calculated according to Formula (1). Statistical evaluation was performed for all the calculated parameters in the form of calculating the standard deviation $s_x$ and the coefficient of variation $v_x$. The tensile test results can be seen in Table 1.

**Table 1.** Results of tensile adhesion $\sigma_{adh}$ (in MPa) and force F (in N), standard deviation $s_x$ (in MPa for $\sigma_{adh}$, in % for ε) and coefficient of variation $v_x$ (in %) when acting on different areas.

| | 125 mm$^2$ | 250 mm$^2$ | 500 mm$^2$ | 1000 mm$^2$ | 2000 mm$^2$ |
|---|---|---|---|---|---|
| $\sigma_{adh}$ | 0.519 | 0.518 | 0.543 | 0.433 | 0.300 |
| $s_x$ | 0.069 | 0.081 | 0.023 | 0.013 | 0.022 |
| $v_x$ | 13.346 | 15.711 | 4.193 | 3.065 | 7.414 |
| F | 64.79 | 128.92 | 270.88 | 432.89 | 600.49 |
| $s_x$ | 8.61 | 20.26 | 11.33 | 13.22 | 44.52 |
| $v_x$ | 13.29 | 15.72 | 4.18 | 3.05 | 7.41 |

[1] The average values are a summary of six measurements conducted for each tested combination.

## 4 Discussion

The aim was to verify the influence of the size of the bonded joint on the overall load-bearing capacity of the bonding technology of sheet metal elements on facades with ETICS. Partial results without thermal insulation have already been published under normal laboratory conditions, where the joint had a load bearing capacity of 0.694 MPa on a target area of 2500 mm$^2$ [3].

The closest to this result were the 125 mm$^2$, 250 mm$^2$ and 500 mm$^2$ targets, but these were also affected by an approximate 24% reduction in tensile adhesion. A significant decreasing trend was seen for the larger targets. This trend can be explained by the significant influence on the cohesive strength by the thermal insulation, which has a strength only one sixth compared to the strength of the acrylic plaster itself.

As the area increases, there is more interaction with the thermal insulation, which extends more deeply and more elongation can be seen.

## 5 Conclusions

Measured tensile adhesive strengths were the highest for the measured surfaces 125 $mm^2$ 0.519 MPa, 250 $mm^2$ 0.518 MPa, 500 $mm^2$ 0.543 MPa and further decreased linearly or slightly exponentially for individual surfaces. The smallest measured tensile adhesion was measured at the largest area of 2000 $mm^2$. On the other side, the measured tensile forces decreased gradually due to increasing area and the difference between the smallest area and the largest area was decreased by 42%.

The following conclusions were drawn based on the results obtained:

- The dependence of force on area is not linear
- The tensile strength is decreasing with increasing area and is limiting towards the weakest layer of ETICS, i.e. thermal insulation with a strength of 0.1 MPa.
- For the overall load-bearing capacity of the joint, a larger amount of smaller joints is preferable due to the limitation of the joint interaction only to the plaster with a high load-bearing capacity, not the thermal insulator

It is obvious that the load-bearing capacities of the individual base layers significantly influence the overall load-bearing capacity of the bonded joint by their interaction. For a complete verification of the load capacity, the shear adhesion test to the substrate must be additionally carried out, which, as expected, should not be affected as much as in the tensile test.[7]

This research has been funded by an Internal Grant Programmes of the Brno University of Technology under No. FAST-J-22-7871 and No. FAST-J-23-8230.

## References

1. J. Ślusarek, B. Orlik-Kożdoń, J. Bochen, T. Muzyczuk, *Impact of the imperfection of thermal insulation on structural changes of thin-layer façade claddings in ETICS*, J. Build. Eng., **32** (2020)
2. L. Olsson, *Rain resistance of façades with façade details: A summary of three field and laboratory studies*, J. Build. Phys. **41**, 521–532, (2018)
3. J. Šlanhof, A. Průcha, B. Nečasová, A. Boháček, *Use of Bonded Joints for Fastening Sheet-Metal Components to Contemporary Facades Fitted with an External Thermal Insulation Composite System with Thin-Layer Acrylic Plaster*, Mat. Proc., **13**, 0-9, (2023)
4. ČSN EN 10025-2, *Hot rolled products of structural steels—Part 2: Technical delivery conditions for non-alloy structural steels*, Czech Standards Institute: Prague, Czech Republic, (2020)
5. P. Davies, L. Sohier, J.-Y. Cognard, A. Bourmaud, D. Choqueuse, E. Rinnert, R. Créac'hcadec, *Influence of adhesive bond line thickness on joint strength*, Int. J. Adhes. Adhes. **29**, 724–736 (2009)
6. ČSN 732577, *Test for Surface Finish Adhesion of Building Structures to the Base. Sorting No. 32691*, Czech Office for Technical Approvals and Measurements: Prague, Czech Republic, 4, (1981)
7. EN 1465, *Adhesives - Determination of Tensile Lap-Shear Strength of Bonded Assemblies, Sorting No. 66 8510*, Czech Office for Standards, Metrology and Testing: Prague, Czech Republic, 12, (2009)

# HBIM: The digital transformation of cultural and existing heritage

*Maria* Grazianova[1*], and *Peter* Mesaros[1]

[1]Technical University of Kosice, Faculty of Civil Engineering, Institute of Construction Technology, Economics and Management, Vysokoskolska 4, 042 00 Kosice, Slovakia

**Abstract**. BIM, or Building Information Modelling, is the future of architecture, engineering, construction and operations. We very often associate this definition with new buildings, while 80% of buildings in 2050 already exist today. A significant part of them are historical buildings. It is important to know how we can use BIM in this context and how this modern design tool serves to model and manage the unpredictable situations and complexities of historic buildings. The dissimilarity, the difficulty of data exchange between interested parties and therefore the difficulty of standardization is in this regard the challenge to be overcome with older buildings. In connection with architectural heritage, information modelling of heritage buildings is referred to under the name of HBIM, which is a built architecture representation system composed of libraries of semantically structured and parameterized objects. The basic process is the processing of the data in our possession, which are all transferred to a BIM-type model. SCAN to BIM is an operation that allows you to connect a point cloud created by laser scanner or photogrammetry to develop historical building management models that can be used in different application areas: accessibility and improvement, archival documentation research, conservation and visualization of cultural heritage, reconstruction and management of heritage, artificial intelligence (AI) and virtual reality (VR).The purpose of the paper is to provide an basic overview of the concept of HBIM. The contribution provides a brief overview of the regulatory framework and an overview of the methodology and workflow in the mentioned area with the appropriate conclusions.

## 1 Introduction

NIBS (National Institutes of Building Science) defines BIM as the "digital representation of the physical and functional characteristics of an object" [1].

The first definition of Heritage Building Information Modelling (HBIM) appeared in 2009 as an information representation system for built buildings consisting of a library of semantically structured and parametric objects.

* Corresponding author: maria.grazianova@tuke.sk

Since its inception, HBIM has created new possibilities for sharing disparate information in a single digital environment and, in the context of cultural heritage, has been configured as a way to increase portability and make data available to a wider community of users for different purposes. The concept of collaborative work is integrated into BIM itself and its principles are an efficient approach to building design and maintenance for effective information management throughout a building's life cycle.

In the field of cultural heritage, HBIM's great potential lies in its ability to combine multidisciplinary knowledge and skills: BIM methods applied to heritage require an interdisciplinary and collaborative approach, starting from reliefs, passing through modeling, historical analysis, database implementation, etc. to continue to engage in specific skills.

HBIM can be understood as a holistic process, as a tool in which information and models on different scales can be kept interconnected in space and time and allows experts with different skills to share work in order to collect all the information from the analysis phase to meet maximum knowledge in support of conservation, dissemination and enhancement.

The workflow begins with the survey as a database and ends with the development of management models of the historic building that can be used in different application areas: accessibility and enhancement, archival document research, conservation and artificial intelligence.

Thanks to the integration and linking of different knowledge and skills, it is possible to reuse information obtained directly or extracted from in-depth research to provide a comprehensive and all-encompassing knowledge framework in order to improve accessibility, fruition but also to increase the intrinsic historical and documentary value of cultural heritage.

## 2 Literature review

### 2.1 Overview of BIM technology on existing buildings

A substantial part of building in Europe is older than 50 years with many buildings in use today that are hundreds of years old. More than 40% of our residential buildings have been constructed before the 1960s when energy building regulations were very limited. Countries with the largest components of older buildings include the UK, Denmark, Sweden, France, Czech Republic and Bulgaria. A large boom in construction in 1961-1990 is also evident through analysis where the housing stock, with a few exceptions, more than doubles in this period [2].

Developments in information and communication technologies (ICT) have led to a significant renewal in recent years in different sectors such as the design, enhancement, restoration and conservation of historic buildings and are still today, implying the need to apply new approaches to Heritage.

Traditional techniques of graphic representation and building management are constantly revisited with new knowledge and updated through the appropriate use of new generation tools to transform the process of information processing into digital data.

BIM was introduced to facilitate the building process and the management of orders by professionals, without neglecting the fundamental contribution provided by this process in simplifying the exchange of data during the design process. Its success is internationally recognized, but the fact remains that this constantly evolving method has been designed fundamentally for the design of the new.

This aspect represents a great limitation of BIM, especially in a historical period in which the theme of safeguarding historical and built heritage has assumed great importance. In the wake of these limitations and in the attempt one day to overcome them completely, we are now witnessing continuous experimentation of BIM approaches on existing heritage, with the aim not only of documenting the state of the art, but also with a view to future maintenance and design of buildings.

Nowadays, Building Information Modelling is at the center of scientific debate and has achieved important milestones, first in the field of new buildings and then in that of existing buildings.

Unlike newly built projects, BIM application in the heritage sector is a relatively new field of academic research. The challenge with historic buildings is that the elements are not similar and therefore difficult to standardize. Most elements are unique, implying that many have irregular geometries and shapes along their length.

Heritage buildings are still managed using the traditional method that does not provide heritage data in a single digital database but through a multi-format system where different parties are involved. The multi-format system is characterized by various formats such as PDF documents, word reports, Excel spreadsheets, etc., which are hardly linked and lack integration. It is also supported by 2D representation, therefore lacking 3D visualization, making it difficult for managers to make critical decisions.

Retrieving and updating information from such a system is challenging. In addition, the exchange of data between stakeholders is difficult because the various formats create a lack of collaboration and a communication gap. The adoption of such a traditional method creates many problems during the different building processes (management, reuse and FM)

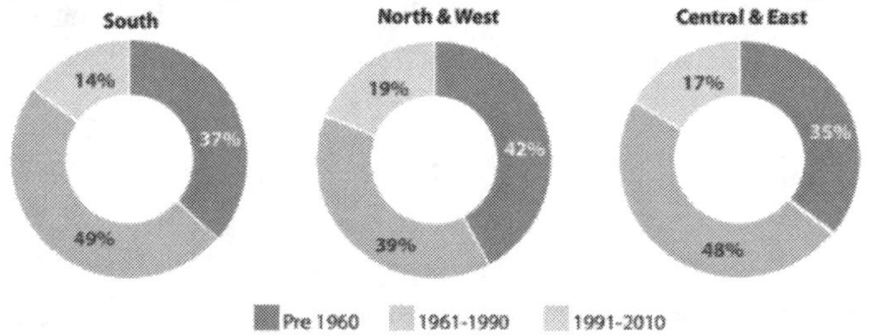


**Fig 1.** Age categorisation of housing stock in Europe. Source: BPHE survey

## 2.2 Heritage Building Information Modelling (HBIM)

In recent years we have seen experimentation with the application of approaches such as BIM to existing assets and in this regard the acronym HBIM is used to describe this particular area of interest.

The term was coined in 2009 by Professor Maurice Murphy of the Dublin Institute of Technology, whose meaning emerged a few years later in an article in the journal INSPR Photogrammetry and Remote Sensing: "Historic Building Information Modelling (HBIM) is a novel solution whereby interactive parametric objects representing architectural elements are constructed from historic data, these elements (including detail behind the scan surface) are accurately mapped onto a point cloud or image based survey " [3].

From this definition it is clear that BIM, usually used to create projects from scratch, has expanded its typical scope by exploring existing markets. A new challenge, in fact, is represented by the possibility of modeling existing building components in the territory by creating parametric digital models containing different types of information.

In the last decade, the scientific community has recognized HBIM as the real modeling process from relevant data, with the aim of creating libraries of parametric objects and combining them with informative data to enrich them. The main purpose of the BIM methodology applied to the heritage that already exists is the acquisition of a digital twin [4] to be used as a model and database to insert geometric and technical information useful for acquiring awareness and knowledge of the building in question.

BIM, applied to built heritage, greatly increases the intrinsic potential of the system itself because it increases its use in the creation of existing models not only as their digital and geometric restitution, but above all as intelligent collectors of data and information, parametric and with a well-defined semantics. With HBIM a sort of reverse engineering process is applied to existing buildings, the constructive evolutions are studied and, in the modeling phase of the architectural artefact, the procedures that led to the construction of that on the false (imaginary?) line of the final goal building organism are studied.

The BIM models of the Architectural Heritage (AH) are made up of intelligent objects that, despite the uniqueness of the artifact, can be re-used and adapted to other architectures that present constructive and architectural style similarities. In addition, the elements of the parametric libraries of HBIM models contain real and extremely heterogeneous information on the graphic documentation, the type and the main construction characteristics of the construction period to which the building belongs, information that can be updated, replaced and added. Ultimately, the HBIM paradigm is able to produce real "data collectors" populated by both geometric and non-geometric information related to various topics: historical documents, monitoring data, structural information, conservation or maintenance status.

## 2.3 The methodology and workflow of the HBIM

The workflow from data acquisition to the HBIM three-dimensional model is now consolidated and consists of a sequence of key operational phases in the process for the realization of models for the management of the historical architectural heritage [5].

The design activity for the recovery or restoration of an existing structure, which is based on the BIM methodology, starts from the in-depth knowledge of the artifact: the history and evolution must be investigated, the constructive stratifications that determine the current state of the art, reaching a high level of knowledge of the building in its entirety and complexity.

The historical investigation and the relief activity lead to the knowledge of architectural, structural, geometric and material characteristics of the building system to be preserved in the design of the recovery activity. Upstream, it is therefore essential to undertake an accurate and in-depth analysis and historical investigation that, through consultation of documents and documents, historical sources, archival documents, drawings and description of the techniques and materials used, allow to reconstruct the entire life of the artifact.

This must be accompanied by the relief and measurement activity on the site, which takes place through the instruments and technologies offered and made available based on the investment and the economic budget of the project. For this reason, the first phase of knowledge, data collection and awareness, is one of the main ones. The information collected lays the foundations that allow diagnostic analysis to be carried out on the building, to identify and define the recovery interventions to be implemented and the subsequent maintenance activities. The technologies adopted for the collection of data and information applied in the HBIM methodology can be multiple: photogrammetry and laser scanner detection, which in turn allow to associate geometric information to images, in such a way as to provide a virtual model highly faithful to reality.

The next step to be undertaken is the processing of the data collected as a result of the relief campaign. In the use of laser scanner and photogrammetry, usually, following a post-production and data processing activity, which also requires the insertion of georeferenced coordinates, we arrive at the realization of a point cloud that constitutes the basis for the modeling of the artifact, defining in detail volumetry, dimensions and colorimetric information. The processing of survey data allows to obtain a point cloud from which it is possible to create meshes that lead to a first definition of 3D model.

The next step is that of the actual digitization and realization of the three-dimensional and parametric BIM model, which relies and reflects on what has been detected in the previous phases. The information and data entered concern historical information, analysis of degradation or deformation, materials used and level of reliability of the data present.

In this way it is possible to create a complete model, which collects all the data relating to the history of the building and which will therefore allow to produce all the necessary design drawings: from technical drawings to orthographic projections to the production of intervention schedules. Obviously, of relative importance is to be aware of the purpose for which the BIM model of the artifact is made, inserting and collecting more detailed data based on the recovery or restoration intervention that is planned to be performed.

HBIM, therefore, explains a working method that allows you to design the existing from the point of view of modelling, or a model is created that simulates the construction of the building and that presents most of its characteristics. The application of the methodology to the existing heritage provides the possibility of optimizing its management, maintenance, and protection, allowing the monitoring of degradation, the planning of restoration interventions, the simulation of catastrophic events.

## 2.4 HBIM and the regulatory framework

In the last ten years, the use of Building Information Modeling (BIM) has been discussed and experimented, which was born and developed in the sectors of new buildings, infrastructures and industrial production, thanks to the evolution of digital tools and methods typical of Industry 4.0.

In December 2018, ISO 19650 was adopted, in its parts 1 (Concept and principles) and 2 (Delivery phase of assets), which through the mechanism of direct adoption of the Vienna Agreement becomes a European (EN) and national standard for each member state during 2019.

ISO 19650, we can say that concludes, with the first international standard of "principle" (in the vein of ISO 9000-quality, 14000-environment, 55000-asset management, 31000-risk, etc.) a first phase, "historical", of the regulatory world (and not) on BIM and digitization of the construction sector. A first phase much more oriented to IT aspects, to its beginnings (ISO STEP 10303-11-21), and then increasingly aimed at information management and processes.

In parallel, national standards have been developed as a corollary, which are now being revised to align with the concepts introduced with the ISO 19650 series and make them operational in their own country. So we have the English PAS 1192 - process, the Italian series UNI 11337 - application, the German DIN 91392 - CDE and the French AFNOR PRXPP07-150 - products.

ISO 10650, in its main characters (part 1), maintains the approach and concepts that have been consolidated over the years throughout the world "BIM": Information Specifications, Information Management Plan, etc. [6].

The absolute reference point in Europe with regard to the regulatory introduction of BIM is "European Union Public Procurement Directive" 2014/24 of 26 February 2014. The "European Union Public Procurement Directive", voted in January 2014 by the European Parliament and subsequently adopted, invites EU Member States, by 2016, to "encourage, specify or impose" through dedicated legislative measures the use of BIM, as a reference standard, for all projects and works with public funding [7].

However, within the scope of the history, procedural guidelines have not yet been drawn up that establish a shared HBIM methodology covering all phases of the process, starting from the survey and diagnostics to the design and use phase [8].

A problematic aspect that brings with it a BIM model referring to a historic building is the reliability of the measurements and information reported. In fact, research and survey activities do not always provide complete answers that allow an accurate and safe reconstruction of the artefact. For this reason, the issue of the reliability of the model and its transparency is fundamental in order to provide a precise awareness of the building to those who will then have to intervene. The indication of the reliability and interpretative level of the information in the model therefore becomes a parameter of great importance and interest in the context of HBIM.

These issues related to the reliability of the data, were introduced by referring to the concept of transparency of the model of the state of the art already starting from 2009 with the London Charter and then, subsequently integrated in 2012 in the Principles of Seville, although these were mainly aimed at the world of archeology. However, the foundations of these arguments were laid as early as 2003 with UNESCO's Charter on the Conservation of Digital Heritage.

The objective of the London Charter is to establish methodological principles to ensure correct digital display, dissemination and digital communication within the Cultural Heritage sector. The Principles of Seville, on the other hand, represent a practical implementation that lays the foundations of these issues within the world of archaeology, trying to make them implemented.

The objective is to define the process by which the image, model or information has been traced, disseminated and made available, we want to reach a scientific and unambiguous definition linked to the reliability and truthfulness of the data, whether of a geometric or informative nature.

In particular, the London Charter speaks of "intellectual transparency" referring to the result viewable by computer that reports information that is transmitted to users, while, in the Seville Principles, "scientific transparency" is introduced, with respect to which the conclusions reached through computer visualization by researchers and professionals must be able to be confirmed, refuted or modified by other experts over time.

For this reason, in these documents the concept of para-data is introduced, i.e. metadata concerning information of understanding and interpretation due to human processes that are collected within databases and have the purpose of interpreting the critical reading of an artifact performed by experts.

It is evident that, for the BIM methodology, referring to artifacts of historical and artistic interest, the theme of transparency and reliability of the visualization of the digital model is of considerable importance, both as regards the dimensional geometric aspect and the information and knowledge of the artifact.

## 3 Conclusions

BIM proposes a new working model: integrated, interactive, able to contain in a single platform both the individual project specifications and the contribution of all the subjects involved; but above all with a new control over the temporal component. The inconsistency between the project and the actual state has often been one of the major causes of the extension of construction times and the increase in planned costs.

The application of this methodology, certainly advantageous for the design of new buildings, can be equally valid also on existing buildings since its true potential is linked to the definition of a new collective working mode in which the different professional technicians, clients and users interact in order to limit changes and surprises.

It is no longer a question, therefore, of creating descriptive plan drawings in a more or less reliable way of the data contained in them, but it is a matter of creating three-dimensional models that allow to associate the geometries with the information necessary to correctly describe the architectural, structural and plant elements that must be managed and updated daily by the FM and by all the subjects involved in the process.

Beyond the technical or performance specifications of the software, it is this change of perspective compared to the past that makes BIM an interesting and probably irreversible process.

"Tangible cultural heritage, especially real estate (monuments, archaeological sites, and so on), is the main topic of application in the new approaches. A digital 3D model structured as part of the process of improving the architectural heritage is an urgent need today. In addition, the digital 3D model must be converted into a crucial framework for understanding and monitoring documentation, thereby creating a source of data (graphic and semantic) suitable for assisting in conservation, restoration and reconstruction projects." [9]

## References

1. G. M. Di Giuda, V. Villa - *Il BIM. Guida completa al Building Information Modeling per committenti, architetti, ingegneri, gestori immobiliari e imprese*, Hoepli, Italian edition, (2016)
2. B.Atanasiu, C. Despret, M. Economidou, J. Maio, I. Nolte, O. Rapf, *EUROPE'S BUILDINGS UNDER THE MICROSCOPE. A country-by-country review of the energy performance of buildings.* (Published by Buildings Performance Institute Europe (BPIE) in October 2011)
3. M. Murphy, E. McGovern, S. Pavia, *Historic Building Information Modelling – Adding intelligence to laser and image-based surveys of European classical architecture.* ISPRS J. Photogram. Remote Sensing, 2013, DOI: 10.1016/j.isprsjprs.2012.11.006 (2013)
4. R. Plana *"Il digital twin consente di costruire una copia virtuale dell'impianto reale, ma anche di una singola macchina, in grado di replicare in tutto e per tutto il funzionamento reale o di verificarne tutte le possibili alternative"*, (2017)
5. R. Quattrini et al. *'From tls to hbim. high quality semantically-aware 3d modelling of complex architecture'*, in International Archives of the Photogrammetry, Remote Sensing and Spatial Information Sciences - ISPRS Archives. DOI: 10.5194/isprsarchives-XL-5-W4- 367- (2015)
6. https://www.ingenio-web.it/articoli/standard-bim-il-mondo-dopo-la-iso 19650/
7. https://www.teamsystem.com/construction/magazine-edilizia/bim/la-normativa-europea
8. https://www.ingenio-web.it/articoli/hbim-nei-progetti-di-recupero-e-restauro-gestione-informativa-e-automatizzata-dei-processi-di-diagnostica/
9. H. Penttilä, et al. *Building information modelling of modern historic buildings. In Wiesbaden, F., ed. Predicting the Future*, 25th eCAADe Conference Proceedings, Frankfurt, (2007)
10. R. Ferzoco *HBIM e gestione della progettazione tramite la metodologia BIM Rifunzionalizzazione e riqualificazione dell'edificio Parella (Santuario di San Giovanni d'Andorno, Campiglia Cervo - BI)*, Thesis Politecnico di Torino Universidad Politécnica de Madrid, Master's Degree Course in Building Engineering, Italy, (2022)
11. R. Mammoli *PARADIGMI TECNOLOGICI PER UN NUOVO APPROCCIO NELLA DIGITAL TRANSFORMATION DEL PATRIMONIO CULTURALE BIM per una semantica dell'architettura storica.* Thesis, Marche Polytechnic University,PhD

School in Engineering Science,s Curriculum in Civil Engineering, Construction and Architecture, Italy, (2020)

12. L. Ori *Ottimizzazione della metodologia HBIM applicata al progetto di restauro architettonico: il caso studio di Villa Victorine a Pisa*. Thesis, UNIVERSITY OF PISA School of Engineering – D.E.S.T.e.C. Master's Degree Course in Building Engineering Architecture, Italy, (2017)
13. F.Capparelli *Soluzioni BIM per l'architettura e l'archeologia: il caso del Kyrenia Shipwreck Museum, un percorso di 25 secoli*. Thesis, School of Architecture Single-cycle master's degree in architecture, Italy, (2019)
14. F.Brruku, C. Cigliutti *HBIM, Realtà Virtuale e Aumentata a supporto della condivisione del modello BIM Social del Santuario del Trompone.* Thesis, TURIN POLYTECHNIC, Master's Degree Course in Architecture, Construction and City, Italy, (2019)

MATEC Web of Conferences **385**, 01041 (2023) https://doi.org/10.1051/matecconf/202338501041
*Young Scientist 2023*

# Environmental assessment of wooden roof compositions

*Jana* Budajova[1*]

[1]Institute of Architectural Engineering, Faculty of Civil Engineering, Technical University of Kosice, Vysokoskolska 4, 042 00 Kosice, Slovakia

**Abstract.** The construction industry has high energy consumption and environmental impact, such as greenhouse gases, acidification, and the like, leading to concerns about the environmental impact of buildings on climate change. Sustainable and ecological construction requires innovation in materials and their durability to support a low-carbon and efficient city. The construction industry needs measures to reduce emissions, and buildings are assessed using methods such as Life Cycle Assessment, which allows for a holistic view and supports sustainable development. Timber roofs are crucial in the fight against the environmental crisis, as buildings have a significant impact on lower greenhouse gas emissions.

## 1 Introduction

The construction industry is characterized by high energy consumption and large production of emissions such as greenhouse gases, acidification, eutrophication, and others [1]. It is obvious that the increasing pressure on the environment due to human activity is beginning to cause bad effects on the environment. Climate change and loss of biodiversity are becoming increasingly unbearable [2]. To create a healthy and sustainable environment, changes in society are necessary. Making these changes and adjustments is a worldwide challenge to mitigate global warming. One of the main contributors to negative environmental impacts is the construction industry. The largest emissions of greenhouse gases are produced either during the extraction of materials, transport to the construction site or because of the energy consumed during the operation of the building and because of the materials and energy required for the construction of the building and their disposal after the demolition of the building [3]. Indeed, buildings account for approximately 40% of global energy-related $CO_2$ emissions [4]. Life Cycle Assessment (LCA) is a technique for assessing the impact of a product on the environment in all phases of the life cycle - i.e., from the extraction of raw materials through the processing of materials, production, distribution, use, repairs, maintenance to disposal or recycling [5]. The goal of LCA is to compare the range of effects of environmental impacts assigned to goods and services to improve processes, promote economy, and provide a solid basis for qualified decisions [6]. The term life cycle refers to the idea that an objective and comprehensive assessment requires an assessment of

* Corresponding author: jana.budajova@tuke.sk

the production of raw materials, production, distribution, use and disposal, including "intermediate transport" and all other actions necessary for the existence of the product. This article is focused on the analysis of the composition of wooden roofs using environmental indicators.

## 2 Material and methods

The LCA method was used to evaluate the composition of roofs with a border from the cradle to the gate with choices according to the EN 15978 standard [7]. System boundaries and modules found in this study are shown in Fig. 1 [8]. The characteristics of the examined roof compositions are described in Table 1. The environmental impacts were calculated per functional unit [FU] of 1 $m^2$ of usable area and the estimated lifespan is 50 years. One Click LCA software [9] was used to determine environmental indicators, life cycle and circularity of materials. Environmental impact indicators such as Global warming potential (GWP) expressed as kg $CO_{2eq}$, Biogenic carbon storage (BS) expressed as kg $CO_{2eqbio}$, Ozone depletion potential (ODP) expressed in kg $CFC11_{eq}$, Acidification potential (AP) expressed in kg $SO_{2eq}$, Eutrophication potential (EP) expressed in kg $PO_{4eq}$, Photochemical ozone creation potential (POCP) expressed in kg $C_2H_{4eq}$, Abiotic depletion potential (ADP-elements) for non-fossil sources expressed in kg $Sb_{eq}$ and Abiotic depletion potential (ADP-fossil fuels) for fossil resources expressed in MJ. The results of the LCA analysis was subjected to a multicriteria analysis using the MCA7 tool [10] to select the appropriate composition of the wooden roof.

| Product stage | | | Construction stage | | Use stage | | | | | | | End of life stage | | | | Benefits and loads beyond system boundary | | |
|---|---|---|---|---|---|---|---|---|---|---|---|---|---|---|---|---|---|---|
| Raw material extraction | Transport | Manufacturing | Transport | Assembly | Use | Maintenance | Repair | Replace | Refurbishment | Operational energy | Operational water | Deconstruction | Transport | Waste processing | Disposal | Reuse | Renovation | Recycling |
| A1 | A2 | A3 | A4 | A5 | B1 | B2 | B3 | B4 | B5 | B6 | B7 | C1 | C2 | C3 | C4 | D | D | D |

**Fig. 1.** System boundary (From the cradle to the gate with options).

**Table 1.** Characteristics of building materials and compositions (C1-C20).

| No | Material | Thickness | No | Material | Thickness |
|---|---|---|---|---|---|
| C1 | Gypsum board 2x10mm | 0.020 | C2 | Plasterboard 2x12.5mm | 0.025 |
| | Wood fiber insulation | 0.100 | | Hemp insulation with PE | 0.080 |
| | OSB board 3 | 0.022 | | Wooden formwork | 0.022 |
| | Wooden ins./KVH | 0.260 | | Vapor barrier (PE) | --- |
| | Wood fiber insulation | 0.060 | | Hemp insulation /KVH | 0.220 |
| | Wood fiber insulation | 0.035 | | Wooden ins. /KVH | 0.140 |
| | Ventilated gap - 60x60 | 0.060 | | Wood fiber insulation | 0.060 |
| | Contraltos 50x30 | 0.030 | | DHF wood fiber | 0.015 |
| | Roofing - burnt tiles | --- | | Ventilated gap - 60x80 | 0.080 |
| C3 | Internal plaster - gypsum | 0.010 | | OSB board 2x15mm | 0.030 |
| | Wood fiber pressed board | 0.015 | | Waterproofing+geotextile | --- |
| | Wood fiber insulation | 0.050 | | Expanded clay | 0.030 |

| | | | | | |
|---|---|---|---|---|---|
| | OSB 3 board | 0,022 | | Substrate | 0.060 |
| | Linen insulation /KVH | 0.220 | | Gypsum board 2x10 mm | 0.020 |
| | Linen ins. Between KVH | 0.160 | | Sheep's wool | 0.060 |
| | DHF wood fiber board | 0.015 | | OSB 3 board | 0.018 |
| | Ventilated gap - 60x80 | 0,080 | | Cellulose fiber b./KVH | 0.220 |
| | Contralats 50x30 | 0.030 | **C4** | Cellulose fiber b./KVH | 0.160 |
| | Roofing - burnt tiles | --- | | DHF wood fiber board | 0.015 |
| | Wooden cladding | 0.018 | | Ventilated gap - 60x80 | 0.080 |
| | Mineral insulation | 0.080 | | OSB board | 0.015 |
| | OSB board | 0.015 | | Insurance waterproofing | --- |
| | Vapor barrier Jutafol NAL | --- | | Titanium zinc sheet | --- |
| | Cellulose between KVH | 0.260 | | Clay plaster on a reed mat | 0.015 |
| **C5** | OSB board | 0.015 | | Plasterboard | 0.015 |
| | Wood fiber insulation | 0.100 | | Hemp insulation with PE | 0.100 |
| | Diffusion film | --- | | OSB 3 board | 0.022 |
| | Ventilated gap - 60x80 | 0.080 | | Sheep wool/KVH 60x200 | 0.200 |
| | Wooden formwork | 0.027 | **C6** | Sheep wool/KVH | 0.140 |
| | Waterproofing + geotextile | --- | | Wood fiber insulation | 0.050 |
| | Gravel | 0.060 | | DHF wood fiber | 0.015 |
| | Gypsum board 2x12.5mm | 0.025 | | Ventilated gap - wood | 0.060 |
| | Blown cellulose | 0.140 | | Wooden formwork | 0.027 |
| | OSB 3 board | 0.022 | | Waterproofing+ geotextile | --- |
| | Cellulose fiber b./KVH | 0.280 | | Substrate | 0.060 |
| **C7** | Wood fiber insulation | 0.050 | | Clay plaster on a reed mat | 0.015 |
| | Ventilated gap - 60x80 | 0.080 | | Wood fiber pressed board | 0.015 |
| | OSB board 2x15mm | 0.030 | | Sheep's wool | 0.100 |
| | Waterproofing + geotextile | --- | | OSB 3 board | 0.022 |
| | Gravel | 0.060 | **C8** | Straw /KVH 100x280 | 0.280 |
| | Gypsum board 2x10mm | 0.020 | | Wood fiber insulation | 0120 |
| | Wood fiber insulation | 0.140 | | Diffusion foil | --- |
| | OSB 3 board | 0.015 | | Ventilated gap - 60x60 | 0.060 |
| | Cellulose wood fiber/ BSH | 0.280 | | Contralats 50x30 | 0.030 |
| | Wood fiber insulation | 0.035 | | Wooden shingles - double | --- |
| **C9** | Ventilated gap - 60x80 | 0.080 | | Clay plaster on a reed mat | 0.015 |
| | OSB board 2x15mm | 0.030 | | OSB 3 board | 0.030 |
| | Waterproofing + geotextile | --- | | Hemp insulation/BSH | 0.260 |
| | Drainage layer-ceramsite | 0.030 | | Hemp insulation/KVH | 0.160 |
| | Filter fabric | --- | | DHF wood fiber diffusion | 0.015 |
| | Substrate | 0.080 | **C10** | Ventilated gap | 0.080 |
| | Gypsum board | 0.015 | | OSB board 2x15mm | 0.030 |
| | Sheep's wool | 0.080 | | Waterproofing +geotextile | --- |
| | Intelligent vapor barrier | --- | | Drainage layer - gravel | 0.030 |
| | OSB board | 0.018 | | Filter fabric | --- |
| | Linen insulation | 0.200 | | Substrate | 0.080 |
| | Linen insulation /KVH | 0.140 | | BSH profile 100x220 | 0.095 |
| **C11** | DHF diffusion fiber board | 0.013 | | Plasterboard 2x12.5mm | 0.025 |
| | Ventilated gap - 60x60 | 0.060 | | Wood fiber insulation | 0.100 |
| | OSB board 2x15mm | 0.030 | | OSB 3 board | 0.022 |
| | Waterproofing + geotextile | --- | **C12** | Wood fiber ins./KVH | 0.280 |
| | Drainage layer (nop film) | 0.020 | | Wood fiber insulation | 0.050 |
| | Filter fabric | --- | | Ventilated gap - 60x60 | 0.060 |
| | Substrate | 0.080 | | Wooden formwork (rare) | 0.024 |
| **C13** | Gypsum board 2x10mm | 0.020 | | OSB board | 0.015 |
| | Wood fiber insulation | 0.060 | | Waterproofing +geotextile | --- |

| | | | | | |
|---|---|---|---|---|---|
| | OSB 3 board | 0.022 | | Gravel | 0.080 |
| | Linen insulation/ I profiles | 0.360 | **C14** | Plasterboard 2x12.5mm | 0.025 |
| | Diffusion film | --- | | Wood fiber ins./KVH | 0.060 |
| | Ventilated gap - 60x60 | 0.060 | | OSB board | 0.013 |
| | Contralats 50x30 | 0,030 | | Vapor barrier (Al) | --- |
| | Roofing - burnt tiles | --- | | Wood fiber ins./I profiles | 0.360 |
| **C15** | Fireproof plasterboard | 0.025 | | Wood fiber insulation | 0.035 |
| | Hemp ins. with PE/KVH | 0.080 | | Ventilated gap - 60x80 | 0.080 |
| | OSB 3 board | 0.022 | | Wooden formwork | 0.024 |
| | Cellulose wood/ I profile | 0.360 | | Insurance waterproofing | --- |
| | Diffusion film | --- | | Sheet metal covering | --- |
| | Ventilated gap - 60x60 | 0.060 | **C16** | Internal plaster - lime | 0.010 |
| | Contralats 50x30 | 0.030 | | OSB 3 board | 0.022 |
| | Roofing - concrete tiles | 0.020 | | Linen insulation/I profiles | 0.400 |
| **C17** | Wooden cladding | 0.024 | | Diffusion foil | --- |
| | Vapor barrier (PE) | --- | | Ventilated gap | 0.060 |
| | Cork dust/I profiles | 0.400 | | Contralats 50x35 | 0.035 |
| | Wood fiber insulation | 0.035 | | Wooden shingles - double | --- |
| | Ventilated gap - 60x60 | 0.060 | **C18** | Gypsum board 2x10mm | 0.020 |
| | Contralats 50x30 | 0.030 | | Hemp insulation with PE | 0.060 |
| | Roofing - burnt tiles | | | OSB 3 board | 0.022 |
| **C19** | Lay plaster on a reed mat | 0.015 | | Hemp ins. /PE/I profile | 0.360 |
| | Plasterboard | 0.015 | | DHF wood fiber | 0.013 |
| | Linen ins. with PE/KVH | 0.080 | | Diffusion foil | --- |
| | OSB 3 board | 0.022 | | Ventilated gap - 60x80 | 0.080 |
| | Straw between I profiles | 0.400 | | Wooden formwork | 0.024 |
| | Wood fiber insulation | 0.050 | | Insurance waterproofing | --- |
| | Ventilated gap - 60x60 | 0.060 | | Titanium zinc sheet | --- |
| | Contralats 50x35 | 0.035 | **C20** | Plasterboard 2x12.5mm | 0.025 |
| | Wooden shingles - double | --- | | Cellulose fiber boards | 0.060 |
| | | | | OSB board | 0.015 |
| | | | | Intelligent vapor barrier | --- |
| | | | | Blown cellulose/I profiles | 0.360 |
| | | | | DHF 2x15 mm | 0.030 |
| | | | | Ventilated gap - 60x60 | 0.060 |
| | | | | Contralats 50x30 | 0.030 |
| | | | | Roofing - burnt tiles | --- |

## 3 Results and discussion

Table 2 shows the results for the investigated impact categories for the twenty designed wooden roof compositions. Fig. 2 shows the results for GWP expressed as kg of $CO_{2eq}$ emissions. According to the results, roof composition C17 (116.7 kg $CO_{2eq}$) achieves the worst results and composition C7 (39.2 kg $CO_{2eq}$) achieves the best result. Fig. 3 shows biogenic carbon BS. The best result is achieved by composition C4 (92.8 kg $CO_{2eq}$bio) and composition C17 (281.4 kg $CO_{2eq}$bio) has the worst impact on the environment. Fig. 4 shows the results of ODP. The best result is achieved by compositions C9 and C10 (2.45E-06 and 2.3E-06 kg $CFC11_{eq}$) and the worst result by C3 and C16 (1.9E-05 and 2.08E-05 kg $CFC11_{eq}$). In terms of $SO_{2eq}$ emissions, C17 has the worst result (0.49 kg $SO_{2eq}$) and C1 has the best result (0.15 kg $SO_{2eq}$) in AP (Fig. 5). The results for EP shown in Fig. 6 show that C1 achieves the best result (0.03 kg $PO_{4eq}$) and C16 the worst result (0.09 kg $PO_{4eq}$). The best result in terms of POCP (kg $C_2H_{4eq}$) is achieved by composition C15 and C18 with a value of 0.02 kg $C_2H_{4eq}$, and the worst result is achieved by composition C8 and C19 (0.07 kg

$C_2H_{4eq}$) (Fig. 7). Fig. 8 presents results for abiotic depletion potential (ADP-elements) for non-fossil sources. Composition C14 achieves the best result (0.02 kg $Sb_{eq}$), composition C4 and C18 the worst result (0.4 kg $Sb_{eq}$). And finally, the abiotic depletion potential (ADP-fossil fuels) for fossil resources is shown in Fig. 9. Composition C15 achieves the best result (711 MJ) and composition C17 the worst result (2058 MJ).

**Table 2.** Results of impact categories for roof compositions (C1-C20).

| | **GWP** *kg* $CO_{2eq}$ | **BS** *kg* $CO_{2eq}$ *bio* | **ODP** *kg* $CFC11_{eq}$ | **AP** *kg* $SO_{2eq}$ | **EP** *kg* $PO_{4eq}$ | **POCP** *kg* $C_2H_{4eq}$ | **ADP-elements** *kg* $Sb_{eq}$ | **ADP-fossil fuels** *MJ* |
|---|---|---|---|---|---|---|---|---|
| C1 | 62.9 | 105.33 | 3.0E-06 | 0.15 | 0.03 | 0.022 | 0.031 | 1056 |
| C2 | 77.3 | 140.3 | 4.8E-06 | 0.24 | 0.07 | 0.022 | 0.033 | 1103 |
| C3 | 57.6 | 204.5 | 1.9E-05 | 0.24 | 0.08 | 0.030 | 0.027 | 1010 |
| C4 | 50.8 | 92.8 | 5.8E-06 | 0.25 | 0.06 | 0.022 | 0.418 | 830 |
| C5 | 63.9 | 225.8 | 7.3E-06 | 0.29 | 0.08 | 0.025 | 0.033 | 1050 |
| C6 | 66.5 | 137.5 | 1.7E-05 | 0.22 | 0.06 | 0.027 | 0.031 | 890 |
| C7 | 39.2 | 129.6 | 2.7E-06 | 0.17 | 0.04 | 0.022 | 0.042 | 827 |
| C8 | 71.1 | 231.4 | 8.4E-06 | 0.29 | 0.08 | 0.067 | 0.029 | 1330 |
| C9 | 66.1 | 156.6 | 2.5E-06 | 0.22 | 0.05 | 0.026 | 0.039 | 1227 |
| C10 | 46.2 | 113.2 | 2.3E-06 | 0.17 | 0.06 | 0.017 | 0.040 | 794 |
| C11 | 43.8 | 103.0 | 1.5E-05 | 0.21 | 0.06 | 0.026 | 0.035 | 836 |
| C12 | 52.2 | 129.6 | 3.4E-06 | 0.20 | 0.05 | 0.021 | 0.033 | 989 |
| C13 | 60.1 | 139.2 | 1.7E-05 | 0.21 | 0.06 | 0.022 | 0.025 | 924 |
| C14 | 65.30 | 152.0 | 6.3E-06 | 0.21 | 0.05 | 0.021 | 0.023 | 989 |
| C15 | 59.0 | 155.6 | 7.5E-06 | 0.23 | 0.06 | 0.016 | 0.028 | 711 |
| C16 | 66.38 | 197.4 | 2.1E-05 | 0.31 | 0.09 | 0.039 | 0.026 | 988 |
| C17 | 116.7 | 281.4 | 1.7E-05 | 0.49 | 0.09 | 0.036 | 0.024 | 2058 |
| C18 | 50.6 | 153.3 | 6.8E-06 | 0.26 | 0.09 | 0.017 | 0.402 | 781 |
| C19 | 69.7 | 229.1 | 1.2E-05 | 0.33 | 0.09 | 0.070 | 0.030 | 1230 |
| C20 | 68.6 | 180.0 | 7.3E-06 | 0.27 | 0.07 | 0.019 | 0.028 | 774 |

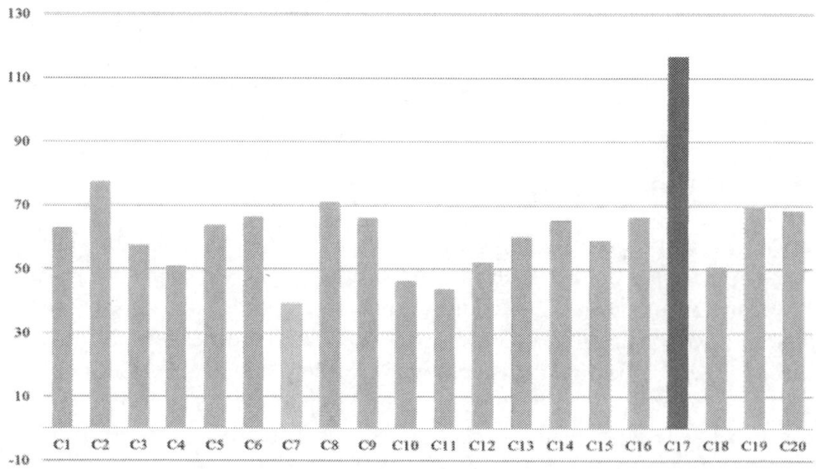


**Fig. 2.** GWP for roof compositions - kg $CO_{2eq}$.

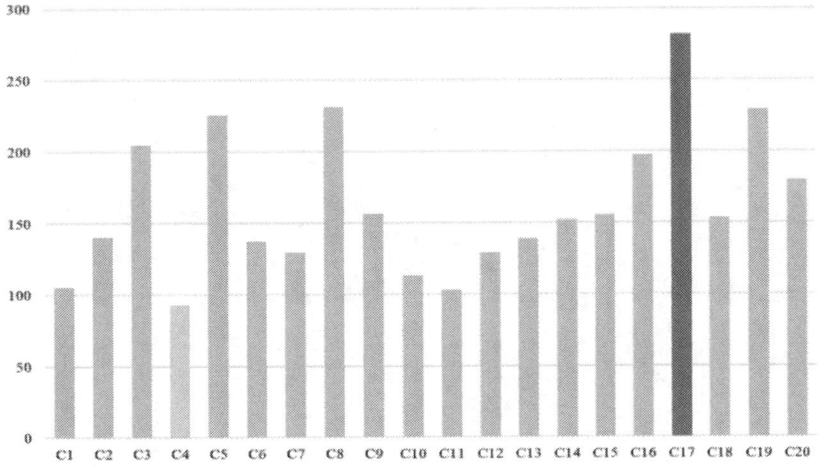


**Fig. 3.** BS for roof compositions - kg $CO_{2eq\ bio}$.

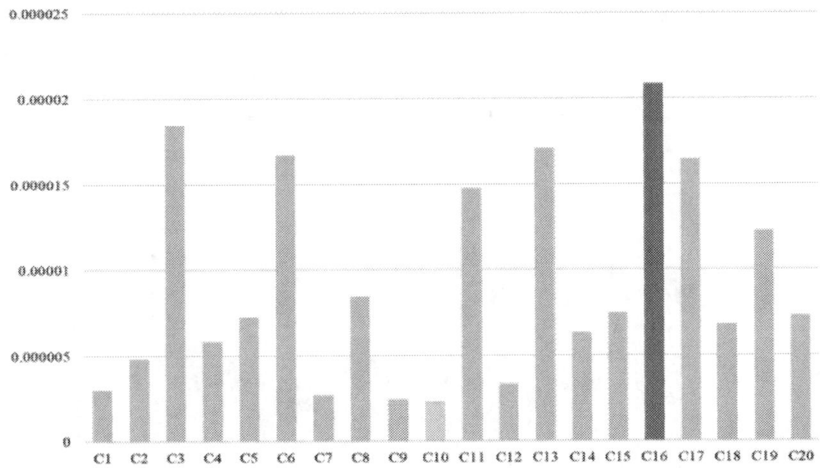


**Fig. 4.** ODP for roof compositions - kg $CFC11_{eq}$.

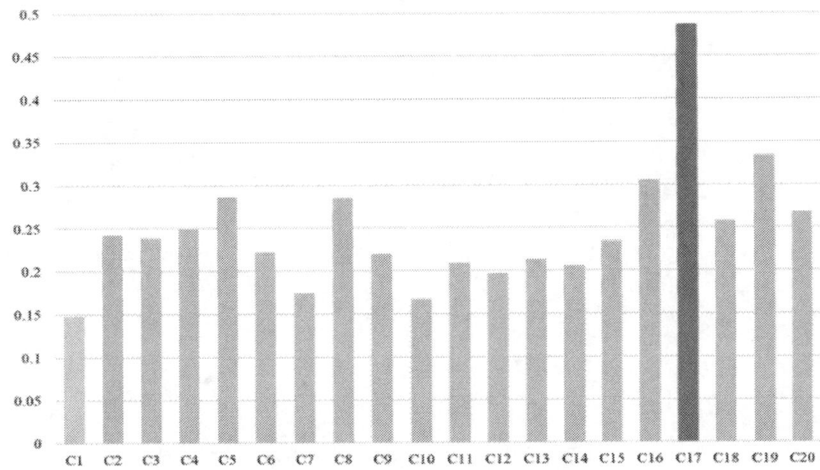


**Fig. 5.** AP for roof compositions - kg $SO_{2eq}$.

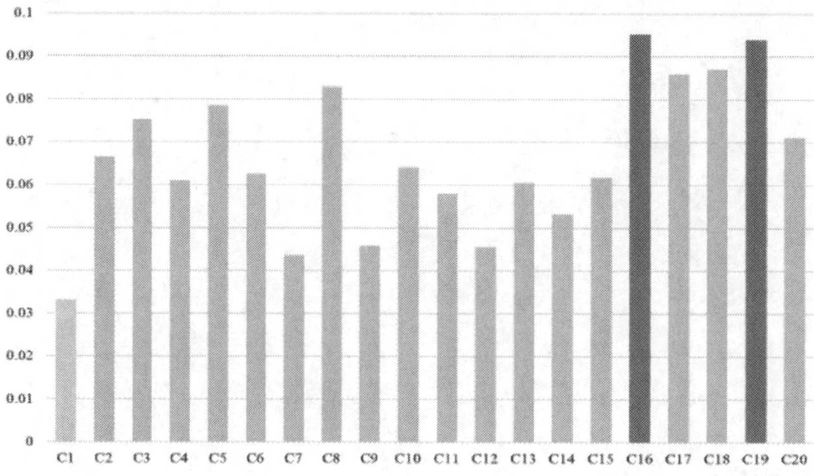


**Fig. 6.** EP for roof compositions - kg $PO_{4eq}$.

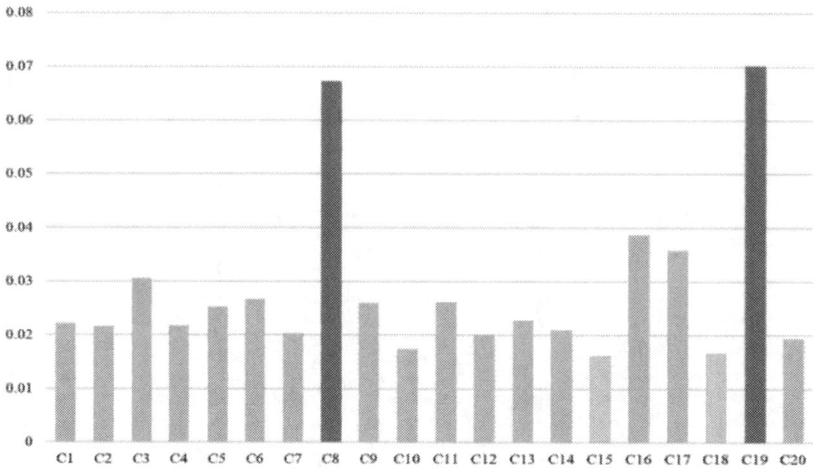


**Fig. 7.** POCP for roof compositions - kg $C_2H_{4eq}$.

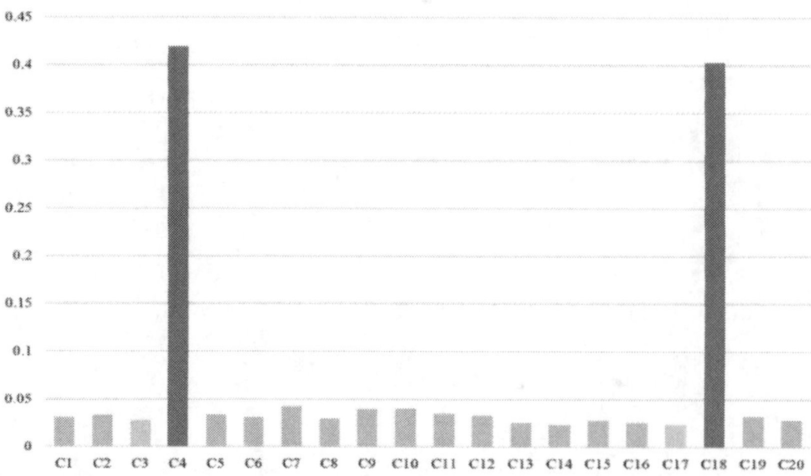


**Fig. 8.** ADP-elements for roof compositions - kg $Sb_{eq}$.

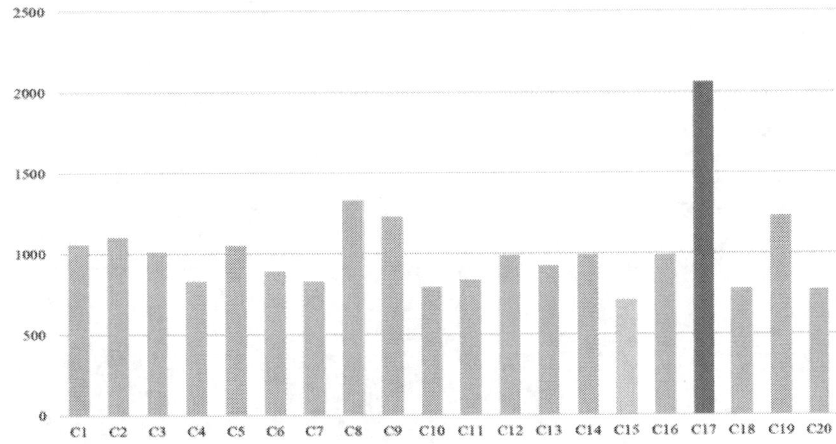


**Fig. 9.** ADP-fossil fuels for roof compositions - MJ.

Table 3 presents the largest material contributors to GWP within the cradle-to-gate boundary (A1-A3) and their percentage of roof composition.

**Table 3.** Materials with the largest share of emissions in the GWP category (C1-C20).

| | Compositions | | | | | | | | | |
|---|---|---|---|---|---|---|---|---|---|---|
| **Material** | *C1* | *C2* | *C3* | *C4* | *C5* | *C6* | *C7* | *C8* | *C9* | *C10* |
| Wood fiber insulation, loose, L = 0.038 W/mK | 12 kg $CO_{2e}$ 24% | | | | 14 kg $CO_{2e}$ 29% | | | | | |
| Clay roofing tile, 2 - 4 kg/piece, 40 - 60 kg/m2 | 10 kg $CO_{2e}$ 21% | | 10 kg $CO_{2e}$ 22% | | | | | | | |
| Substrate for green roof system | | 17 kg $CO_{2e}$ 31% | | | | 17 kg $CO_{2e}$ 37% | | | | |
| Wood fiber ins., L = 0.036 W/mK | | 6.4 kg $CO_{2e}$ 12% | | 2.8 kg $CO_{2e}$ 9% | | | 5.8 kg $CO_{2e}$ 25% | 17 kg $CO_{2e}$ 31% | 25 kg $CO_{2e}$ 48% | |
| Wooden lamella floors on a wooden frame | | | 14 kg $CO_{2e}$ 30% | | 14 kg $CO_{2e}$ 28% | | | | | |
| Insulation from hemp, flax and cotton fibers | | | 7.3 kg $CO_{2e}$ 16% | | | 8.3 kg $CO_{2e}$ 17% | | | | 5.7 kg $CO_{2e}$ 25% |
| OSB board | | | | 6.7 kg $CO_{2e}$ 22% | | | 11 kg $CO_{2e}$ 36% | | 9.2 kg $CO_{2e}$ 17% | 12 kg $CO_{2e}$ 37% |
| Wooden shingles | | | | | | | | 23 kg $CO_{2e}$ 40% | | |

| | Compositions | | | | | | | | | |
|---|---|---|---|---|---|---|---|---|---|---|
| **Material** | *C11* | *C12* | *C13* | *C14* | *C15* | *C16* | *C17* | *C18* | *C19* | *C20* |
| Wood fiber insulation, loose, L = 0.038 W/mK | | | | | | | 55 kg $CO_{2e}$ 64% | | | |
| Clay roofing tile, 2 - 4 kg/piece, 40 - 60 kg/m2 | | | 10kg $CO_{2e}$ 23% | | | | 10 kg $CO_{2e}$ 12% | | | 10 kg $CO_{2e}$ 22% |
| Substrate for green roof system | | | | | | | | | | |
| Wood fiber ins., L = 0.036 W/mK | | 20 kg $CO_{2e}$ 48% | | 21 kg $CO_{2e}$ 44% | | | | | | |
| Wooden lamella floors on a wooden frame | | | 14 kg $CO_{2e}$ 32% | 14 kg $CO_{2e}$ 29% | 14 kg $CO_{2e}$ 32% | 14 kg $CO_{2e}$ 27% | 14 kg $CO_{2e}$ 16% | 14 kg $CO_{2e}$ 34% | 14 kg $CO_{2e}$ 25% | 14 kg $CO_{2e}$ 29% |
| Insulation from hemp, flax and cotton fibers | 6.5 kg $CO_{2e}$ 20% | | 6.9 kg CO2e 16% | | | 7.7 kg $CO_{2e}$ 15% | | 8.4 kg $CO_{2e}$ 20% | | |
| OSB board | 9.8 kg $CO_{2e}$ 30% | 7.5 kg $CO_{2e}$ 18% | | | 4.5 kg $CO_{2e}$ 10% | | | 4.5 kg $CO_{2e}$ 11% | 4.5 kg $CO_{2e}$ 8.2% | |
| Wooden shingles | | | | | | 23 kg $CO_{2e}$ 44% | | | 23 kg $CO_{2e}$ 42% | |

Eight materials with the largest share of emissions in the GWP category. Two materials for each roof composition: first (green); second (green); third (green)

**Table 4.** Results of multicriteria analysis (C1-C20).

| No. | variant | CDA | variant | IPA | variant | WSA | variant | TOPSIS |
|---|---|---|---|---|---|---|---|---|
| 1 | C7 | 2.9826 | C7 | 0.0652 | C7 | 0.9348 | C7 | 0.9396 |
| 2 | C10 | 3.8595 | C10 | 0.1108 | C10 | 0.8892 | C10 | 0.8725 |
| 3 | C12 | 5.8023 | C12 | 0.1422 | C12 | 0.8578 | C12 | 0.8674 |
| 4 | C1 | 7.8054 | C1 | 0.161 | C1 | 0.839 | C1 | 0.8036 |
| 5 | C4 | 9.1154 | C9 | 0.2012 | C9 | 0.7988 | C4 | 0.7929 |
| 6 | C9 | 9.4385 | C4 | 0.2512 | C4 | 0.7488 | C9 | 0.7775 |
| 7 | C18 | 9.7019 | C15 | 0.2561 | C15 | 0.7439 | C18 | 0.7327 |
| 8 | C15 | 9.8582 | C14 | 0.2585 | C14 | 0.7415 | C15 | 0.7242 |
| 9 | C14 | 10.0845 | C18 | 0.2832 | C18 | 0.7168 | C14 | 0.7236 |
| 10 | C5 | 13.0385 | C20 | 0.316 | C20 | 0.684 | C5 | 0.6769 |
| 11 | C20 | 13.9396 | C2 | 0.3161 | C2 | 0.6839 | C20 | 0.6622 |
| 12 | C11 | 15.161 | C5 | 0.3184 | C5 | 0.6816 | C2 | 0.6613 |
| 13 | C2 | 15.4798 | C11 | 0.3383 | C11 | 0.6617 | C11 | 0.5957 |
| 14 | C8 | 18.7901 | C13 | 0.4412 | C13 | 0.5588 | C8 | 0.5889 |
| 15 | C13 | 19.6697 | C8 | 0.4452 | C8 | 0.5548 | C19 | 0.4936 |
| 16 | C19 | 21.2068 | C6 | 0.4718 | C6 | 0.5282 | C13 | 0.4866 |
| 17 | C3 | 21.7779 | C3 | 0.4841 | C3 | 0.5159 | C6 | 0.461 |
| 18 | C6 | 22.1985 | C19 | 0.5367 | C19 | 0.4633 | C3 | 0.4607 |
| 19 | C16 | 27.8898 | C16 | 0.6176 | C16 | 0.3824 | C16 | 0.3727 |
| 20 | C17 | 30.1684 | C17 | 0.7921 | C17 | 0.2079 | C17 | 0.1867 |

## 4 Conclusion

Roof structures play a key role in the energy efficiency of buildings and in the prevention of heat loss. The modernisation and renovation of roof structures can make a significant contribution to reducing energy costs for heating and cooling and increasing the overall comfort of living. It is also important to select the right material and construction for roofing to minimize the need for maintenance and increase the overall service life of the structure. There are various programmes and institutions in the European Union that aim to promote the renovation of buildings and improve their energy efficiency. At Member State level, there are specific regulations and standards for the energy sources used in the construction and renovation of buildings. In some countries, funding is also available to support the modernisation of buildings. Nowadays, the renovation of buildings is seen as a positive investment for the future and a step towards environmental protection and thermal comfort. It is important for us to realise that every step we take towards improving energy management and protecting nature has a positive impact on our overall quality of life. Roof structures are one of the key elements in the renovation of buildings and the right choice of materials and construction can have a significant impact on the overall efficiency and quality of the building. This study presents the results of the evaluation of timber roof compositions from an environmental aspect. Based on the assessment of the proposed compositions, it can be concluded that the best rating was achieved by the C7 composition with KVH timbers, wood fibre insulation, a ventilated gap and an aggregate capping layer. The results of the environmental indicators show that the use of wood fibre and cellulose insulation in the compositions is up to 66% lower for GWP, 88% for ODP, 69% for AP, 65% for EP, 77% for POCP, 94% for ADP-elements and 65% for ADP-fossil fuels compared to the compositions with sheep wool, cork, or linen insulation. The KVH composition with wood-fibre insulation also performs the best in the multi-criteria analysis including all LCA categories. Further research work will focus on the in-depth analysis and comparison of buildings and their structures in terms of life cycle costs, circularity, and circular economy.

## References

1. Y. Pamu, V. S. S. Kumar, M. A. Shakir, H. Ubbana, Life Cycle Assessment of a building using Open-LCA software. Mater. Today **52**. 1968-1978, (2022)
2. R. Watson, I. Baste, A. Larigauderie, P. Leadley, U. Pascual, B. Baptiste, H. Mooney. Summary for policymakers of the global assessment report on biodiversity and ecosystem services of the Intergovernmental Science-Policy Platform on Biodiversity and Ecosystem Services. IPBES Secretariat: Bonn, Germany. 22-47 (2019)
3. M. W. Ryberg, P. K. Ohms, E. T. Møller, Lading Comparative life cycle assessment of four buildings in Greenland. Building and Environment. 204, 108130 (2021)
4. IEA, U. N. Global status report for buildings and construction 2019. UN Enviroment programme, 224 (2019)
5. Števo, S. Zero house – fiction, but good business
6. National Risk Management Research Laboratory,Life Cycle Assessment: Principles and Practice,Cincinnati, Ohio 45268
   Available on: www.epa.gov/nrmrl/lcaccess/pdfs/chapter1_frontmatter_lca101.pdf
7. EN 15978:2011 Sustainability of construction works. Assessment of environmental performance of buildings. Calculation method

8. EN 15804+A2:2019 Sustainability of construction works – Environmental product declarations - Core rules for the product category of construction products
9. One Click LCA. Available on: https://www.oneclicklca.com/?utm_source=google&utm_medium=cpc&utm_campaign=EEU%202019%20Search&gclid=Cj0KCQjw39uYBhCLARIsAD_SzMSRSYF7cnhyEbMoj2DQYbNCK7wtVBjtrB4blHQTFRN6lRxdE9Fm-poaAg-WEALw_wcB (25.04.2023)
10. P. Korviny. MCA7. Program for multicriteria decision analysis. Ostrava. p. 27 (2003)

# The effect of mesh density and element type on the analysis of 3D objects

*Michala* Weissova[1*], *Vincent* Kvocak[2], *Daniel* Dubecky[3], and *Patricia* Vanova[3]

[1]Technical University of Kosice, Faculty of Civil Engineering, Institute of Structural Engineering and Transportation Structures, 040 01 Kosice, Slovakia
[2]Technical University of Kosice, Faculty of Civil Engineering, Institute of Structural Engineering and Transportation Structures, Department of Load Bearing Structures and Intelligent Adaptive Structures, Vysokoskolska 4, 040 01 Kosice, Slovakia
[3]Technical University of Kosice, Faculty of Civil Engineering, Center for Research and Innovation in Construction, Park Komenskeho 10A, 042 00 Kosice, Slovakia

**Abstract.** This paper focuses on the usage of mesh density and element type in the analysis of 3D objects, which are crucial components of 3D object Finite Element Method (FEM) analysis. The document is based on theoretical principles and will be supported by practical numerical examples using the FEM software Abaqus. In order to conduct an accurate FEM analysis, it is vital to employ an appropriate mesh that is easily convertible for calculations and free from imperfections. The suitability of the mesh depends on two key parameters: density, which is determined by the material being used, and element type, such as tetrahedrons, hexahedrons, polyhedrons, and others. Different types of elements and mesh densities in FEM software often involve multiple settings that primarily impact the accuracy of the model, as well as the calculation time and convergence quality. This article is divided into three sections: the first part focuses on the theoretical foundations, the second part presents a practical example, and the third part offers conclusions and comparisons.

## 1 Introduction

Preparing the 3D mesh is a crucial initial step in obtaining accurate results from a model. This paper will discuss the various types of finite elements and their proper configuration, including density and size, in static models. The correct adjustment of finite elements significantly affects the calculation time. The presented knowledge will be illustrated through a selected example where the analysis will be conducted using Abaqus software. The fundamental types of 3D finite elements include tetrahedrons, hexahedrons, and triangular prisms, which can take on linear, quadratic, or cubic forms, among others. (Fig. 1) [1,2]

* Corresponding author: michala.weissova@tuke.sk

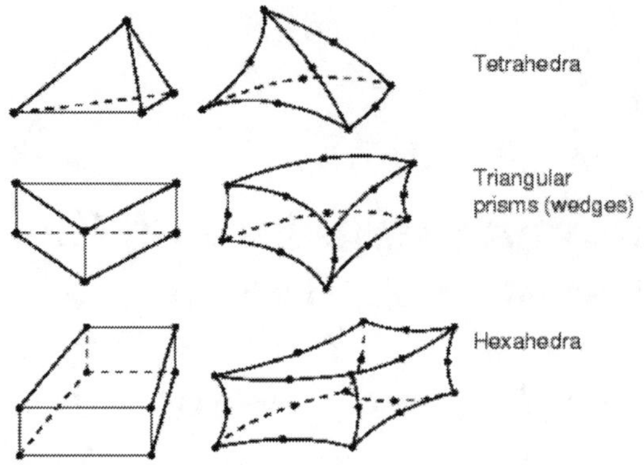


**Fig. 1.** Types of 3D elements in the mesh.

Apart from their size, finite elements also possess a density that is influenced by the material employed in the model. The size of the finite elements within the model is determined by factors such as the material density, the size and geometry of the problem being solved, and the stress-strain diagram. These aspects are interconnected and directly proportional. By maintaining the correct balance among these parameters, the calculation time can be reduced. [1,3]

## 2 FEM model

To showcase previous knowledge and experience, a simple example of a tensile test model has been chosen to determine the strength of the steel and generate the stress-strain diagram. (Fig. 2)

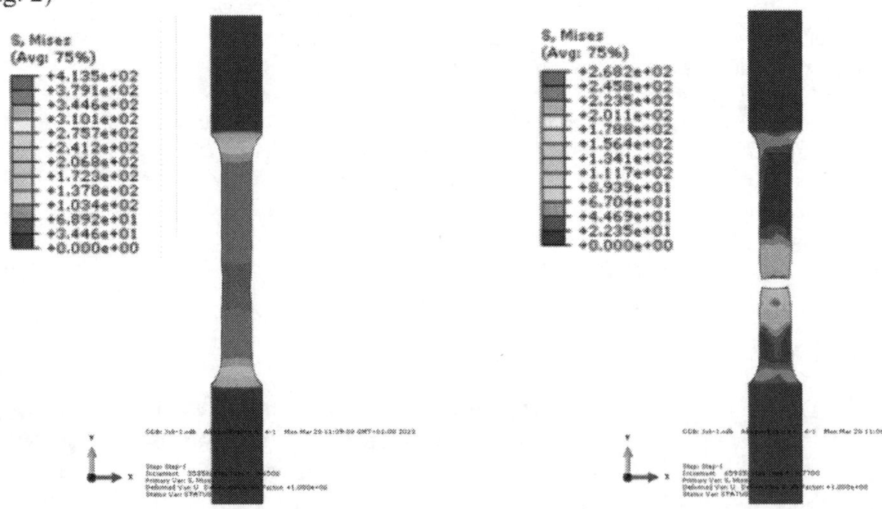


**Fig. 2.** The results of the tensile test.

The material properties, geometry, and loading conditions of the sample used in the model correspond to the results obtained from an experiment conducted at the Research and Innovation Centre for Construction. The experiment yielded the stress-strain diagram and material characteristics of the steel, which are represented by the model in the software. (Fig. 3).

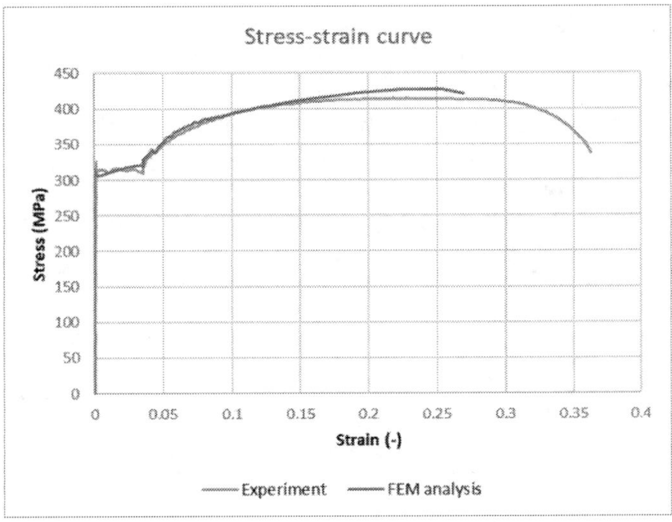


**Fig. 3.** Stress-strain diagram from FEM analysis and laboratory experiment.

Based on the previous figure, it is evident that the FEM model is accurate, allowing us to shift our focus towards assessing the mesh quality and the duration of the model calculation.

## 3 3D Mesh

### 3.1 Types of elements in the 3D mesh

The choice of mesh type and size significantly impacts the calculation process and, consequently, the obtained results. In the Abaqus software, the element shapes correspond appropriately to the overall model geometry. When employing the finite element method, calculations tend to converge more effectively when simpler element shapes are selected, although this may not always be feasible. Hexahedrons are elements that facilitate faster convergence and yield more accurate results. On the other hand, tetrahedrons are better suited for capturing complex problem geometries, but their convergence can be affected by element distortion. This distortion arises from the non-perpendicular walls of tetrahedrons, making their geometry more intricate compared to that of hexahedrons. (Fig. 4)

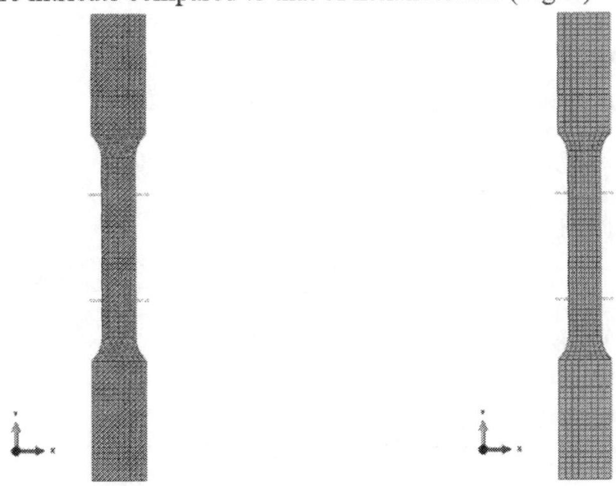

**Fig. 4.** The samples in FEM models.

Both samples share the same mesh density and other conditions. The only distinction between the samples lies in the type of finite element used: hexahedrons on the right and tetrahedrons on the left in Figure 4.

- Sample with final elements in the form of hexahedrons:
  - Element count: 704
  - Calculation time: 8.8 sec
- Sample with final elements in the form of tetrahedrons:
  - Element count: 8450
  - Calculation time: 13711.5 sec

### 3.2 Element size in the 3D mesh

The final calculation time is affected by the size of the elements in the mesh. The overall convergence time of a calculation step is determined by the size of the smallest element within the entire model. Therefore, it is crucial to minimize significant differences in element sizes to ensure efficient convergence.

$$\Delta t = L^e (\rho/E)^{1/2} \tag{1}$$

The previous formula represents the relationship between the material characteristics, the element size and the calculation time where $L^e$ is the size of the element, $\rho$ is the material density, $E$ is Young's modulus and $\Delta t$ is the increment stable time.

$$L^e = (VOLUME)^{1/3} \tag{2}$$

Based on these relationships, we can adjust the input values in Abaqus to shorten the calculation time and not affect the results. (Fig.5)

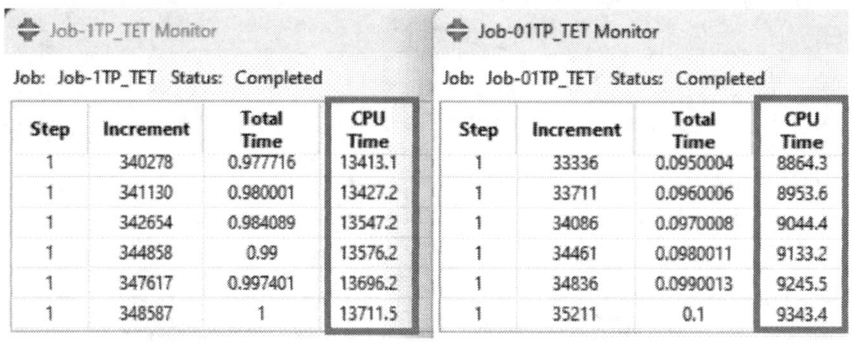
Job-1TP_TET Monitor

Job: Job-1TP_TET Status: Completed

| Step | Increment | Total Time | CPU Time |
|---|---|---|---|
| 1 | 340278 | 0.977716 | 13413.1 |
| 1 | 341130 | 0.980001 | 13427.2 |
| 1 | 342654 | 0.984089 | 13547.2 |
| 1 | 344858 | 0.99 | 13576.2 |
| 1 | 347617 | 0.997401 | 13696.2 |
| 1 | 348587 | 1 | 13711.5 |

Job-01TP_TET Monitor

Job: Job-01TP_TET Status: Completed

| Step | Increment | Total Time | CPU Time |
|---|---|---|---|
| 1 | 33336 | 0.0950004 | 8864.3 |
| 1 | 33711 | 0.0960006 | 8953.6 |
| 1 | 34086 | 0.0970008 | 9044.4 |
| 1 | 34461 | 0.0980011 | 9133.2 |
| 1 | 34836 | 0.0990013 | 9245.5 |
| 1 | 35211 | 0.1 | 9343.4 |

**Fig. 5.** The comparison of the calculation time.

### 3.3 The calculation time

The calculation time is impacted by various factors, including the complexity of the model geometry, the type of elements employed, their size, and their density. The material density incorporated in the model contributes to the overall calculation time and can be adjusted if necessary. Altering the material density may influence both the accuracy and the duration of the calculation. In models, the Mass Scaling function can be utilized to automatically increase the material density. The impact on the results will be minimal, contingent upon the chosen modeling approach.

The calculation time can be influenced by adjusting the utilization of computer logic processors and computer memory, with these settings being constrained by computer parameters.

In addition to modifying material density, it is also possible to adjust the calculation period within the range of 0.0 to 1.0. This adjustment affects both the total calculation time and the obtained results. Choosing a lower value from this interval reduces the convergence time for individual increments, leading to variations in results, especially within the plastic region of the graph. Figure 6 illustrates a comparison of certain model changes described above.

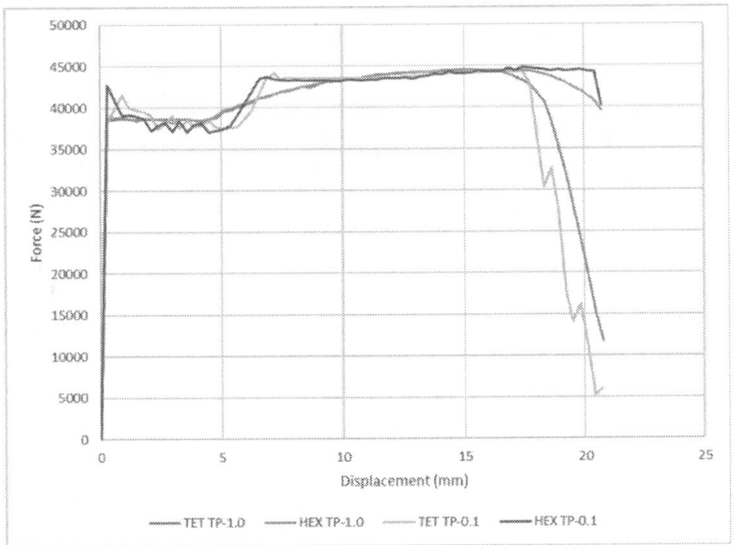


**Fig. 6.** Results of tensile test from the different FEM models.

Figure 6 displays a graph depicting the relationship between displacement and force at a reference point chosen on the site of the tensile test sample during the actual experiment. The image presents a comparison of the test results obtained using different types of mesh and calculation periods. The graph showcases the results of four models, with each individual curve representing:

- TET TP-1.0 – tetrahedron elements, time period 1.0
- HEX TP-1.0 – hexahedron elements, time period 1.0
- TET TP-0.1 – tetrahedron elements, time period 0.1
- HEX TP-0.1 – hexahedron elements, time period 0.1

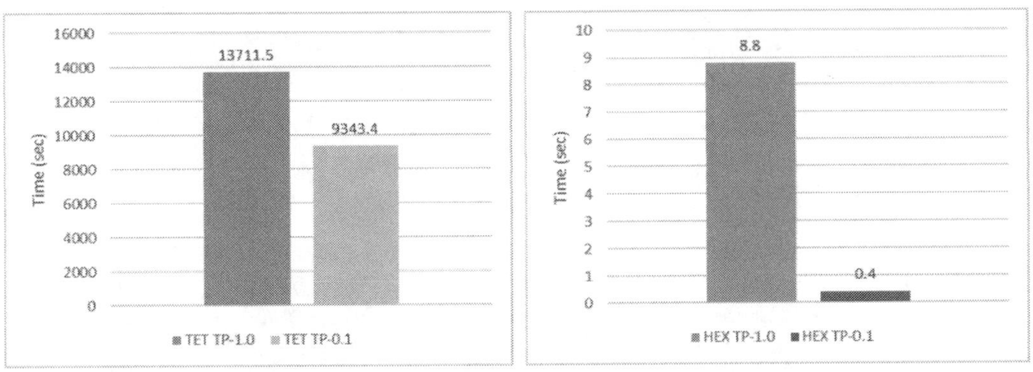


**Fig. 7.** Comparison of time in different types of samples.

## 4 Conclusion

The accuracy and efficiency of the analytical model are influenced by the 3D mesh, as well as the proper adjustment of load conditions, material characteristics, and geometry. The software compiles stiffness matrices and equations based on the model's mesh, which are essential for the overall analysis.

This paper has demonstrated how mesh settings impact both the calculation time and the accuracy of the results. The analytical models clearly show that changing the element type and the calculation period leads to differences in results, with a significant impact on the total calculation time.

The conclusion drawn from these findings is that when selecting a 3D mesh, it is important to provide correct and appropriate inputs in order to strike a balance between the model's accuracy and the calculation time.

The presented paper was supported by the project: VEGA 1/0172/20 "Stress and deformation analysis of load-bearing components made out of steel, glass and composite materials" of the Scientific Grant Agency of the Ministry of Education, Science, Research and Sport of the Slovak Republic.

## References

1. K.P. Shaik, D.K. Vaddi, and R.R. Bsv, Experimental Investigation of Glass Fibre Reinforced Polymer Laminates (SCHOLARS' PRESS, Beau Bassin, Mauritius, 2020).
2. H. Ataei and M. Mamaghani, Finite Element Analysis: Applications and Solved Problems Using Abaqus® (United States, 2018).
3. M. Stroeven, Discrete Numerical Modelling of Composite Materials: Application to Cementitious Materials. Doctoral Thesis. (Meinema BV, Delft, Netherlands, 1999), p. 238.
4. Kolář, V., Němec, I., & Kanický, V. (1997). FEM principy a praxe metody konečných prvků (M. Mojzesová, Ed.; 1st ed.). Computer Press.
5. Oguejiofor, E. C., & Hosain, M. U. (1997). Numerical analysis of push-out specimens with perfobond rib connectors. Computers and Structures, 62(4), 617–624. https://doi.org/10.1016/S0045-7949(96)00270-2
6. Vanova, P., Dubecky, D., & Kvocak, V. (2021). Push-Out Test and FEM Analysis of Continuous Shear Connector. IOP Conference Series: Materials Science and Engineering, 1203(3), 32080. https://doi.org/10.1088/1757-899x/1203/3/032080

# Effect of particle size distribution on shear strength of soil

*Hasan* Eteraf[1]*, *Balazs* Kovacs[1], *Viktoria* Mikita[1], and *Zahra* Delshad[1]

[1]University of Miskolc, H-3515 Miskolc, Hungary

**Abstract.** Soil grain size distribution has a significant impact on the mechanical properties of geotechnical materials. This study presents a laboratory investigation into the effects of fine content and grading factors on shear parameters. A series of shear tests and triaxial tests on sandy soils were performed for this aim. Finally, the outcomes of both strategies were compared and assessed. It was discovered that altering the grain size distribution results in changing the values of shearing parameters. Furthermore, the outcomes of different laboratory tests are not the same.

## 1 Introduction

Since the weight of all buildings is eventually passed to it, soil, as the most plentiful and readily available building element, plays a unique role in geotechnical matters. Although mixed soil is found more frequently in nature than pure soil, it has less geotechnical information and its qualities are always changing due to variances in the relative amounts of components that make up the mixed soil. We frequently meet soil made of sand–clay or sand–silt mixes while dealing with a site for a building.

To gain a thorough understanding of these places, researchers must conduct field and laboratory experiments to determine their behaviour. For sands, several investigations have shown that the angle of internal friction obtained from plane strain testing differs from that obtained from triaxial compression experiments. Studies comparing the results of angle of internal friction from direct shear tests and triaxial compression testing for granular soils are also available in the literature. While there are a lot of studies on pure sands, there aren't many studies on sand-fine particle mixtures [1-3]. Besides that, in terms of soil qualities, there hasn't been much research into the impact of grain size distribution on shearing behaviour at the soil–structure interface. Indeed, the grain size distribution of the soil has a substantial influence on the mechanical behaviours of geotechnical materials, as stated in the literature [4]. Shear tests should be undertaken to determine the features of stress-strain and volume change in soils subjected to load. By using a shear test, it can be expressed that how soils attain the stage of plastic failure as the stress applied to them increases [5].

To measure the shear strength properties of soils, standard testing procedures include triaxial compression and shear box tests. Because it is straightforward and practicable to

* Corresponding author: hghassan@uni-miskolc.hu

apply to disturbed and undisturbed soils, the shear box test is widely used. The sample is distorted at a regulated pace until failure occurs in this very simple test, which directly evaluates shear forces. The sample is made to yield via a shear plane at varied shearing rates by the testing apparatus [6]. In this study, a series of shear tests and triaxial tests on six distinct grain size distributions for clayey and silty sand soils were conducted. Each sample's maximum, mean, and lowest grain sizes, as well as its relative density, were held constant. Both methods of tests yielded shear strength metrics such as cohesiveness and internal friction angle, and the results were compared.

## 2 Literature review

The goal of this research is to determine the impact of particle size on soil shear strength and compare the results of various types of experiments. According to Islam et al. [7], particle size has a significant impact on the strength behavior of granular materials. The fabric is altered by the size of the particles in the granular mass, which is responsible for the variance in strength behavior. In terms of the influence of particle size on shear strength, previous research has produced mixed results. Kirkpatrick [8] used tests on two cohesionless materials to investigate the impact of particle size. The results revealed that when particle size increases, the friction angle decreases, which is consistent with the findings of Marschi et al. [9] and Marsal [10]. Zelasko et al. [11] investigated three sands and discovered that as the mean particle diameter increases, the friction angle decreases somewhat. In the meantime, several research support the opposing viewpoint.

In the case of fine-grained soils, such as the bottom of the circular failure surface under an embankment, it is usually essential to analyze or conduct comparison research. Kulhawy et al. [12] hypothesized that fine-grained soils behave similarly to sands under direct shear and triaxial shear. The same effective stress shear strength characteristics were obtained from experiments conducted on eight specimens of Boulder Clay using the direct shear and triaxial devices, according to Skempton. Casagrande and Poulos provided the findings of CD direct shear and CD triaxial tests performed on compacted specimens of a lean clay, demonstrating that the two tests yielded almost the same shear strength envelope [13]. Abdel-Ghaffar (1990) collated findings from studies that used undisturbed samples of the same soil to perform direct shear and triaxial testing. He came to the conclusion that the effective stress friction angle and cohesion intercept for the direct shear and triaxial devices are comparable [14].

In this study, the effect of shear rates has also been investigated.so it is better to consideration the previous works regarding that. To investigate the effects of shear rate on internal friction angle and cohesiveness, Thermann et al. [15] employed a direct shear apparatus and applied varied shear rates to silty sand soils. At four different shear rates, Saito et al. [16] performed direct shear experiments on silica sands containing illite and bentonite. The effective residual internal friction angle of the silica sands remained practically constant at 34° for all shear speeds, according to these researchers. Shear box studies on coarse grained granular backfill soils were conducted by Nakao et al. [17], who discovered that peak and residual internal friction angles varied in response to different shear rates.

## 3 Methodology

The investigated soil of our study was taken from the Hejopapi region located in northern Hungary. The soil was sieved and remixed into different soils compound with different course contents. Figure 1 shows the grain size distribution of samples. The soil specimen

https://doi.org/10.1051/matecconf/202338501043

was prepared in agreement with the testing standard. The initial density, water content, compaction process, and other sample state parameters were given special attention because the outcomes are influenced by all of these factors. A series of direct shear box tests were performed to characterize the shear strength of the soil. The properties of different tested soils are listed in Table 1.

**Table 1.** Properties of samples.

| Sample No. | Water content $w$ (%) | Average diameter $D_{50}$ (mm) | Fines content% |
|---|---|---|---|
| | | | Clay+silt |
| #1 | 15 | 0.002 | 50 |
| #2 | 15 | 0.2 | 45 |
| #3 | 15 | 0.26 | 40 |
| #4 | 15 | 0.3 | 35 |
| #5 | 15 | 0.33 | 30 |
| #6 | 15 | 0.36 | 25 |

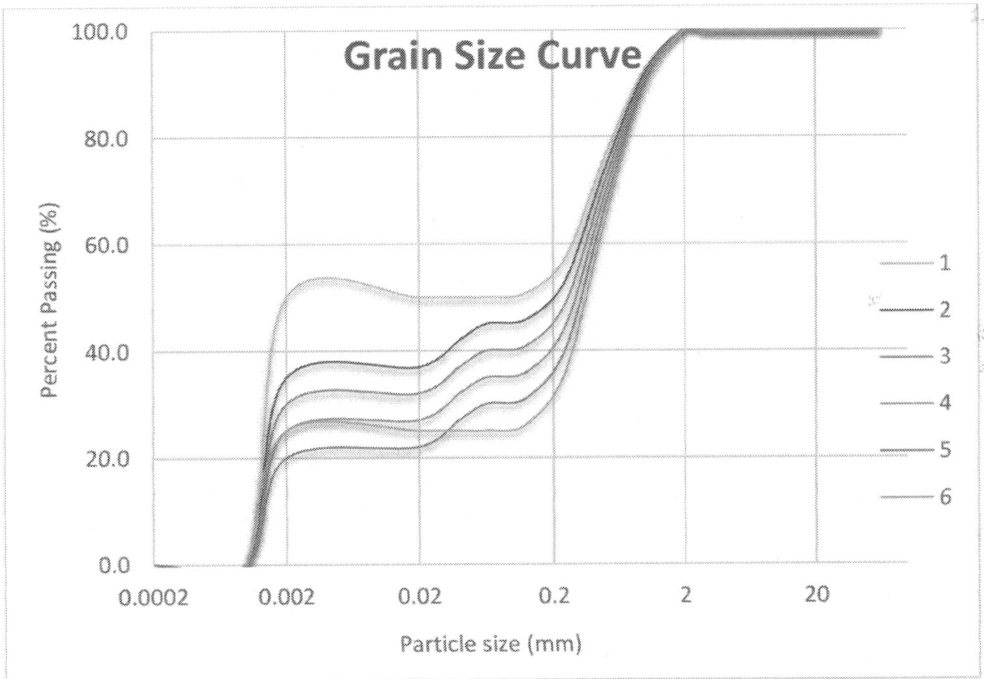


**Fig 1.** Grain size distribution of samples.

The specimens were compacted in three equal layers in the shear box using a rubber hammer for the direct shear test, with a water content of 15%. After applying vertical pressure to the specimen in the shear box of 100 kPa, 200 kPa, and 300 kPa, horizontal thrust was applied to the top box until soil collapse. The horizontal loading rate in the experiments was set at 0.05 and 0.5 mm/min as a slow and rapid rate for comparative purposes. The rate of displacement was determined using ASTM D3080-90 (ASTM, 1990) and Bishop and Henkel's findings (**1957**) [18]. A linear fit through the origin in the plot of peak or maximum shear stress against normal stress defines the peak angle of shearing resistance. In addition, the tests revealed the cohesiveness of each sample. Furthermore, the materials were put to drained triaxial compression testing. The experiments were conducted with effective confining forces in place. During the consolidation and shearing phases, filter paper side drains were used to shorten the test duration. Before testing, the samples were soaked by applying back pressure. To derive the cohesion and internal friction angle from triaxial test, the p-q diagram has been used. The p-q diagram is a method of simplifying the analysis of triaxial and other stress data which are commonly used in soil mechanics. It can be used in a variety of applications and solve a range of problems. Using this method, you won't have to solve two equations with two unknowns. Thus, the results obtained by triaxial test and direct shear test for sand with different fine content were compared and analyzed. In addition, the effect of shear rate has been investigated.

## 4 Result and discussion

The effects of the Median particle diameter and fines content on the shear strength parameters of the soil by using different methods were discussed in present section. Using a linear fit across the origin in the plot of shear stress versus normal stress, the internal friction angle and cohesion were computed for the direct shear tests. It should be taken into consideration that the shear rate applied for these tests was 0.05 mm/min. Figure 2 to 7 shows the result of direct shear tests for sample number 1 to 6 respectively.

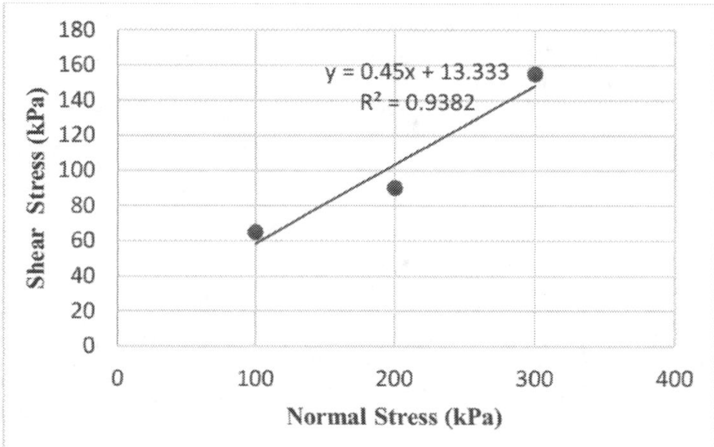


**Fig 2.** Shear stress versus normal stress for sample #1.

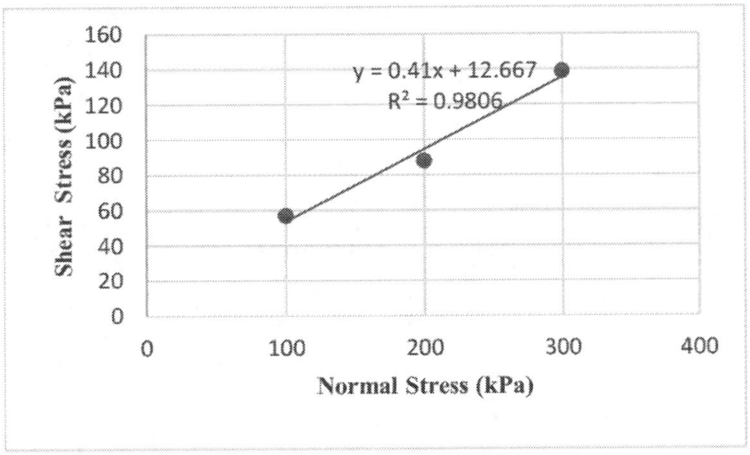


**Fig 3.** Shear stress versus normal stress for sample #2.

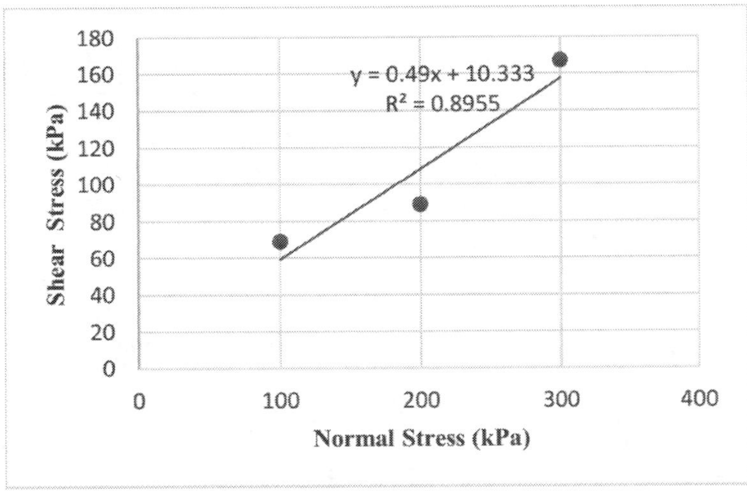


**Fig 4.** Shear stress versus normal stress for sample #3.

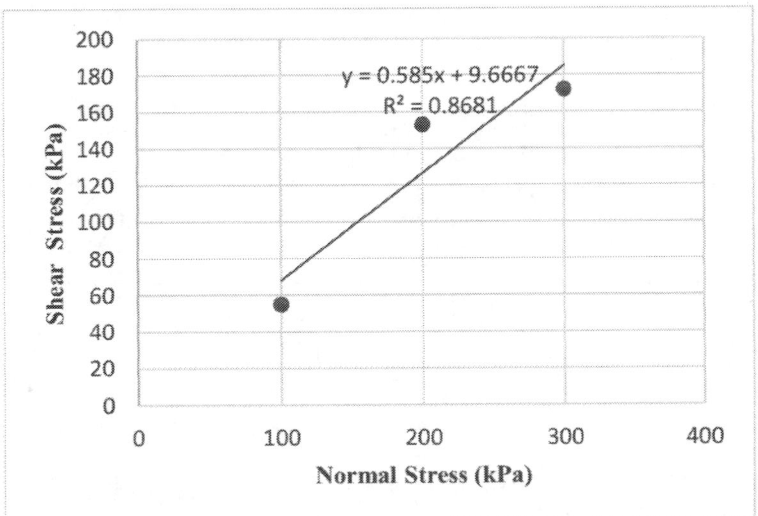


**Fig 5.** Shear stress versus normal stress for sample #4.

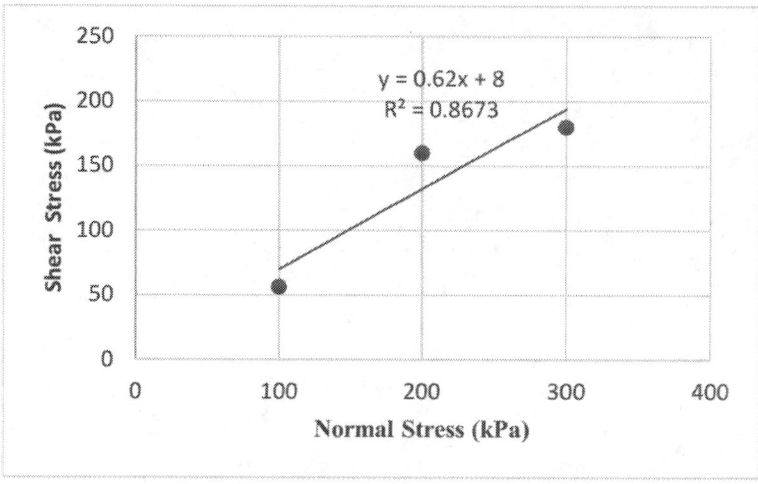


**Fig 6.** Shear stress versus normal stress for sample #5.

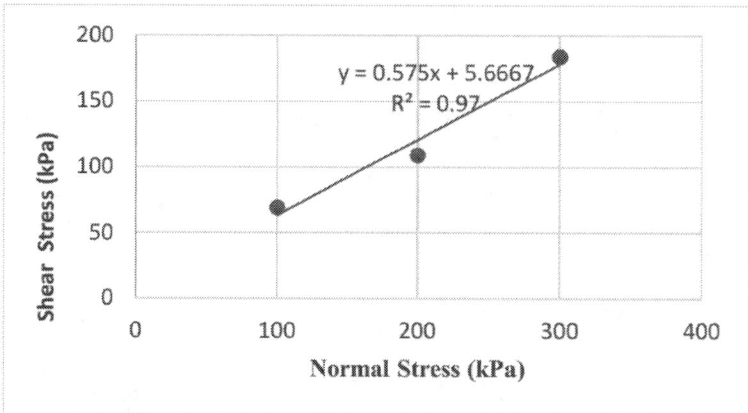


**Fig 7.** Shear stress versus normal stress for sample #6.

Table 2 shows all the values of cohesion and internal friction angle obtained from the diagrams. It can be seen that the angles of shearing resistance of the samples vary from 22 to 32 degree. The range of cohesion is also claimed to be between 5 and 14 kPa. The results also shows that generally by increase in median diameter of particles, the values of cohesion decrease and the amount of internal friction angle increase.

**Table 2.** Results of direct shear tests.

| Sample No. | Cohesion (kPa) | Internal friction angle (degree) |
|---|---|---|
| #1 | 13.1 | 24.23 |
| #2 | 12.7 | 22.29 |
| #3 | 10.2 | 26.10 |
| #4 | 9.5 | 30.32 |
| #5 | 8 | 31.79 |
| #6 | 5.1 | 29.88 |

All samples were subjected to a triaxial test to improve accuracy and compare shear strength data acquired using different techniques. The data were retrieved in a p-q diagram for simplicity, and the quantity of cohesion and internal friction angle were then derived from them. As an example, for specimen number one, Figure 8 illustrates a p-q diagram. The same procedure was followed for all samples, with the findings listed in Table 3.

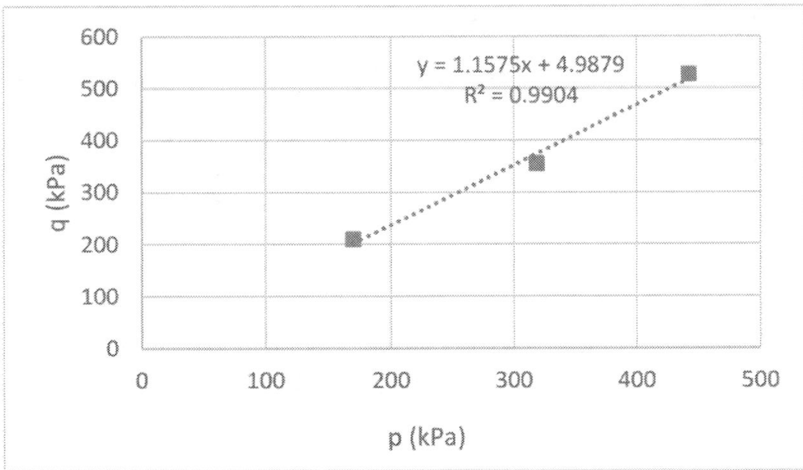


**Fig 8.** p-q diagram of a triaxial test for sample number 1.

**Table 3.** Results of direct shear tests.

| Sample No. | Cohesion(kP) | Internal friction angle (Degree) |
|---|---|---|
| #1 | 15.4 | 29.2 |
| #2 | 13.8 | 25.7 |
| #3 | 10 | 27.9 |
| #4 | 9.8 | 32.4 |
| #5 | 7.3 | 33.1 |
| #6 | 4.1 | 33.8 |

Figures 9 and 10 illustrate a better comparison to notice the differences in the results. In terms of increasing the average dispersion rate of grains, this graph compares the cohesion and internal friction angles acquired from two tests. According to the results of a series of direct shear and triaxial tests on different sands, the angle of internal friction values obtained from triaxial tests are about 1–5° higher than that obtained from direct shear tests. In general, it appears that the difference is greater in soils with a higher median particle diameter. It can be referred to amount of fine which each sample contain. Regarding cohesion parameter, while there is a significant difference, there are no discernible tendencies. As it was mentioned before, to have a consideration about the effect of shear rate parameter on the results, the specimen was tested with a faster rate in direct shear test. The applied rate for rapid test was 0.5 mm/min. Table 4 shows the effect of shear rate on shearing characteristics under both testing circumstances for sample numbers 1 to 6.

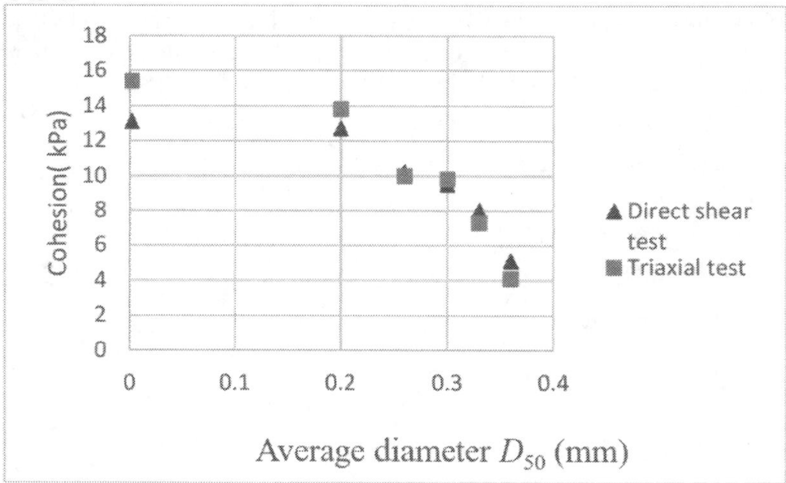


**Fig 9.** Comparison of cohesion obtained by both methods.

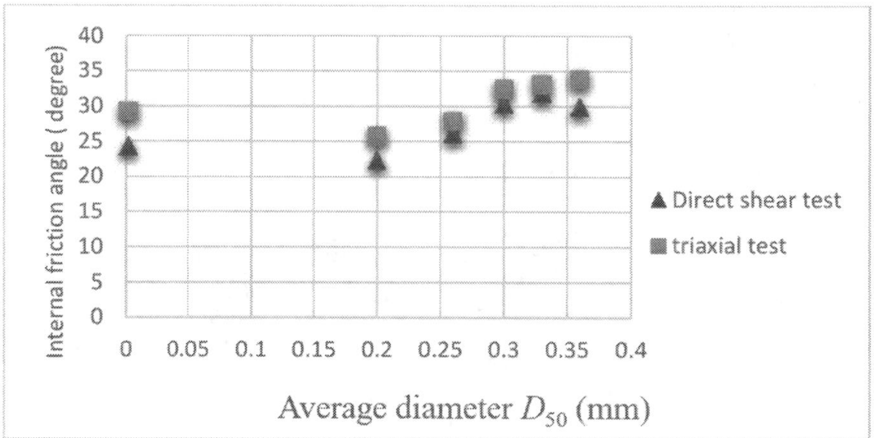


**Fig 10.** Comparison of internal friction angle obtained by both methods.

**Table 4.** Results obtained by direct shear tests with different shear rates.

| Sample No. | Cohesion(kPa) | | Internal friction angle (degree) | |
|---|---|---|---|---|
| | Shear rate = 0.05mm/min | Shear rate = 0.5mm/min | Shear rate= 0.05mm/min | Shear rate= 0.5mm/min |
| #1 | 13.1 | 14.2 | 24.23 | 28.33 |
| #2 | 12.7 | 13.9 | 22.29 | 26.24 |
| #3 | 10.2 | 11.1 | 26.10 | 28.07 |
| #4 | 9.5 | 9.1 | 30.32 | 31.42 |
| #5 | 8 | 8.2 | 31.79 | 33.78 |
| #6 | 5.1 | 5.3 | 29.88 | 34.2 |

Internal friction angle and cohesion parameters are both affected by shear rate. When shear rate was increased, the internal friction angle increased further. There is also a rise in cohesion values, which may be neglected. Furthermore, the results show that the values of the internal friction angle obtained by the high-speed rate direct shear test are almost similar to the values acquired from the triaxial tests.

## 5 Conclusion

A series of shear tests and triaxial tests on sandy soils were conducted in this study. The soils have different fine contents. The results obtained by both types of tests were compared and the effect of median particle diameter was investigated. The results of direct shear test shows that generally by increase in median diameter of particles, the values of cohesion decrease and the amount of internal friction angle increase. According to the results of triaxial tests, the angle of internal friction values obtained from triaxial tests are about 1–5° higher than that obtained from direct shear tests. In general, it appears that the difference is greater in soils with a higher median particle diameter. It can be referred to amount of fine which each sample contain. In addition, it was found out that internal friction angle and cohesion parameters are both affected by shear rate which is the same as that research done in the past. By increasing in shear rates, higher values of internal friction angle were observed. those values are almost similar to the values acquired from the triaxial tests.

## References

1. A. Krim, A. Arab, M. Chemam, A. Brahim, M. Sadek, & I. Shahrour. *Experimental study on the liquefaction resistance of sand–clay mixtures: Effect of clay content and grading characteristics*. Marine Georesources & Geotechnology, **37(2)**, 129-141 (2019).
2. S. Goodarzi, & H. Shahnazari. *Strength enhancement of geotextile-reinforced carbonate sand*. Geotextiles and Geomembranes, **47(2)**, 128-139 (2019).
3. H. Venkateswarlu, K. N. Ujjawal, & A. Hegde. *Laboratory and numerical investigation of machine foundations reinforced with geogrids and geocells*. Geotextiles and Geomembranes, **46(6)**, 882-896 (2018).
4. H. L. Wang, Y. J. Cui, F. Lamas-Lopez, N. Calon, G. Saussine, J. C. Dupla, ... & R. P. Chen. *Investigation on the mechanical behavior of track-bed materials at various contents of coarse grains*. Construction and Building Materials, **164**, 228-237 (2018).
5. D. Kim, & S. Ha. *Effects of particle size on the shear behavior of coarse-grained soils reinforced with geogrid. Materials*, **7(2)**, 963-979 (2014).
6. M. Beren, I. Cobanoglu, S. B. Çelik, & Ö. Ündül. *Shear Rate Effect on Strength Characteristics of Sandy Soils*. Soil Mechanics and Foundation Engineering, **57(4)**, 281-287 (2020).
7. M. N. Islam, A. Siddika, M. B. Hossain, A. Rahman, and M. A. Asad. *Effect of particle size on the shear strength of sand*. Australian Geomechanics, **46(3)**, September 2011.
8. W. M. Kirkpatrick. *Effects of grain size and grading on the shearing behavior of granular materials*. Proceedings of the sixth International Conference on Soil Mechanics and Foundation Engineering (pp. 273-277), September 1965.
9. N. D. Marschi, C. K. Chan, & H. B. Seed. *Evaluation of properties of rockfill materials*. Journal of the Soil Mechanics and Foundations Division, **98(1)**, 95-114 (1972).

10. R. J. Marsal. *Mechanical properties of rockfill.* In Embankment-Dam Engineering, R.C. Hirschfeld and S. J.

11. J. S. Zelasko, R. J. Krizek, & T. B. Edil. *Shear Behavior of Sands as a Function of Grain Characteristics.*, **1**, 55-64 (1975).
12. F. H. Kulhawy, & P. W. Mayne. *Manual on estimating soil properties for foundation design* (No. EPRI-EL-6800). Electric Power Research Inst., Palo Alto, CA (USA); Cornell Univ., Ithaca, NY (USA). Geotechnical Engineering Group (1990).
13. B. A. Castellanos, & T. L. Brandon. *A comparison between the shear strength measured with direct shear and triaxial devices on undisturbed and remolded soils.* Proceedings of the 18th International Conference on Soil Mechanics and Geotechnical Engineering, Paris (Vol. 2013, pp. 317-320) (2013).
14. M. E. D. M. Abdel-Ghaffar. *The meaning and practical significance of the cohesion intercept in soil mechanics.* Doctoral dissertation, University of Illinois at Urbana-Champaign (1990).
15. K. Thermann, C. Gau, & J. Tiedemann. *Shear strength parameters from direct shear tests–influencing factors and their significance.* The Geological Society of London (Vol. 484, pp. 1-12) (2006).
16. R. Saito, H. Fukuoka, & K. Sassa. *Experimental study on the rate effect on the shear strength.* Disaster Mitigation of Debris Flows, Slope Failures and Landslides, 421-427 (2006).
17. T. Nakao, & S. Fityus. *Direct shear testing of a marginal material using a large shear box.* Geotechnical Testing Journal, **31(5)**, 393-403 (2008).
18. ASTM D 3080: *Standard test method for direct shear test of soils under consolidated drained conditions*, ASTM International (2012).

# AUTHOR INDEX